Silvanus Phillips Thompson

The Electromagnet, And Electromagnetic Mechanisms

Silvanus Phillips Thompson

The Electromagnet, And Electromagnetic Mechanisms

ISBN/EAN: 9783744675529

Printed in Europe, USA, Canada, Australia, Japan

Cover: Foto ©berggeist007 / pixelio.de

More available books at **www.hansebooks.com**

Finsbury Technical Manuals.

EDITOR OF THE SERIES,

PROFESSOR SILVANUS P. THOMPSON,

D.Sc., B.A., F.R.S., M.I.E.E., &c.

Finsbury Technical Manuals.

THE
ELECTROMAGNET,

AND

ELECTROMAGNETIC MECHANISM.

BY

SILVANUS P. THOMPSON,

D.SC., B.A., F.R.S.,

Principal of, and Professor of Physics in, the City and Guilds of London Technical College, Finsbury;
Late Professor of Experimental Physics in University College, Bristol;
Vice-President of the Physical Society of London;
Member of Council of the Institution of Electrical Engineers;
Membre du Bureau de la Société Française de Physique;
Membre de la Société Internationale des Électriciens;
Honorary Member of the Physical Society of Frankfort-on-the-Main.

E. & F. N. SPON, 125, STRAND, LONDON.

NEW YORK: 12, CORTLANDT STREET.

1891.

PREFACE.

THE electromagnet in a practical form was first publicly exhibited on May 23rd, 1825, by William Sturgeon, on the occasion of reading a paper which is to be found in the volume of the *Transactions of the Society of Arts* for that year. For this invention we may rightfully claim the very highest place. Electrical engineering embraces many branches, but most of these are concerned with electro-magnets. The dynamo for generating electric currents, the motor for transforming their energy back into work, the arc lamp, the electric bell, the telephone, the recent electro-magnet machinery for coal mining and for the separation of ore, and many other electro-mechanical contrivances, come within the purview of the electrical engineer. In every one of these, and in many more of the useful applications of electricity, the central organ is *an electromagnet*. By means of this simple and familiar contrivance—an iron core surrounded by a copper wire coil—mechanical actions are produced at will, at a distance, under control, by the agency of electric currents. These mechanical actions are known to vary with the mass, form, and quality of the iron core, the quantity and disposition of the copper wire wound upon it, the quantity of electric current circulating around it, the form, quality, and distance of the iron armature upon which it acts. But the laws which govern the mechanical action in relation to these various matters are by no means well known, and, indeed, several of them have long been a matter of dispute. Gradually, however, that which has been vague and indeterminate becomes clear and precise. The laws of the steady circulation of electric currents, at one time altogether obscure, were cleared up by the discovery of the famous law of Ohm. Their extension to the case of rapidly interrupted currents, such as are used in telegraphic working, was discovered by von Helmholtz; whilst to Maxwell is due their further extension to alternating, or, as they are sometimes called, undulatory currents. All this was purely electric work. But the law of the electromagnet was still undiscovered; the magnetic part of the problem was still buried in obscurity. The only exact reasoning about magnetism dealt with problems of another kind; it was couched in language of a misleading character; for the practical problems connected with the electromagnet it was worse than useless. The doctrine of two magnetic fluids distributed over the end surfaces of magnets, had, under the sanction of the great names of Coulomb, of Poisson, and of Laplace, unfortunately become recognized as an accepted part of science, along with the law of inverse squares. How greatly the progress of electromagnetic science has been impeded and retarded by the weight of these great names it is impossible now to gauge. We now know that for all purposes, save only those whose value lies in the domain of abstract mathematics, the doctrine of the two magnetic fluids is false and misleading. We know that magnetism, so far from residing on the end or surface of the magnet, is a property resident throughout the mass; that the internal not the external magnetization is the important fact to be considered; that the so-called free magnetism on the surface is, as it were, an accidental phenomenon; that the magnet is really most highly magnetized at those parts where there is least surface magnetization; finally, that the doctrine of surface distribution of fluids is absolutely incompetent to afford a basis of calculation such as is required by the electrical engineer. He requires rules to enable him not only to predict the lifting power of a given electromagnet, but also to guide him in designing and constructing electromagnets of special forms suitable for the various cases that arise in his practice. He wants in one place a strong electromagnet to hold on to its armature like a limpet to its native rock; in another case he desires a magnet having a very long range of attraction, and wants a rule to guide him to the best design; in another he wants a special form having the most rapid action attainable; in yet another he must sacrifice everything else to attain maximum action with minimum weight. Toward the solution of such practical problems as these the old theory of magnetism offered not the slightest aid. Its array of mathematical symbols was a mockery. It was as though an engineer asking for rules to enable him to design the cylinder and piston of an engine were confronted with receipts how to estimate the cost of painting it.

Gradually, however, new light dawned. It became customary, in spite of the mathematicians, to regard the magnetism of a magnet as something that traverses or circulates around a definite path, flowing more freely through such substances as iron, than through other relatively non-magnetic materials. Analogies between the flow of electricity in an electrically conducting circuit, and the passage of magnetic lines of force through circuits possessing magnetic conductivity, forced themselves upon the minds of experimenters, and compelled a mode of thought quite other than the previously accepted. So far back as 1821, Cumming[*] experimented on magnetic conductivity. The idea of a magnetic circuit was more or less familiar to Ritchie,[†] Sturgeon,[‡] Dove,[§] Dub,[‖] and De la Rive,[¶] the last-named of whom explicitly uses the phrase, "a closed magnetic circuit." Joule[**] found the maximum power of an electromagnet to be proportional to "the least sectional area of the entire magnetic circuit," and he considered the resistance to induction

---

[*] *Camb. Phil. Trans.*, April 2, 1821.
[†] *Phil. Mag.*, series iii. vol. iii. p. 122.
[‡] *Ann. of Electr.*, xii. p. 217.
[§] *Pogg. Ann.*, xxix. p. 462, 1833. See also *Pogg. Ann.*, xliii. p. 517, 1838.
[‖] Dub, *Elektromagnetismus* (ed. 1861), p. 401; and *Pogg. Ann.*, xc. p. 440, 1853.
[¶] De la Rive. *Treatise on Electricity* (Walker's translation), vol. i. p. 292.
[**] *Ann. of Electr.*, iv. 59, 1839; v. 195, 1841; and *Scientific Papers*, pp. 8, 34, 35, 36.

as proportional to the length of the magnetic circuit. Indeed, there are to be found, scattered in Joule's writings on the subject of magnetism, some five or six sentences, which, if collected together, constitute a very full statement of the whole matter. Faraday,[*] considered that he had proved that each magnetic line of force constitutes a closed curve; that the path of these closed curves depended on the magnetic conductivity of the masses disposed in proximity; that the lines of magnetic force were strictly analogous to the lines of electric flow in an electric circuit. He spoke of a magnet surrounded by air being like unto a voltaic battery immersed in water or other electrolyte. He even saw the existence of a power, analogous to that of electromotive force in electric circuits, though the name, "magnetomotive force," is of more recent origin. The notion of magnetic conductivity is to be found in Maxwell's great treatise (vol. ii. p. 51), but is only briefly mentioned. Rowland,[†] in 1873, expressly adopted the reasoning and language of Faraday's method in the working out of some new results on magnetic permeability, and pointed out that the flow of magnetic lines of force through a bar could be subjected to exact calculation; the elementary law, he says, "is similar to the law of Ohm." According to Rowland, the "magnetising force of helix" was to be divided by the "resistance to the lines of force;" a calculation for magnetic circuits which every electrician will recognize as precisely as Ohm's law for electric circuits. He applied the calculations to determine the permeability of certain specimens of iron, steel, and nickel. In 1882,[‡] and again in 1883, Mr. R. H. M. Bosanquet[§] brought out at greater length a similar argument, employing the extremely

---

[*] *Experimental Researches*, vol. iii. art. 3117, 3228, 3230, 3260, 3271, 3276, 3294, and 3361.

[†] *Phil. Mag.*, series iv. vol. xlvi. August 1873, "On Magnetic Permeability and the Maximum of Magnetism of Iron, Steel, and Nickel."

[‡] *Proc. Royal Soc.*, xxiv. p. 445, December 1882.

[§] *Phil. Mag.*, series v. vol. xv. p. 205, March, 1883. *On Magneto-Motive Force.* Also *ibid.*, vol. xix. February, 1885, and *Proc. Roy. Soc.*, No. 223, 1883. See also *Electrician*, xiv. p. 291, February 14th, 1885.

apt term "magnetomotive force," to connote the force tending to drive the magnetic lines of induction through the "magnetic resistance," or "reluctance" (to use a more modern term), of the circuit. In these papers the calculations are reduced to a system, and deal not only with the specific properties of iron, but with problems arising out of the shape of the iron. Bosanquet shows how to calculate the several resistances, or "reluctances," of the separate parts of the circuit, and then add them together to obtain the total resistance, or "reluctance," of the magnetic circuit.

Prior to this, however, the principle of the magnetic circuit had been seized upon by Lord Elphinstone and Mr. Vincent, who proposed to apply it in the construction of the dynamo-electric machines. On two occasions* they communicated to the Royal Society the results of experiments to show that the same exciting current would evoke a larger amount of magnetism in a given iron structure, if that iron structure formed a closed magnetic circuit, than if it were otherwise disposed.

In recent years the notion of the magnetic circuit has been vigorously taken up by the designers of dynamo-machines, who indeed base the calculation of their designs upon this all-important principle. Having this, they need no laws of inverse squares of distances, no magnetic moments, none of the elaborate expressions for surface distribution of magnetism, none of the ancient paraphernalia of the last century. The simple law of the magnetic circuit, and a knowledge of the properties of iron, is practically all they need. About four years ago, much was done by Mr. Gisbert Kapp† and by Drs. J. and E. Hopkinson‡ in the application of these considerations to the design of dynamo-machines, which previously had been a matter of empirical practice.

* *Proc. Roy. Soc.*, xxix. p. 292, 1879, and xxx. p. 287, 1880. See *Electrical Review*, viii. p. 134, 1880.

† *The Electrician*, vols. xiv. xv. and xvi. 1885-6; also *Proc. Inst. Civil Engineers*, lxxxiii. 1885-6; and *Journ. Soc. Telegr. Engineers*, xv. 524, 1886.

‡ *Phil. Trans.*, 1886, pt. i. p. 331; and *The Electrician*, xviii. pp. 39, 63, 86, 1886.

To this end the formulæ of Professor Forbes* for calculating magnetic leakage, and the researches of Professors Ayrton and Perry† on magnetic shunts, contributed a not unimportant share. As the result of the advances made at that time, the subject of dynamo design was reduced to an exact science.

It is the aim and object of the present work to show how the same considerations which have been applied with such great success to the subject of the design of dynamo-electric machines may be applied to the study of the electromagnet. The theory and practice of the design and construction of electromagnets will thus be placed, once for all, upon a rational basis. Definite rules will be laid down for the guidance of the constructor, directing him as to the proper dimensions and form of iron to be chosen, and as to the proper size and amount of copper wire to be wound upon it in order to produce any desired result.

In Chapter I. is given a historical account of the invention. This is followed by a chapter dealing with general considerations respecting the uses and forms of electromagnets, and electromagnetic phenomena in general. This is followed in Chapter III. by a discussion of the magnetic properties of iron and steel and other materials; some account being added of the methods used for determining the magnetic permeability of various brands of iron at different degrees of saturation. Tabular information is given as to the results found by different observers. In connection with the magnetic properties of iron, the phenomenon of magnetic hysteresis is also described and discussed. In Chapter IV. the principle of the magnetic circuit is discussed, with numerical examples, and a number of experimental data respecting the performance of electromagnets are adduced, in particular those bearing upon the tractive power of electromagnets. The law of traction between an electromagnet and its armature is then laid down, followed by the rules for predetermining the iron cores and copper coils required to

---

* *Journ. Soc. Telegr. Engineers*, xv. 555, 1886.
† *Ibid.*, xv. 530, 1886.

give any prescribed tractive force. In Chapter V. comes the extension of the calculation of the magnetic circuit to those cases where there is an air-gap between the poles of the magnet and the armature ; and where, in consequence, there is leakage of the magnetic lines from pole to pole. Chapter VI. is devoted to the rules for calculating the winding of the copper coils ; and the limiting relation between the magnetizing power of the coil and the heating effect of the current in it is explained. After this comes a detailed discussion, in Chapter VII., of the special varieties of form that must be given to electromagnets in order to adapt them to special services. Those which are designed for maximum traction, for quickest action, for longest range, for greatest economy when used in continuous daily service, for working in series with constant current, for use in parallel at constant pressure, and those for use with alternate currents, are separately considered.

Toward the close of the book some account is given of the various forms of electromagnetic mechanism which have arisen in connection with the invention of the electromagnet. In Chapter VIII. the plunger and coil is specially considered as constituting a species of electromagnet adapted for a long range of motion. Chapter IX. is devoted to electric mechanism, and in it sundry modes of mechanically securing long range for electromagnets, and of equalizing their pull over the range of motion of the armature, are also described. In the development of this subject some analogies between sundry electro-mechanical movements and the corresponding pieces of ordinary mechanism are traced out. Chapter X. deals with electromagnetic vibrators ; Chapter XI. with alternate-current mechanisms ; Chapter XII. with motors ; and Chapter XIII. with electromagnetic machine tools. Chapter XIV. is occupied by a consideration of the various modes of preventing or minimising the sparks which occur in the circuits in which electromagnets are used. Chapter XV., relating to the use of the electromagnet in surgery, has been mainly contributed by the Author's brother, Dr. J. Tatham

Thompson, of Cardiff. Chapter XVI., which concludes the book, deals with permanent magnets of steel.

The work now presented to the public is an amplification of the Cantor Lectures delivered by the Author in 1890, before the *Society of Arts*. To the Council of that Society his thanks are due for the permission to reproduce much of the text and many of the cuts. It has been thought well to retain in many passages the direct form of address, as to an audience, rather than recast the matter in purely descriptive terms. The chapter on electromagnetic mechanism constituted the topic of the Author's presidental discourse to the Junior Engineering Society.

The Author's grateful thanks are due to Professor R. Mullineux Walmsley, for assistance in revising proofs; and to Mr. Eustace Thomas, one of the demonstrators in the department of Electrical Engineering in the Technical College, Finsbury, for much help in the preparation of the original lectures.

Finally, the Author has to acknowledge the reception accorded to his Cantor Lectures, both in this country and in the United States, and to express the hope that in the present more extended form, his labours will prove of service to those who are occupied in the electrical industries, as well as to those who follow science for its own sake.

CITY AND GUILDS' TECHNICAL COLLEGE, FINSBURY.
July 1891.

CONTENTS.

	PAGE
PREFACE	v

CHAPTER I.
HISTORICAL INTRODUCTION 1

CHAPTER II.
GENERALITIES CONCERNING ELECTROMAGNETS AND ELECTROMAGNETISM. TYPICAL FORMS OF ELECTROMAGNETS. MATERIALS OF CONSTRUCTION 35

CHAPTER III.
PROPERTIES OF IRON 65

CHAPTER IV.
PRINCIPLE OF THE MAGNETIC CIRCUIT. THE LAW OF TRACTION. DESIGN OF ELECTROMAGNETS FOR MAXIMUM TRACTION 112

CHAPTER V.
EXTENSION OF THE LAW OF THE MAGNETIC CIRCUIT TO CASES OF ATTRACTION OF AN ARMATURE AT A DISTANCE. CALCULATION OF MAGNETIC LEAKAGE 156

CHAPTER VI.
RULES FOR WINDING COPPER WIRE COILS 191

CHAPTER VII.
SPECIAL DESIGNS. RAPID-ACTING ELECTROMAGNETS. RELAYS AND CHRONOGRAPHS 211

CHAPTER VIII.

Coil-and-Plunger 240

CHAPTER IX.

Electromagnetic Mechanism 275

CHAPTER X.

Electromagnetic Vibrators and Pendulums 318

CHAPTER XI.

Alternate-current Electromagnets 331

CHAPTER XII.

Electromagnetic Motors 350

CHAPTER XIII.

Electromagnetic Machine Tools 359

CHAPTER XIV.

Modes of preventing Sparking 363

CHAPTER XV.

The Electromagnet in Surgery 374

CHAPTER XVI.

Permanent Magnets 381

Appendix A.—William Sturgeon 412
Appendix B.—Electric and Magnetic Units 419
Appendix C.—Calculation of Excitation, Leakage, Etc. 427

LIST OF ILLUSTRATIONS.

FIG.		PAGE
1.	Sturgeon's First Electromagnet	3
2.	Side view of same	3
3.	Sturgeon's Straight-bar Electromagnet	4
4.	Sturgeon's Lecture-table Electromagnet	8
5.	Henry's Electromagnet	14
6.	Henry's Experimental Electromagnet	17
7.	Joule's Electromagnet	21
8.	Joule's Cylindrical Electromagnet	24
9.	Roberts' Electromagnet	26
10.	Joule's Zigzag Electromagnet	26
11.	Faraday's Electromagnet at the Royal Institution	29
12.	S. P. Thompson's Electromagnet	31
13.	Electromagnet made of Two Cannons	32
14.	Diagram of Magnetic Lines of Force at Mouth of Cannon	33
15.	Circulation of Current around a Two-pole Electromagnet	37
16.	Diagram illustrating Relation of Magnetizing Circuit and Resulting Magnetic Force	37
17.	Lines of Force running through Bar Magnet	41
18.	Filing-figure of the Bar Magnet	43
19.	Magnetizing Coil wound around a Magnetic Circuit	45
20.	Action of Magnetic Field on Conductor carrying Current	47
21.	Rotation of Conductor of Current around Magnetic Pole	48
22.	Bar Electromagnet	50
23.	Typical Two-pole Electromagnet	51
24.	Club-foot Electromagnet	51
25.	Horse-shoe Electromagnet with one Coil on Yoke	51
26.	Iron-clad Electromagnet	52
27.	Annular Iron-clad Electromagnet	53
28.	Ruhmkorff's Electromagnet	54
29.	Coil-and-Plunger	55
39.	Stopped Coil-and-Plunger (Bonelli's Electromagnet)	55
31.	Ring Electromagnet with Consequent Poles	56
32.	Circular Electromagnet	57
33.	Faggot of Electromagnets	57
34.	Spiral Electromagnet	58

FIG.		PAGE
35.	Curves of Magnetization of different Magnetic Materials	68
36.	Ring Method of Measuring Permeability (Rowland's Arrangement)	69
37.	Bosanquet's Data of Magnetic Properties of Iron and Steel Rings	72
38.	Hopkinson's Divided Bar Method of Measuring Magnetic Permeability	73
39.	Curves of Magnetization of Iron	75
40.	Curves of Magnetic Properties of Annealed Wrought Iron, plotted from Table IV.	77
41.	Curves of Permeability	78
42.	The Permeameter	81
43.	Curve of Magnetization of Magnetic Circuit with Air-gap	85
44.	Magnetization of Soft Iron Rods of Various Lengths	87
45.	Ewing's Curves for Effect of Joints	90
46.	Magnetization of Mild Steel at Various Temperatures	93
47.	Relation between Permeability (in weak field) and Temperature in Hard Steel	94
48.	Cycle of Magnetic Operations of Annealed Steel Wire	97
49.	Work Done in Increasing Magnetization, and Restored in Decreasing	99
50.	Magnetic Cycle for Annealed Wrought Iron	100
51.	Hysteresis in Wrought Iron and in Steel	101
52.	Magnetic Flux	104
53.	Bosanquet's Verification of the Law of Traction	120
54.	Stumpy Electromagnet	126
55.	Experiment on Rounding Ends	132
56.	Experiment of Detaching Armature	133
57.	Contrasted Effect of Flat and Pointed Poles	136
58.	Ayrton's Apparatus for Measuring Surface Distribution of Permanent Magnetism	138
59.	Dub's Experiments with Pole-pieces	143
60.	Dub's Deflexion Experiment	144
61.	Deflecting a Steel Magnet, Pole-piece on Near End	145
62.	Deflecting a Steel Magnet, Pole-piece on Distant End	145
63.	Electromagnet and Iron Jacket	148
64.	Experiment with Tubular Core and Iron Ring	149
65.	Experiment with Iron Disk on Pole of the Electromagnet	150
66.	Exploring Polar Distribution with small Iron Bar	151
67.	Pole of Electromagnet explored by Vom Kolke	153
68.	Iron Ball attached to Edge of Polar Face	154
69.	Apparatus to illustrate the Law of Inverse Squares	159
70.	Deflexion of Needle caused by Bar Magnet Broadside-on	161
71.	Closed Magnetic Circuit	163
72.	Divided Magnetic Circuit	163
73.	Electromagnet with Armature in Contact	164

List of Illustrations.   xvii

FIG.		PAGE
74.	Electromagnet with narrow Air-gaps	165
75.	Electromagnet with wider Gaps	166
76.	Electromagnet without Armature	167
77.	Experiment on Leakage of Electromagnet	169
78.	Curves of Magnetization, plotted from preceding	172
79.	Curves of Flow of Magnetic Lines in Air from one Cylindrical Pole to another	175
80.	Diagram of Leakage Reluctances	176
81.	Von Feilitzsch's Curves of Magnetization of Rods of Various Diameters	180
82.	Von Feilitzsch's Curves of Magnetization of Tubes	184
83.	Hughes's Electromagnet	187
84.	Du Moncel's Experiment on Armatures	189
85.	Three-branch Electromagnet	207
86.	Electromagnet, with Multiple Cores	208
87.	Multiple Cores and Single Core	208
88.	Experiment with Permanent Magnet	211
89.	Club-footed Electromagnet	217
90.	Jensen's Electric Bell	218
91.	Curves of Rise of Currents	223
92.	Curves of Rise of Current, with Different Groupings of Battery	228
93.	Electromagnets of Relay and their Effects	230
94.	P.O. Relay, "A" pattern	231
95.	P.O. Relay, "B" pattern	232
96.	P.O. Relay, "C" pattern	232
97.	Siemens' Relay	233
98.	Cores of Exchange Company's Electromagnets	334
99.	Electromagnet of Marcel Deprez's Chronograph, No. 1, full size	237
100.	Polarized Electromagnet for Deprez's Chronograph	237
101.	Electromagnet of Smith's Chronograph	238
102.	Curve of Loss of Magnetism	239
103.	Experiment with Coil-and-Plunger	241
104.	Vertical Coil-and-Plunger	242
105.	Hjörth's Electromagnetic Mechanism	243
106.	Action of Single Coil on Point-pole in Axis	246
107.	Action along Axis of Single Coil	246
108.	Action of Tubular Coil	248
109.	Diagram of Force and Work of Coil-and-Plunger	251
110.	Von Feilitzsch's Experiment on Plungers of Iron and Steel	258
111.	Bruger's Experiments on Coils and Plungers	259
112.	Bruger's Experiments, using Currents of Various Strengths	260
113.	Hollow *versus* Solid Cores	263
114.	Iron-clad Coil	264
115.	Two forms of Differential Coil-and-Plunger	265

List of Illustrations.

FIG.		PAGE
116.	Curve of Forces in Mechanism of Pilsen Arc Lamp	266
117.	Differential Plungers of the Brockie-Pell Arc Lamp	266
118.	Menges' Arc Lamp	267
119.	Electromagnet of Brush Arc Lamp	270
120.	Plunger Electromagnet of Stevens and Hardy	270
121.	Mechanism of Kennedy's Arc Lamp	271
122.	Holroyd Smith's Electromagnet	272
123.	Roloff's Electromagnet	272
124.	Ayrton and Perry's Tubular Ironclad Electromagnet	273
125.	Gaiser's Long-range Armature	273
126.	E. Davy's Mode of Controlling Armature by a Spring	279
127.	Callaud's Equalizer	280
128.	Robert Houdin's Equalizer	280
129.	Mechanism of Duboscq's Arc Lamp	281
130.	Froment's Equalizer with Stanhope Lever	282
131.	Froment's Equalizer depending on Oblique Approach	283
132.	Use of Shaped Polar Extension	283
133.	Use of Pierced Armature and Coned Pole-piece	284
134.	Oblique Approach between Electromagnet and Mass of Iron	285
135.	Hinged Electromagnets	286
136.	Interlocking Electromagnets	287
137.	Plunger Core made in Separate Joints	288
138.	Repulsion between two Parallel Cores	288
139.	Electromagnetic Mechanism, working by Repulsion	288
140.	Electromagnetic Pop-gun	289
141.	Magnetic Elongation of Double Core of Iron	290
142.	Polarized Mechanism of Sturgeon's Electromagnetic Telegraph	293
143.	Mechanism of Polarized Trembling Bell	294
144.	Abdank's Polarized Bell	295
145.	Hughes's Electromagnet Mechanism	296
146.	Bain's Moving-coil Mechanism	298
147.	Doubrava's Mechanism with Sliding Coils	299
148.	D'Arsonval's Telephonic Receiver	300
149.	Ader's Telephonic Receiver	300
150.	Mechanism of Evershed's Ampere-meters	302
151.	Shaped Iron Armature between Poles of Electromagnet	302
152.	Curved Plunger Core and Tubular Coil	302
153.	Siemens' Form of Pivoted Armature	303
154.	Waterhouse's Form of Pivoted Armature	303
155.	Telephonic Receiver with Magnetic Shunt	303
156.	D'Arlincourt's Relay	305
157.	Magnetic Circuit of d'Arlincourt's Relay	306
158.	Magnetic Circuit of d'Arlincourt's Relay, after current is cut off	306
159.	Electromagnetic Adherence of Wheel to Rail	307
160.	Nicklès' Magnetic Friction-gear	308

List of Illustrations.

FIG.		PAGE
161.	Magnetic Separator ..	309
162.	Wynne and Raworth Electromagnetic Clutch ..	311
163.	Forbes and Timmis Electromagnetic Railway Brake ..	312
164.	Colombet's Mechanism	313
165.	Mechanism of Wagener's Electric Bell	314
166.	Mechanism of Draw-up Indicator	315
167.	Thorpe's Semaphore Indicator	316
168.	Moseley's Indicator..	316
169.	Polarized Indicator Movement	317
169a.	Tripolar Electromagnet from Gent's Indicator ..	317
170.	Mechanism of Ordinary Electric Trembling Bell	319
171.	Short-circuit Bell Mechanism	322
172.	Vibrating Break of Induction Coil	323
173.	Spottiswoode's Rapid Break	324
174.	Lacour's Apparatus ..	325
175.	Elisha Gray's Vibrator	328
176.	Langdon-Davies' Rate-governor	330
177.	Laminated Iron Cores	332
178.	Two Waves in Similar Phase	334
179.	Two Alternate Currents in Opposite Phases	335
180.	Two Alternate Currents in Quadrature	335
181.	Action of Alternate Current Electromagnet on Copper Ring ..	337
182.	Effect on Sheet of Copper of turning Electromagnet on and off	339
183.	Repulsion of a Copper Ring	340
184.	Copper Ring tethered to Table, floating above the Alternate Current Electromagnet	341
185.	Lateral Repulsion of Copper Ring	341
186.	Displacement of Copper Ring towards Neutral Zone of Electromagnet	342
187.	Deflection of Pivoted Ring by Alternate Current Electromagnets	343
188.	Rotation produced by Shading half the Pole	344
189.	Rotation due to Dissymmetrical Induction of Eddy-currents ..	344
190.	Rankin Kennedy's Electromagnet for Heating purposes	348
191.	Ritchie's Motor	351
192.	Jacobi's Motor	352
193.	Page's Double-beam Engine	352
194.	Froment's Motor	353
195.	Bourbouze's Engine..	354
196.	Roux's Motor	354
197.	Immisch's Motor	356
198.	Field-magnet of Alternate-current Machine	356
199.	Field-magnet of Mordey's Alternator	357
200.	Marcel Deprez's Electromagnetic Hammer	360
201.	Rowan's Electromagnetic Riveter	362

List of Illustrations.

FIG.		PAGE
202.	Short-circuit working	370
203.	Differential winding ..	371
204.	Multiple wire winding	371
205.	Compensating the Self-Induction of an Electromagnet	373
206.	Special Electromagnet for Ophthalmic Surgery ..	376
207.	Van Wetteren's Magnet	405
208.	Unipolar Magnets ..	409
209.	Compensating Magnet	409
210.	Astatically Balanced Magnet	410
211.	Lebailliff's Sideroscope	410
212.	Nobili's Pair ..	410
213.	Vertical Astatic Pair	410

THE ELECTROMAGNET.

CHAPTER I.

HISTORICAL INTRODUCTION.

THE effect which an electric current, flowing in a wire, can exercise upon a neighbouring compass needle was discovered by Oersted in 1820.[*] This first announcement of the possession of magnetic properties by an electric current was followed speedily by the researches of Ampère,[†] Arago,[‡] Davy,[§] and by the devices of several other experimenters, including De la Rive's [||] floating battery and coil, Schweigger's [¶] multiplier, Cumming's [**] galvanometer, Faraday's [††] apparatus for rotation of a permanent magnet, Marsh's [‡‡] vibrating pendulum, and Barlow's [§§] rotating star-wheel. But it was not until 1825 that the electromagnet was invented. Arago announced, on 25th September, 1820, that a copper wire uniting the poles of a voltaic cell, and consequently traversed by an electric current, could attract iron filings to itself laterally. In the same communication he described how he had succeeded in communicating permanent magnetism to steel needles laid at right angles to the copper wire, and how, on showing this experiment to Ampère, the latter had suggested that the magnetizing action would be more intense

---

[*] See Thomson's *Annals of Philosophy*, Oct. 1820 ; see translation of original paper in *Journ. Soc. Telegr. Engineers*, v. p. 464, 1876.
[†] *Ann. de Chim. et de Physique*, xv. pp. 59 and 170, 1820.
[‡] *Ibid.*, xv. p. 93, 1820.   [§] *Phil. Trans.*, 1821.
[||] *Bibliothèque Universelle*, March 1821.
[¶] *Ibid.*   [**] *Camb. Phil. Trans.*, 1821.
[††] *Quarterly Journal of Science*, Sept. 1821.
[‡‡] Barlow's *Magnetic Attractions*, 2nd edition, 1823.   [§§] *Ibid.*

if for the straight copper wire there were substituted one wrapped in a helix, in the centre of which the steel needle might be placed. This suggestion was at once carried out by the two philosophers. "A copper wire wound in a helix was terminated by two rectilinear portions which could be adapted, at will, to the opposite poles of a powerful horizontal voltaic pile; a steel needle wrapped up in paper was introduced into the helix." "Now, after some minutes' sojourn in the helix, the steel needle had received a sufficiently strong dose of magnetism." Arago then wound upon a little glass tube some short helices, each about $2\frac{1}{4}$ inches long, coiled alternately right-handedly and left-handedly, and found that on introducing into the glass tube a steel wire, he was able to produce "consequent poles" at the places where the winding was reversed. Ampère, on October 23rd, 1820, read a memoir, claiming that these facts confirmed his theory of magnetic actions. Davy had, also, in 1820, surrounded with temporary coils of wire the steel needles upon which he was experimenting, and had shown that the flow of electricity around the coil could confer magnetic power upon the steel needles. From these experiments it was a grand step forward to the discovery that a core of soft iron, surrounded by its own appropriate coil of copper, could be made to act not only as a powerful magnet, but as a magnet whose power could be turned on or off at will, could be augmented to any desired degree, and could be set into action and controlled from a practically unlimited distance.

The electromagnet, in the form which can first claim recognition for these qualities, was devised by William Sturgeon,[*] and is described by him in the paper which he contributed to the Society of Arts in 1825, accompanying a set of improved apparatus for electromagnetic experiments.[†] Amongst this set of apparatus are two electromagnets, one of horse-shoe shape (Figs. 1 and 2), and one a straight bar (Fig. 3). It will be seen that the former figures represent an electromagnet consisting of a bent iron

---

[*] See Appendix A for a biographical notice of William Sturgeon.
[†] *Trans. Society of Arts*, 1825, xliii. p. 38.

rod about 1 foot long and ½ inch in diameter, varnished over and then coiled with a single left-handed spiral of stout uncovered copper wire of eighteen turns. This coil was found appropriate to the particular battery which Sturgeon preferred, namely, a single cell containing a spirally enrolled pair of zinc and copper plates of large area (about 130 square inches) immersed in acid; which cell having small internal

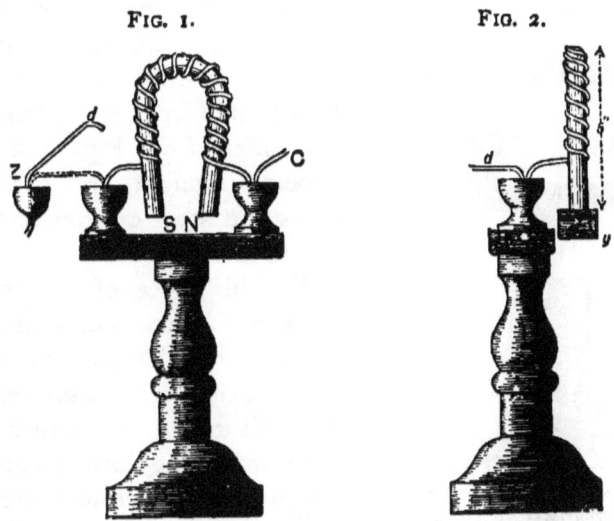

Fig. 1.   Fig. 2.

Sturgeon's First Electromagnet.

resistance, would yield a large quantity of current when connected to a circuit of small resistance. The ends of the copper wire were brought out sideways and bent down so as to dip into two deep connecting cups, marked Z and C, fixed upon a wooden stand. These cups, which were of wood, served as supports to hold up the electromagnet, and, having mercury in them, served also to make good electrical connexion. In Fig. 2 the magnet is seen sideways, supporting a bar of iron, y. The circuit was completed to the battery through a connecting wire, d, which could be lifted out of the cup, Z, so breaking circuit when desired, and allowing the weight to drop. Sturgeon added, in his explanatory remarks, that the poles, N and S, of the magnet will be reversed if you wrap the copper wire about the rod as a right-handed screw, instead of a left-handed one, or, more simply, by reversing the connexions with the battery, by causing the wire that dips into the Z cup to dip into the C cup, and *vice versâ*. This electromagnet was capable of supporting 9 lb. when thus excited.

Fig. 3 shows another arrangement to fit on the same stand. "This arrangement communicates magnetism to hardened steel bars as soon as they are put in, and renders soft iron within it magnetic during the time of action; it only differs from Figs. 1 and 2 in being straight, and thereby allows the steel or iron bars to slide in and out."

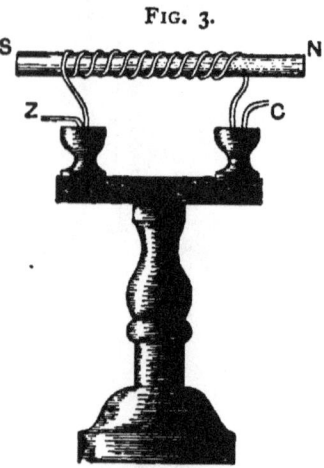

FIG. 3.

STURGEON'S STRAIGHT BAR ELECTROMAGNET.

For this piece of apparatus and other adjuncts accompanying it, all of which are described in the Society's *Transactions* for 1825, Sturgeon, in return for the award of the Society's medal and premium, deposited the apparatus in the museum of the Society, which therefore might be supposed to be the proud possessor of the first electromagnet ever constructed. Alas for the vanity of human affairs, the Society's museum of apparatus has long been dispersed, this priceless relic having been either made over to the now defunct Patent Office Museum, or otherwise disposed of.

Sturgeon's first electromagnet, the core of which weighed only about 7 oz., was able to sustain a load of 9 lb., or about twenty times its own weight. At the time it was considered a truly remarkable performance. Its single layer of stout copper wire was well adapted to the single cell battery employed. The same weight of copper in the form of a fine wire would have produced no better result. Subsequently, in the hands of Joule, the same electromagnet sustained a load of 50 lb., or about 114 times its own weight. Writing in 1832 about his apparatus of 1825, Sturgeon used the following magniloquent language :—

"When first I showed that the magnetic energies of a galvanic conducting wire are more conspicuously exhibited by exercising them on soft iron than on hard steel, my experiments were limited to small masses—generally to a few inches of rod iron about half an inch in diameter. Some of those pieces were employed while straight, and others were bent into the form of a horse-shoe magnet, each piece being encompassed by a spiral conductor of copper wire. The magnetic energies developed by these simple arrangements are of a very distinguished and exalted character, as is conspicuously manifested by the suspension of a considerable weight at the poles during the period of excitation by the electric influence.

"An unparalleled transiliency of magnetic action is also displayed in soft iron, by an instantaneous transition from a state of total inactivity to that of vigorous polarity, and also by a simultaneous reciprocity of polarity in the extremities of the bar—versatilities in this branch of physics for the display of which soft iron is pre-eminently qualified, and which, by the agency of electricity, become demonstrable with the celerity of thought, and illustrated by experiments the most splendid in magnetics. It is, moreover, abundantly manifested by ample experiments; that galvanic electricity exercises a superlative degree of excitation on the latent magnetism of soft iron, and calls forth its recondite powers with astonishing promptitude, to an intensity of action far surpassing anything which can be accomplished by any known application of the most vigorous permanent magnet, or by any other mode of experimenting hitherto discovered. It has been observed, however, by experimenting on different pieces selected from various sources, that, notwithstanding the greatest care be observed in preparing them of a uniform figure and dimensions, there appears a considerable difference in the susceptibility which they individually possess of developing the magnet powers, much of which depends upon the manner of treatment at the forge, as well as upon the natural character of the iron itself.*

---

* "I have made a number of experiments on small pieces, from the results of which it appears that much hammering is highly detrimental to the development of magnetism in soft iron, whether the exciting cause be galvanic or any other. And although good annealing is always essential, and facilitates to a considerable extent the display of polarity, that process is very far from restoring to the

"The superlative intensity of electromagnets, and the facility and promptitude with which their energies can be brought into play, are qualifications admirably adapted for their introduction into a variety of arrangements in which powerful magnets so essentially operate, and perform a distinguished part in the production of electromagnetic rotations; whilst the versatilities of polarity of which they are susceptible are eminently calculated to give a pleasing diversity in the exhibition of that highly interesting class of phenomena, and lead to the production of others inimitable by any other means."*

Sturgeon's further work during the next three years is best described in his own words:—

"It does not appear that any very extensive experiments were attempted to improve the lifting powers of electromagnets, from the time that my experiments were published in the *Transactions of the Society of Arts*, &c., for 1825, till the latter part of 1828. Mr. Watkins, philosophical instrument maker, Charing Cross, had, however, made them of much larger size than any which I had employed, but I am not aware to what extent he pursued the experiment.

"In the year 1828, Professor Moll, of Utrecht, being on a visit to London, purchased of Mr. Watkins an electromagnet weighing about 5 lb., at that time I believe the largest which had been made. It was of round iron, about one inch in diameter, and furnished with a single copper wire twisted round it eighty-three times. When this magnet was excited by a large galvanic surface, it supported about 75 lb. Professor Moll afterwards prepared another electromagnet, which, when bent, was $12\frac{1}{2}$ inches high, $2\frac{1}{2}$ inches in diameter, and weighed about 26 lb.; prepared like the former with a single spiral conducting wire. With an acting galvanic surface of 11 square feet, this magnet would support 154 lb., but would not lift an anvil which weighed 200 lb.

"The largest electromagnet which I have yet (1832) exhibited in my lectures weighs about 16 lb. It is formed of a small bar of

---

iron that degree of susceptibility which it frequently loses by the operation of the hammer. Cylindric rod iron of small dimensions may very easily be bent into the required form without any hammering whatever; and I have found that small electromagnets made in this way display the magnetic powers in a very exalted degree."

* Sturgeon's *Scientific Researches*, p. 113.

soft iron, 1½ inch across each side; the cross piece, which joins the poles, is from the same rod of iron, and about 3¾ inches long. Twenty separate strands of copper wire, each strand about 50 feet in length, are coiled round the iron, one above another, from pole to pole, and separated from each other by intervening cases of silk : the first coil is only the thickness of one ply of silk from the iron; the twentieth, or outermost, about half an inch from it. By this mean the wires are completely insulated from each other without the trouble of covering them with thread or varnish. The ends of wire project about 2 feet for the convenience of connection. With one of my small cylindrical batteries, exposing about 150 square inches of total surface, this electromagnet supports 400 lb. I have tried it with a larger battery, but its energies do not seem to be so materially exalted as might have been expected by increasing the extent of galvanic surface. Much depends upon a proper acid solution; good nitric or nitrous acid, with about six or eight times its quantity of water, answers very well. With a new battery of the above dimensions and a strong solution of salt and water at a temperature of 190° Fahr., the electromagnet supported between 70 and 80 lb. when the first seventeen coils only were in the circuit. With the three exterior coils alone in the circuit, it would just support the lifter, or cross piece. When the temperature of the solution was between 40° and 50°, the magnetic force excited was comparatively very feeble. With the innermost coil alone and a strong acid solution, this electromagnet supports about 100 lb.; with the four outermost wires about 250 lb. It improves in power with every additional coil until about the twelfth, but not perceptibly any further; therefore the remaining eight coils appear to be useless, although the last three, independently of the innermost seventeen, and at the distance of half an inch from the iron, produce in it a lifting power of 75 lb.

"Mr. Marsh has fitted up a bar of iron much larger than mine, with a similar distribution of the conducting wires to that devised and so successfully employed by Professor Henry. Mr. Marsh's electromagnet will support about 560 lb. when excited by a galvanic battery similar to mine. These two, I believe, are the most powerful electromagnets yet produced in this country.

"A small electromagnet, which I also employ on the lecture-table, and the manner of its suspension, is represented by Fig. 3, Plate VI. (Fig. 4 in this work).

"The magnet is of cylindric rod iron, and weighs 4 ounces : its poles are about a quarter of an inch asunder. It is furnished with six coils of wire in the same manner as the large electromagnet before described, and will support upwards of 50 lb.

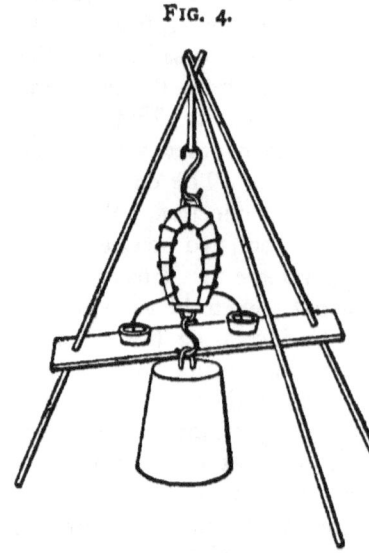

Fig. 4.

Sturgeon's Lecture-table Electromagnet.

"I find a triangular gin very convenient for the suspension of the magnet in these experiments. A stage, A, A, of thin board, supporting two wooden dishes, C and Z, is fastened, at a proper height, to two of the legs of the gin. Mercury is placed in these vessels, and the dependent amalgamated extremities of the conducting wires dip into it—one into each portion.

"The vessels are sufficiently wide to admit of considerable motion of the wires in the mercury without interrupting the contact, which is sometimes occasioned by the swinging of the magnet and attached weight. The circuit is completed by other wires, which connect the battery with these two portions of mercury. When the weight is supported as in the figure, if an interruption be made by removing either of the connecting wires, the weight instantaneously drops on the table. The large magnet I suspend in the same way on a larger gin; the weights which it supports are placed one after another on a square board, suspended by means of a cord at each corner from a hook in the cross piece, which joins the poles of the magnet.

"With a new battery, and a solution of salt and water at a temperature of 190° Fahr., the small electromagnet, Fig. 3, supports 16 lb."

In 1840, after Sturgeon had removed to Manchester, where he assumed the management of the "Victoria Gallery of Practical Science," he continued his work, and in the seventh memoir in his series of *Researches* he wrote as follows:—

"The electromagnet belonging to this Institution is made of a cylindrical bar of soft iron, bent into the form of a horse-shoe magnet, having the two branches parallel to each other, and at the distance of 4·5 inches. The diameter of the iron is 2·75 inches, it is 18 inches long when bent. It is surrounded by fourteen coils of copper wire, seven on each branch. The wire which constitutes the coils is $\frac{1}{17}$th of an inch in diameter, and in each coil there are about 70 feet of wire. They are united in the usual way with branch wires, for the purpose of conducting the currents from the battery. The magnet was made by Mr. Nesbit. . . . The greatest weight sustained by the magnet in these experiments is 12¾ cwt., or 1386 lb., which was accomplished by sixteen pairs of plates, in four groups of four pairs in series each. The lifting power by nineteen pairs in series was considerably less than by ten pairs in series; and but very little greater than that given by one cell or one pair only. This is somewhat remarkable, and shows how easily we may be led to waste the magnetic powers of batteries by an injudicious arrangement of its elements." *

It was not until three years after Sturgeon's invention that any notice of it was taken in Germany. In 1828, Pohl† showed a small electromagnet in Berlin. In 1830, Pfaff‡ described one of Sturgeon's, made by Watkins, which he had seen on a recent visit to London.

At the date of Sturgeon's work the laws governing the flow of electric currents in wires were still obscure. Ohm's epoch-making enunciation of the law of the electric circuit appeared in *Poggendorff's Annalen* in the very year of Sturgeon's discovery, 1825, though his complete book was published only in 1827, and his work, translated by Dr. Francis into English, appeared (in Taylor's *Scientific Memoirs*, vol. ii.), only in 1841. Without the guidance of Ohm's law it was not strange that even the most able experimenters should not understand the relations between battery and circuit which would give them the best effects. These had to be found by the painful method of trial and failure. Pre-eminent amongst those who tried was Professor Joseph Henry, then of the Albany Institute, in New York, later of Princeton, New

---

* Sturgeon's *Scientific Researches*, p. 188.
† See Von Feilitzsch's *Lehre vom galvanischem Strome*, p. 95.
‡ *Schweigger's Journal*, lviii. p. 273, 1830.

Jersey, who not only tried, but succeeded in effecting an important improvement. In 1828, led on by a study of the "multiplier" (or galvanometer,) he proposed to apply to electromagnetic apparatus the device of winding them with a spiral coil of wire "closely turned on itself," the wire being of copper from $\frac{1}{40}$th to $\frac{1}{25}$th of an inch in diameter, covered with silk. In 1831 he thus describes* the results of his experiments:—

"A round piece of iron, about $\frac{1}{4}$ inch in diameter, was bent into the usual form of a horse-shoe, and instead of loosely coiling around it a few feet of wire, as is usually described, it was tightly wound with 35 feet of wire, covered with silk, so as to form about 400 turns; a pair of small galvanic plates, which could be dipped into a tumbler of diluted acid, was soldered to the ends of the wire, and the whole mounted on a stand. With these small plates the horse-shoe became much more powerfully magnetic than another of the same size and wound in the same manner, by the application of a battery composed of twenty-eight plates of copper and zinc, each 8 inches square. Another convenient form of this apparatus was contrived by winding a straight bar of iron, 9 inches long, with 35 feet of wire, and supporting it horizontally on a small cup of copper containing a cylinder of zinc; when this cup, which served the double purpose of a stand and the galvanic element, was filled with dilute acid, the bar became a portable electromagnet. These articles were exhibited to the Institute in March, 1829. The idea afterwards occurred to me, that a sufficient quantity of galvanism was furnished by the two small plates to develop, by means of the coil, a much greater magnetic power in a larger piece of iron. To test this, a cylindrical bar of iron, half an inch in diameter, and about 10 inches long, was bent into the shape of a horse-shoe, and wound with 30 feet of wire; with a pair of plates containing only $2\frac{1}{8}$ square inches of zinc, it lifted 15 lb. avoirdupois. At the same time, a very material improvement in the formation of the coil suggested itself to me on reading a more detailed account of Professor Schweigger's galvanometer, and which was also tested with complete success upon the same horse-shoe; it consisted in using several strands of wire, each covered with silk, instead of one. Agreeably to this construction, a second wire, of the same length as the first, was wound over it, and the ends soldered to the zinc and

* *Silliman's American Journal of Science*, Jan. 1831, xix. p. 400.

copper in such a manner that the galvanic current might circulate in the same direction in both, or, in other words, that the two wires might act as one; the effect by this addition was doubled, as the horse-shoe, with the same plates before used, now suppported 28 lb.

"With a pair of plates 4 inches by 6 inches, it lifted 39 lb., or more than fifty times its own weight.

"These experiments conclusively proved that a great development of magnetism could be effected by a very small galvanic element, and also that the power of the coil was materially increased by multiplying the number of wires without increasing the length of each." *

Not content with these results, Professor Henry pushed forward on the line he had thus struck out. He was keenly desirous to ascertain how large a magnetic force he could produce when using only currents of such a degree of smallness as could be transmitted through the comparatively thin copper wires such as bell-hangers use. During the year 1830 he made great progress in this direction, as the following extracts show:—

"In order to determine to what extent the coil could be applied in developing magnetism in soft iron, and also to ascertain, if possible, the most proper length of the wires to be used, a series of experiments were instituted jointly by Dr. Philip Ten Eyck and myself. For this purpose 1060 feet (a little more than one-fifth of a mile) of copper wire of the kind called bell-wire, ·045 of an inch in diameter, were stretched several times across the large room of the Academy.

"*Experiment* 1.—A galvanic current from a single pair of plates of copper and zinc 2 inches square was passed through the whole length of the wire, and the effect on a galvanometer noted. From the mean of several observations the deflexion of the needle was 15°.

"*Experiment* 2.—A current from the same plates was passed through half the above length, or 530 feet of wire; the deflexion in this instance was 21°.

"By a reference to a trigonometrical table, it will be seen that the natural tangents of 15° and 21° are very nearly in the ratio of the square roots of 1 and 2, or of the relative lengths of the wires in these two experiments.

* *Scientific Writings of Joseph Henry*, p. 39.

"The length of the wire forming the galvanometer may be neglected, as it was only 8 feet long.

"*Experiment* 3.—The galvanometer was now removed, and the whole length of the wire, attached to the ends of the wire of a small soft iron horse-shoe a $\frac{1}{4}$ inch in diameter, and wound with about 8 feet of copper wire with a galvanic current from the plates used in Experiments 1 and 2. The magnetism was scarcely observable in the horse-shoe.

"*Experiment* 4.—The small plates were removed, and a battery composed of a piece of zinc plate 4 inches by 7 inches, surrounded with copper, was substituted. When this was attached immediately to the ends of the 8 feet of wire wound round the horse-shoe, the weight lifted was $4\frac{1}{2}$ lb.; when the current was passed through the whole length of wire (1060 feet), it lifted about half an ounce.

"*Experiment* 5.—The current was passed through half the length of wire (530 feet) with the same battery; it then lifted 2 oz.

"*Experiment* 6.—Two wires of the same length as in the last experiment were used, so as to form two strands from the zinc and copper of the battery; in this case the weight lifted was 4 oz.

"*Experiment* 7.—The whole length of the wire was attached to a small trough on Mr. Cruickshanks' plan, containing twenty-five double plates, and presenting exactly the same extent of zinc surface to the action of the acid as the battery used in the last experiment. The weight lifted in this case was 8 oz.; when the intervening wire was removed, and the trough attached directly to the ends of the wire surrounding the horse-shoe, it lifted only 7 oz. . . .

"It is possible that the different states of the trough, with respect to dryness, may have exerted some influence on this remarkable result; but that the effect of a current from a trough, if not increased, is but slightly diminished in passing through a long wire is certain. . .

"But be this as it may, the fact that the magnetic action of current from a trough is, at least, not sensibly diminished by passing through a long wire is directly applicable to Mr. Barlow's project of forming an electromagnetic telegraph; and it is also of material consequence in the construction of the galvanic coil. From these experiments it is evident that in forming the coil we may either use one very long wire or several shorter ones, as the circumstances may require; in the first case, our galvanic combinations must consist of a number of plates, so as to give "projectile force"; in the second it must be formed of a single pair.

"In order to test on a large scale the truth of these preliminary results, a bar of soft iron, 2 inches square and 20 inches long, was bent into the form of a horse-shoe, 9½ inches high; the sharp edges of the bar were first a little rounded by the hammer; it weighed 21 lb.; a piece of iron from the same bar, weighing 7 lb., was filed perfectly flat on one surface, for an armature or lifter; the extremities of the legs of the horse-shoe were also truly ground to the surface of the armature: around this horse-shoe 540 feet of copper bell-wire were wound in nine coils of 60 feet each; these coils were not continued around the whole length of the bar, but each strand of wire, according to the principle before mentioned, occupied about 2 inches, and was coiled several times backward and forward over itself; the several ends of the wires were left projecting and all numbered, so that the first and last end of each strand might be readily distinguished. In this manner we formed an experimental magnet on a large scale, with which several combinations of wire could be made by merely uniting the different projecting ends. Thus, if the second end of the first wire be soldered to the first end of the second wire, and so on through all the series, the whole will form a continued coil of one long wire.

"By soldering different ends the whole may be formed into a double coil of half the length, or into a triple coil of one-third the length, &c. The horse-shoe was suspended in a strong rectangular wooden frame, 3 feet 9 inches high and 20 inches wide; an iron bar was fixed below the magnet, so as to act as a lever of the second order; the different weights supported were estimated by a sliding weight, in the same manner as with a common steel-yard (see sketch). In the experiments immediately following (all weights being avoirdupois), a small single battery was used, consisting of two concentric copper cylinders with zinc between them; the whole amount of zinc surface exposed to the acid from both sides of the zinc was two-fifths of a square foot; the battery required only half a pint of dilute acid for its submersion.

"*Experiment* 8.—Each wire of the horse-shoe was soldered to the battery in succession, one at a time; the magnetism developed by each was just sufficient to support the weight of the armature, weighing 7 lb.

"*Experiment* 9.—Two wires, one on each side of the arch of the horse-shoe, were attached; the weight lifted was 145 lb.

Experiment 10.—With two wires, one from each extremity of the legs, the weight lifted was 200 lb.

"*Experiment* 11.—With three wires, one from each extremity of the legs and one from the middle of the arch, the weight supported was 300 lb.

"*Experiment* 12.—With four wires, two from each extremity, the weight lifted was 500 lb. and the armature; when the acid was

FIG. 5.

HENRY'S ELECTROMAGNET.

This figure, copied from the *Scientific American*, December 11, 1880, represents Henry's electromagnet, still preserved in Princeton College. The other apparatus at the foot, including a current-reverser and the ribbon-coil used in the famous experiments on secondary and tertiary currents, were mostly constructed by Henry's own hands.

removed from the zinc, the magnet continued to support for a few minutes 130 lb.

"*Experiment* 13.—With six wires the weight supported was 570 lb.; in all these experiments the wires were soldered to the galvanic element; the connection in no case was formed with mercury.

"*Experiment* 14.—When all the wires (nine in number) were attached, *the maximum weight lifted was* 650 *lb.*, and this astonishing result, it must be remembered, was produced by a battery containing only two-fifths of a square foot of zinc surface, and requiring only half a pint of dilute acid for its submersion.

"*Experiment* 15.—A small battery, formed with a plate of zinc 12 inches long and 6 inches wide, and surrounded by copper, was substituted for the galvanic elements used in the last experiment; the weight lifted in this case was 750 lb.

"*Experiment* 16.—In order to ascertain the effect of a very small galvanic element on this large quantity of iron, a pair of plates, exactly 1 inch square, was attached to all the wires; the weight lifted was 85 lb.

"The following experiments were made with wires of different lengths on the same horse-shoe :—

"*Experiment* 17.—With six wires, each 30 feet long, attached to the galvanic element, the weight lifted was 375 lb.

"*Experiment* 18.—The same wires used in last experiment were united so as to form three coils of 63 feet each; the weight supported was 200 lb. This result agrees nearly with that of Experiment 11, though the same individual wires were not used. From this it appears that six short wires are more powerful than three of double the length.

"*Experiment* 19.—The wires used in Experiment 10, but united so as to form a single coil of 120 feet of wire, lifted 60 lb.; while in Experiment 10 the weight lifted was 200 lb. This is a confirmation of the result in the last experiment. . . .

"In these experiments a fact was observed which appears somewhat surprising; when the large battery was attached, and the armature touching both poles of the magnet, it was capable of supporting more than 700 lb., but when only one pole was in contact it did not support more than 5 or 6 lb., and in this case we never succeeded in making it lift the armature (weighing 7 lb). This fact may perhaps be common to all large magnets, but we have never seen the circumstance noticed of so great a difference between a single pole and both. . . .

"A series of experiments was separately instituted by Dr. Ten Eyck, in order to determine the maximum development of magnetism in a small quantity of soft iron.

"Most of the results given in this paper were witnessed by Dr. L. C. Beck, and to this gentleman we are indebted for several suggestions, and particularly that of substituting cotton well waxed for silk thread, which in these investigations became a very considerable item of expense. He also made a number of experiments with iron bonnet-wires, which, being found in commerce already wound, might possibly be substituted in place of copper. The result was that with very short wire the effect was nearly the same as with copper, but in coils of long wire with a small galvanic element it was not found to answer. Dr. Beck also constructed a horse-shoe of round iron 1 inch in diameter, with four coils on the plan before described. With one wire it lifted 30 lb., with two wires 60 lb., with three wires 85 lb., and with four wires 112 lb. While we were engaged in these investigations, the last number of the *Edinburgh Journal of Science* was received, containing Professor Moll's paper on 'Electromagnetism.' Some of his results are in a degree similar to those here described; his object, however, was different, it being only to induce strong magnetism on soft iron with a powerful galvanic battery. The principal object in these experiments was to produce the greatest magnetic force with the smallest quantity of galvanism. The only effect Professor Moll's paper has had over these investigations has been to hasten their publication; the principle on which they were instituted was known to us nearly two years since, and at that time exhibited to the Albany Institute."*

In the next number of *Silliman's Journal* (April 1831), Professor Henry gave "an account of a large electromagnet, made for the laboratory of Yale College." The core of the magnet weighed $59\frac{1}{2}$ lb.; it was forged under Henry's own direction, and wound by Dr. Ten Eyck. This magnet, wound with twenty-six strands of copper bell-wire of total length of 728 feet, and excited by two cells which exposed nearly $4\frac{7}{9}$ square feet of surface, readily supported on its armature, which weighed 23 lb., a load of 2063 lb.

Writing in 1857 of his earlier experiments, Henry speaks thus† of his ideas respecting the use of additional coils on the magnet and the increase of battery power:—

* *Scientific Writings of Joseph Henry*, i. p. 49.
† *Statement in Relation to the History of the Electromagnetic Telegraph* (from the *Smithsonian Annual Report* for 1857, p. 99); and *Scientific Writings*, ii. p. 435.

"To test these principles on a larger scale, the experimental magnet was constructed which is shown in Fig. 6. In this a number of compound helices was placed on the same bar, their ends left projecting, and so numbered that they could all be united into one long helix, or variously combined in sets of lesser length.

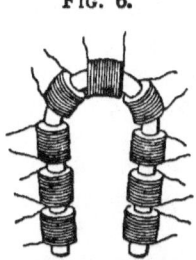

FIG. 6.

HENRY'S EXPERIMENTAL ELECTROMAGNET.

"From a series of experiments with this and other magnets, it was proved that, in order to produce the greatest amount of magnetism from a battery of a single cup, a number of helices is required; but when a compound battery is used, then one long wire must be employed, making many turns around the iron, the length of wire and consequently the number of turns being commensurate with the projectile power of the battery.

"In describing the results of my experiments the terms 'intensity' and 'quantity' magnets were introduced to avoid circumlocution, and were intended to be used merely in a technical sense. By the intensity magnet I designated a piece of soft iron so surrounded with wire that its magnetic power could be called into operation by an intensity battery; and by a quantity magnet, a piece of iron so surrounded by a number of separate coils that its magnetism could be fully developed by a quantity battery.

"I was the first to point out this connexion of the two kinds of the battery with the two forms of the magnet, in my paper in *Silliman's Journal*, January, 1831, and clearly to state that when magnetism was to be developed by means of a compound battery one long coil must be employed, and when the maximum effect was to be produced by a single battery a number of single strands should be used. . . . Neither the electromagnet of Sturgeon nor any electromagnet ever made previous to my investigations was applicable to transmitting power to a distance. . . . The electromagnet made by Sturgeon, and copied by Dana, of New York, was an imperfect quantity magnet,* the feeble power of which was developed by a single battery."

Finally, Henry sums up his own position as follows :—

"1. Previous to my investigations the means of developing

* This criticism is scarcely justified by the facts, since a low-resistance battery is appropriate to a coil of low resistance on the magnet.

magnetism in soft iron were imperfectly understood, and the electromagnet which then existed was inapplicable to transmissions of power to a distance.

"2. I was the first to prove by actual experiment that in order to develop magnetic power at a distance, a galvanic battery of 'intensity' must be employed to project the current through the long conductor, and that a magnet surrounded by many turns of one long wire must be used to receive this current.

"3. I was the first to actually magnetize a piece of iron at a distance, and to call attention to the fact of the applicability of my experiments to the telegraph.

"4. I was the first to actually sound a bell at a distance by means of the electromagnet.

"5. The principles I had developed were applied by Dr. Gale to render Morse's machine effective at a distance."

Though Henry's researches were published in 1831, they were for some years almost unknown in Europe. Until April, 1837, when Henry himself visited Wheatstone at his laboratory at King's College, the latter did not know how to construct an electromagnet that could be worked through a long wire circuit. Cooke, who became the coadjutor of Wheatstone, had originally come to him to consult him,* in February, 1837, about his telegraph and alarum, the electromagnets of which, though they worked well on short circuits refused to work when placed in circuit with even a single mile of wire. Wheatstone's own account† of the matter is extremely explicit :—" Relying on my former experience, I at once told Mr. Cooke that his plan would not and could not act as a telegraph, because sufficient attractive power could not be imparted to an electromagnet interposed in a long circuit; and to convince him of the truth of this assertion, I invited him to King's College to see the repetition of the experiments on which my conclusion was founded. He

---

* See Mr. Latimer Clark's account of Cooke in vol. viii. of *Journal of Society of Telegraph Engineers*, p. 374.

† W. F. Cooke, *The Electric Telegraph: Was it invented by Professor Wheatstone?* 1856–7, pt. ii. p. 87.

came, and after seeing a variety of voltaic magnets, which even with powerful batteries exhibited only slight adhesive attraction, he expressed his disappointment."

After Henry's visit to Wheatstone, the latter altered his tone. He had been using, *faute de mieux*, relay circuits to work the electromagnets of his alarum in a short circuit with a local battery. "These short circuits," he writes, "have lost nearly all their importance and are scarcely worth contending about since *my discovery*" (the italics are our own), "that electromagnets may be so constructed as to produce the required effects by means of the direct current, even in very long circuits." *

We pass on to the researches of the distinguished physicist of Manchester, whose decease we have lately had to deplore, Mr. James Prescott Joule. Sturgeon had removed, as mentioned above, in 1838, to Manchester, where his lectures on electromagnetism excited the attention of many younger men. Amongst them was Joule, who, fired by the work of Sturgeon, made most valuable contributions to the subject. Most of these were published either in Sturgeon's *Annals of Electricity* or in the *Proceedings of the Literary and Philosophical Society of Manchester*, but their most accessible form is the republished volume issued five years ago by the Physical Society of London.

In his earliest investigations he was endeavouring to work out the details of an electric motor. The following is an extract from his own account (*Reprint of Scientific Papers*, p. 7):—

"In the further prosecution of my inquiries, I took six pieces of round bar iron of different diameters and lengths, also a hollow cylinder, $\frac{1}{13}$th of an inch thick in the metal. These were bent in the U-form, so that the shortest distance between the poles of each was half an inch; each was then wound with 10 feet of covered copper wire, $\frac{1}{20}$th of an inch in diameter. Their attractive powers under like currents for a straight steel magnet, $1\frac{1}{2}$ inch long,

* *Ibid.*, p. 95.

suspended horizontally to the beam of a balance, were, at the distance of half an inch, as follows :—

	No. 1. Hollow.	No. 2. Solid.	No. 3. Solid.	No. 4. Solid.	No. 5. Solid.	No. 6. Solid.	No. 7. Solid.
Length round the bend, in inches	6	5½	2⅔	5¼	2½	5¼	2¼
Diameter, in inches	½	½	½	⅜	⅜	¼	¼
Attraction for steel magnet, in grains	7·5	6·3	5·1	5·0	4·1	4·8	3·6
Weight lifted, in ounces	36	52	92	36	52	20	28

"A steel magnet gave an attractive power of 23 grains, while its lifting power was not greater than 60 ounces.

"The above results will not appear surprising if we consider, first, the resistance which iron presents to the induction of magnetism, and, second, how very much the induction is exalted by the completion of the magnetic circuit.

"Nothing can be more striking than the difference between the ratios of lifting to attractive power at a distance in the different magnets. Whilst the steel magnet attracts with a force of 23 grains and lifts 60 ounces, the electromagnet No. 3 attracts with a force of only 5·1 grains, but lifts as much as 92 ounces.

"To make a good electromagnet for lifting purposes :—1st. Its iron, if of considerable bulk, should be compound, of good quality, and well annealed. 2nd. The bulk of the iron should bear a much greater ratio to its length than is generally the case. 3rd. The poles should be ground quite true, and fit flatly and accurately to the armature. 4th. The armature should be equal in thickness to the iron of the magnet.

"*In studying what form of electromagnet is best for attraction from a distance, two things must be considered, viz., the length of the iron and its sectional area.*

"*Now I have always found it disadvantageous to increase the length beyond what is needful for the winding of the covered wire.*"

These results were announced in March, 1839. In May of the same year he propounded a law of the mutual attraction of two electromagnets, as follows :—"*The attractive force of two electromagnets for one another is directly proportional to the square of the electric force to which the iron is exposed* ; or if E denote the electric current, W the length of wire, and M the magnetic attraction, $M = E^2 W^2$." The discrepancies which he himself observed he rightly attributed to the iron becoming saturated magnetically. In March, 1840, he extended this same law to the lifting power of the horse-shoe electromagnet.

In August, 1840, he wrote to the *Annals of Electricity*, on electromagnetic forces, dealing chiefly with some special electromagnets for traction. One of these possessed the form shown in Fig. 7. Both the magnet and the iron keeper were furnished with eye-holes for the purpose of suspension and measurement of the force requisite to detach the keeper. Joule thus writes about the experiments.*

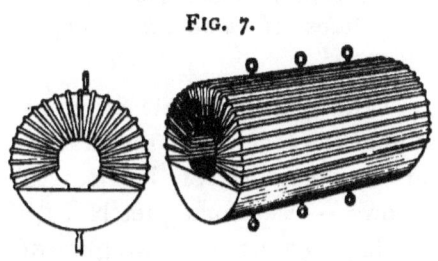

FIG. 7.

JOULE'S ELECTROMAGNET.

"I proceed now to describe my electromagnets, which I constructed of very different sizes in order to develop any curious circumstance which might present itself. A piece of cylindrical wrought iron, 8 inches long, had a hole one inch in diameter bored the whole length of its axis; one side was planed until the hole was exposed sufficiently to separate the thus-formed poles one-third of an inch. Another piece of iron, also 8 inches long, was then planed, and being secured with its face in contact with the other planed surface, the whole was turned into a cylinder 8 inches long, $3\frac{3}{4}$ inches in exterior, and 1 inch interior diameter. The larger piece was then covered with calico and wound with four copper wires covered with silk, each 23 feet long, and $\frac{1}{11}$th of an inch in diameter—a quantity just sufficient to hide the exterior surface and to fill the interior opened hole. . . . The above is designated No. 1; and the rest are numbered in the order of their description.

"I made No. 2 of a bar of $\frac{1}{2}$-inch round iron 2·7 inches long. It was bent into an almost semi-circular shape, and then covered with 7 feet of insulated copper wire, $\frac{1}{20}$th of an inch thick. The

---

* *Scientific Papers*, vol. i. p. 30.

poles are half an inch asunder; and the wire completely fills the space between them.

"A third electromagnet was made of a piece of iron 0·7 inch long, 0·37 inch broad, and 0·15 inch thick. Its edges were reduced to such an extent that the tranverse section was elliptical. It was bent into a semicircular shape, and wound with 19 inches of silked copper wire, $\frac{1}{40}$th of an inch in diameter.

"To procure a still more extensive variety, I constructed what might, from its extreme minuteness, be termed an *elementary electromagnet*. It is the smallest, I believe, ever made, consisting of a bit of iron wire $\frac{1}{4}$ of an inch long, and $\frac{1}{25}$th of an inch in diameter. It was bent into the shape of a sémi-circle, and was wound with three turns of *uninsulated* copper wire $\frac{1}{40}$th of an inch in thickness."

With these magnets experiments were made with various strengths of currents, the tractive forces being measured by an arrangement of levers. The results, briefly, are as follows:—Electromagnet No. 1, the iron of which weighed 15 lb., required a weight of 2090 lb. to detach the keeper. No. 2, the iron of which weighed 1057 grains, required 49 lb. to detach its armature. No. 3, the iron of which weighed 65·3 grains, supported a load of 12 lb., or 1286 times its own weight. No. 4, the weight of which was only half a grain, carried in one instance 1417 grains, or 2834 times its own weight.

"It required much patience to work with an arrangement so minute as this last; and it is probable that I might ultimately have obtained a larger figure than the above, which, however, exhibits a power proportioned to its weight far greater than any on record, and is eleven times that of the celebrated steel magnet which belonged to Sir Isaac Newton.

"It is well known that a steel magnet ought to have a much greater length than breadth or thickness; and Mr. Scoresby has found that when a large number of straight steel magnets are bundled together, the power of each when separated and examined is greatly deteriorated. All this is easily understood, and finds its cause in the attempt of each part of the system to induce upon the other part a contrary magnetism to its own. Still there is no reason why the principle should in all cases be extended from the steel to the electromagnet, since in the latter case a great and commanding inductive power is brought into play to sustain what the former has to support by its own unassisted retentive property. All the preceding experiments support this position; and the following Table gives proof of the obvious and necessary general consequence, *the maximum power of the electromagnet is directly proportional to its least transverse sectional area.* The second column of the Table contains the least sectional area in square inches of the entire magnetic circuit. The maximum power in pounds avoirdupois is recorded in the third; and this, reduced to an inch square of sectional area, is given in the fourth column under the title of specific power.

TABLE I.

Description.		Least Sectional Area.	Maximum Power.	Specific Power.
My own electromagnets	No. 1	10	2090	209
	No. 2	0·196	49	250
	No. 3	0·0436	12	275
	No. 4	0·0012	0·202	162
Mr. J. C. Nesbit's. Length round the curve, 3 ft.; diameter of iron core, 2¾ in.; sectional area, 5·7 in.; ditto of armature, 4·5 in.; weight of iron, about 50 lb.		4·5	1428	317
Professor Henry's. Length round the curve, 20 in.; section, 2 in. square; sharp edges rounded off; weight, 21 lb.		3·94	750	190
Mr. Sturgeon's original. Length round the curve, about 1 ft.; diameter of the round bar, ½ in.		0·196	50	255

"The above examples, are, I think, sufficient to prove the rule I have advanced. No. 1 was probably not fully saturated; otherwise I have no doubt that its power per square inch would have approached 300. Also the specific power of No. 4 is small, because of the difficulty of making a good experiment with it."

These experiments were followed by some to ascertain the effect of the length of the iron of the magnet, which he considered, at least in those cases where the degree of magnetization is considerably below the point of saturation, to offer a directly proportional resistance to magnetization ; a view the justice of which is now, after fifty years, amply confirmed.

In November of the same year further experiments* in the same direction were published. A tube of iron, spirally made and welded, was prepared, planed down as in the preceding case, and fitted to a similarly prepared armature. The hollow cylinder thus formed, shown in Fig. 8, was 2 feet in length, its external diameter was 1·42 inch, its internal being 0·5 inch. The least sectional area was 10¼ square inches. The exciting coil consisted of a single copper rod, covered with tape, bent into a sort of S-shape. This was later replaced by a coil of twenty-one copper wires, each $\frac{1}{25}$ inch in diameter and 32 feet long, bound together by cotton tape. This magnet, excited by a battery of sixteen of Sturgeon's cast-iron cells, each 1 foot square and 1½ inch in interior width, arranged in a series of four, gave a lifting power of 2775 lb.

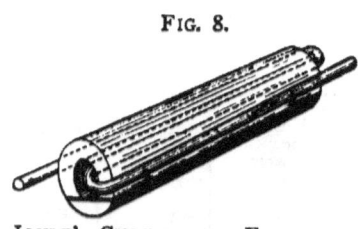

FIG. 8.

JOULE'S CYLINDRICAL ELECTRO-MAGNET.

Joule's work was well worthy of the master from whom he had learned his first lesson in electromagnetism. He showed his devotion not only by writing descriptions of them for Sturgeon's *Annals*, but by exhibiting two of his electromagnets at the Victoria Gallery of Practical Science, of which Sturgeon was director. Others, stimulated into activity by Joule's example, proposed new forms, amongst them being two Manchester gentlemen, Mr. Radford and Mr. Richard Roberts, the latter being a well-known engineer and inventor. Mr. Radford's electromagnet consisted of a flat iron disk, with deep spiral grooves cut in its face, in which were laid the

* *Scientific Papers*, p. 40, and *Annals of Electricity*, vol. v. p. 170.

insulated copper wires. The armature consisted of a plain iron disk of similar size. This form is described in vol. iv. of Sturgeon's *Annals*. Mr. Roberts' form of electromagnet consisted of a rectangular iron block, having straight parallel grooves cut across its face, as in Fig. 9. This was described in vol. vi. of Sturgeon's *Annals*, p. 166. Its face was $6\frac{5}{8}$ inches square, and its thickness $2\frac{7}{16}$ inches. It weighed, with the conducting wire, 35 lb.; and the armature, of the same size and $1\frac{1}{2}$ inch thick, weighed 23 lb. The load sustained by this magnet was no less than 2950 lb. Roberts inferred that a magnet, if made of equal thickness, but 5 feet square, would sustain 100 tons weight. Some of Roberts' apparatus is still preserved in the Museum of Peel Park, Manchester.

FIG. 9.   FIG. 10.

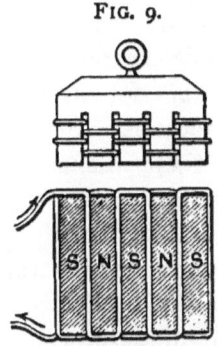 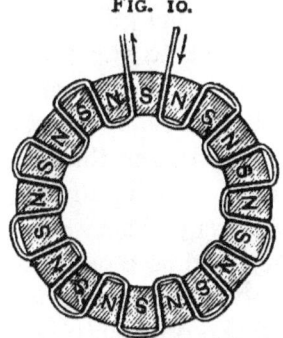

ROBERTS' ELECTROMAGNET.   JOULE'S ZIGZAG ELECTROMAGNET.

On p. 431 of the same volume of the *Annals*, Joule described yet another form of electromagnet, the form of which resembled in general Fig. 10; but which, in actual fact, was built up of twenty-four separate flat pieces of iron bolted to a circular brass ring. The armature was a similar structure, but not wound with wire. The iron of the magnet weighed 7 lb., and that of the armature 4·55 lb. The weight was 2710 lb., when excited by sixteen of Sturgeon's cast-iron cells.

In a subsequent paper on the calorific effects of magneto-electricity,* published in 1843, Joule described another form

* *Scientific Papers*, vol. i. p. 123; and *Phil. Mag.*, ser. 3, vol. xxiii. p. 263, 1843.

of electromagnet of horse-shoe shape, made from a piece of boiler-plate. This was not intended to give great lifting power, and was used as the field-magnet of a motor. In 1852, another powerful electromagnet of horse-shoe form, somewhat similar to the preceding, was constructed by Joule for experiment. He came to the conclusion* that, owing to magnetic saturation setting in, it was improbable that any force of electric current could give a magnetic attraction greater than 200 lb. per square inch. "That is, the greatest weight which could be lifted by an electromagnet formed of a bar of iron 1 inch square, bent into a semi-circular shape, would not exceed 400 lb."

With the researches of Joule may be said to end the first stage of development. The notion of the magnetic circuit, which had thus guided Joule's work did not commend itself at that time to the professors of physical theories; and the practical men, the telegraph engineers, were for the most part content to work by purely empirical methods. Between the practical man and the theoretical man there was, at least on this topic, a great gulf fixed. The theoretical man, arguing as though magnetism consisted in a surface distribution of polarity, and as though the laws of electromagnets were like those of steel magnets, laid down rules not applicable to the cases which occur in practice, and which hindered rather than helped progress. The practical man, finding no help from theory, threw it on one side as misleading and useless. It is true that a few workers made careful observations and formulated into rules the results of their investigations. Amongst these, the principal were Ritchie, Robinson, Müller, Dub, Vom Kolke, and Du Moncel; but their work was little known beyond the pages of the scientific journals wherein their experiments were described.

Formulæ connecting together the dimensions of an electromagnet, the strength of current circulating around it, the number of turns in its coil, and the resulting amount of

---

* *Scientific Papers*, vol. i. p. 362; and *Phil. Mag.*, ser. 4, vol. iii. p. 32.

magnetism excited, were proposed by various physicists. Lenz and Jacobi laid down the rule that the amount of magnetism excited in a given electromagnet is proportional to the current and to the number of turns; a rule that is manifestly incorrect, as it fails to take into account the tendency to magnetic saturation of the iron cores. Formulæ which take saturation into account have been given by Müller, Von Waltenhofen, Lamont, Weber, and Frölich, but are for the most part empirical, and only approximate. Some account of these has been given by the Author elsewhere.* The law of the electromagnet could not indeed be accurately stated by any formula that was not based on the principle of the magnetic circuit. Analogies between the flux of electricity in an electrically-conducting circuit, and the flux of magnetic lines of force through circuits possessing magnetic conductivity are to be found, as was remarked in the Preface to this work, abundantly in the literature of the science.

The work of Rowland, Bosanquet, and others, there summarized, paved the way for a sound method of treating calculations appertaining to magnetic circuits. In 1885 and 1886 Mr. G. Kapp discussed the calculation of the field-magnets of dynamos from this point of view, and gave formulæ. In 1886 Drs. J. and E. Hopkinson communicated to the Royal Society† a very complete and elegant investigation of the problem of the magnetic circuit; their chief point being the *à priori* determination of the characteristic curve of magnetization of the dynamo-machine from the ordinary laws of magnetism, and the known properties of a given specimen of iron. They also investigated experimentally the lateral leakage of magnetic lines from a circuit. Ever since that date calculations of the quantities involved in the magnetic circuits of dynamo-machines have been matters of every day practice. In the Cantor Lectures, delivered by the Author in February, 1890, which form the basis of the present book, the

* S. P. Thompson, *Dynamo-electric Machinery*, 3rd ed., 1888, p. 303.
† *Phil. Trans.*, 1886, pt. i. p. 331.

attempt was made to extend this principle to the varied phenomena of electromagnets in all their forms. In Chapter IV. the method of calculating magnetic circuits is given.

Notable Electromagnets.

No historical account of the electromagnet would be complete which did not refer, however briefly, to some electromagnets that are notable for their unusual dimensions.

Faraday's Electromagnet, still preserved at the Royal Institution, with which so many of his researches and those of Professor Tyndall were made, is depicted in Fig. 11. It is thus described by Faraday.*

"Another magnet which I have had made has the horse-shoe form. The bar of iron is 46 inches in length and 3·75 inches in diameter, and is so bent that the extremities forming the poles are 6 inches from each other; 522 feet of copper wire 0·17 inch in diameter, and covered with tape, are wound round the two straight parts of the bar, forming two coils on these parts, each 16 inches in length, and composed of three layers of wire; the poles are, of course, 6 inches apart, the ends are planed true, and against these move two short bars of soft iron 7 inches long and $2\frac{1}{2}$ by 1 inch thick, which can be adjusted by screws and held at any distance less than 6 inches from each other."

Plücker's Electromagnet, constructed in 1847,† was also of horse-shoe form, but larger. The core was 132 centimetres in length, 10·2 centimetres in diameter, and weighed 84 kilogrammes. The poles were 28·4 centimetres apart. The copper wire, of 14·93 square millimetres sectional area, weighed 35 kilogrammes, and was wound in three layers.

Von Feilitzsch's and Holtz's Electromagnet.—This was constructed‡ in 1880. It differs from all the preceding in possessing a laminated core. Its shape is a horse-shoe, the

---

* *Experimental Researches*, vol. iii. p. 29.
† *Pogg. Ann.*, lxxii. p. 315, 1847; and lxxiii. p. 549, 1848.
‡ *Mittheil. a. d. naturw. Verein v. Neu-Pommern u. Rügen*, Nov. 1880, Taf. iv.

total length of core being about 285 centimetres, of circular section 19·5 centimetres in thickness. The shortest distance between the poles is 40·5 centimetres, and the extreme

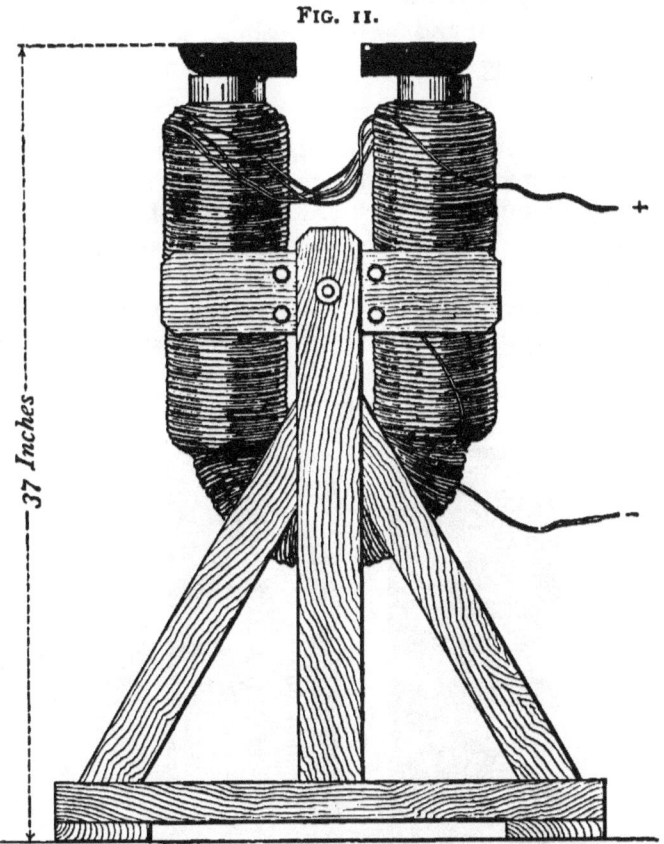

FIG. 11.

FARADAY'S ELECTROMAGNET AT THE ROYAL INSTITUTION.

height from the outer side of the bend to the central point between the flat polar surfaces is 125 centimetres. The core is itself built up of twenty-eight strips of sheet iron 7 millimetres thick, of varying breadth, each separately bent in the forge, lacquered, and put together without any metallic clamping. The whole core weighs 628 kilogrammes. This core, temporarily held together, was imbedded, with its poles upwards, in a bed of cement in a strong oak box on wheels, the limbs projecting to a length of 96 centimetres. The coils are multiple, being composed of fifteen layers of insulated copper strip, and ten layers of wire of 2 millimetres diameter, covered with cotton or wool and well varnished. Arrangements are added for coupling the exciting circuits in several different ways.

Becquerel's Electromagnet.—In 1848 M. Edm. Becquerel constructed an electromagnet, having upright cores 11 cm. in diameter, in total 1 metre in length. Each coil consisted of 910 metres of a copper wire, 2 mm. in diameter, wound in two separate windings, weighing about 25 kilogrammes.

S. P. Thompson's Electromagnet.—In 1883 a large electromagnet (Fig. 12) was constructed for the Author of this book by Mr. Akester,

FIG. 12.

S. P. THOMPSON'S ELECTROMAGNET.

of Glasgow. The ironwork consists of two cylindrical cores of carefully annealed iron of the softest Scotch brand, each 7·3 centimetres in diameter, and 53·8 centimetres high. These are shouldered and inserted into a massive cross-yoke, also of wrought iron. The coils are wound upon removable bobbins of brass, slit lengthways to prevent eddy-currents, and weigh about 75 lb. each. Each coil consists of about 1820 turns of a copper wire 2·67 mm. in diameter, covered with a heavy insulation of indiarubber and tarred hemp braided. It was with this electromagnet that the Author discovered the increase of electric resistance of metals in a strong magnetic field, a discovery announced independently about the same date by M. Righi.

These electromagnets have, however, been far surpassed in recent years by those used as field-magnets in dynamo-machines. The electromagnets thus used in the largest Edison-Hopkinson dynamos, built by Messrs. Mather and Platt, of Manchester, weigh no less than 17 tons, and have a cross-section of 517 square inches of iron. Reckoning (with Joule) the possible magnetic traction at 200 lb. to the square inch, these magnets ought to be capable of sustaining a load of 46 tons.

Two curious experiments carried out in the United States deserve mention here. In 1887, Major W. R. King, of the United States Navy, constructed an enormous electromagnet out of two 15-inch cannons, which were placed side by side and connected together magnetically by piling a number of iron rails across the breech. Each gun was about 15 feet long, and weighed 25 tons. The total cross-section of the rails which served as a yoke was 60 sq. in., and preferably should have been larger. Upon each gun were wound three coils, each about 2 feet long, and having internal and external diameters of about 26 and 40 in. respectively. The wire employed was a stranded cable, having a core of twenty No. 20 gauge copper wires, together with seven smaller wires twisted together and covered with indiarubber and insulating tape to a diameter of four-tenths of an inch. Its length was eight miles, with a total resistance of 19·2 ohms. As an armature, a bundle of fifteen iron plates, each ½ inch thick, 11 in. wide, and 7 feet long, was used. The current was supplied from a dynamo-machine which usually operated

ELECTROMAGNET MADE OF TWO CANNONS.

twenty arc lamps. The whole arrangement is depicted in Fig, 13, p. 32, and in Fig. 14 are given a small plan of the guns, and a diagram of the muzzle of one of them in section, showing the position of the coils, and indicating also the curved paths of the magnetic lines.

FIG. 14.

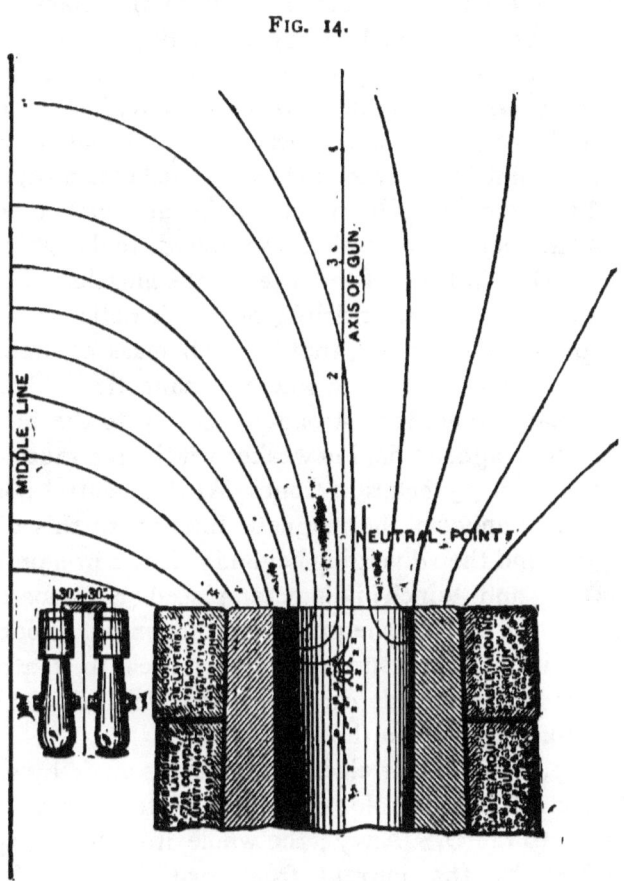

DIAGRAM OF MAGNETIC LINES OF FORCE AT MOUTH OF CANNON.

With this gigantic electromagnet various experiments were made.* A pull of 5 tons at the middle of the armature failed to detach it; and from experiments made with leverage

* See *Electrical World*, xi., p. 27, Jan. 21, 1888.

it appeared that a pull of 20,600 lb., or nearly 10 tons would have been required. Four 15-inch shells of iron, each, weighing 320 lb., were hung in a magnetic chain from one another from the muzzle of one of the guns. By holding in the hand a spike or bit of iron wire, the pull of the magnetic field was felt even 5 or 6 feet away from the poles, and by this means the magnetic lines, given in Fig. 14, p. 33, were marked on a sheet of paper. It was found that if a small piece of iron was held in the axis of the bore, just within the mouth of the cannon, it was repelled strongly outwards; but if held at a point in the same axis some distance away, it was powerfully attracted. The neutral point at which there was neither attraction or repulsion was found to be at a point about $7\frac{1}{2}$ inches in front of the face of the muzzle. This fact illustrates an important principle, of which notice is taken at a later place in this book, that a small mass of iron tends always to move from a place where a magnetic field is weak toward a place where it is stronger. Owing to the hollow of the bore, the magnetic field is weaker within the muzzle than it is at the point $7\frac{1}{2}$ inches in front. At this neutral point the field is nearly uniform, the magnetic lines up to this distance converging, and then diverging beyond. Little iron projectiles 3 inches long and $\frac{1}{10}$ inch in diameter, placed on a smooth piece of board in the axis of the gun, were thrown out about 2 feet from the gun and then suddenly drawn back to it, attaching themselves to the outside rim of the muzzle, and sticking on endways like quills upon the porcupine.

In 1887, a still larger electromagnet was made by winding some electric light cable around the steam-ship *Atlanta* belonging to the U.S. Navy; the whole iron ship being then magnetized by the current from two gramme dynamo-machines. Lieut. Bradley A. Fiske,[*] who carried out these arrangements, hoped thereby to be able to establish a magnetical means of signalling between ships at sea.

[*] See *Electrical World*, Sept. 24, 1887. Also *Electrician*, vol. xix., p. 461, 1887.

CHAPTER II.

GENERALITIES CONCERNING ELECTROMAGNETS AND ELECTROMAGNETISM: TYPICAL FORMS OF ELECTROMAGNETS: MATERIALS OF CONSTRUCTION.

USES IN GENERAL.

REGARDED as a piece of mechanism, an electromagnet may be described as an apparatus for producing a mechanical action at a place distant from the operator who controls it; the means of communication from the operator to the distant point where the electromagnet is being the electric wire. The use of electromagnets may, however, be divided into two main divisions. For certain purposes an electromagnet is required merely for obtaining temporary adhesion or lifting power. It attaches itself to an armature, and cannot be detached so long as the exciting current is maintained, except by the application of a superior opposing pull. The force which an electromagnet thus exerts upon an armature of iron with which it is in direct contact, is always considerably greater than the force with which it can act on an armature at some distance away, and the two cases must be carefully distinguished. *Traction* of an armature in contact, and *attraction* of an armature at a distance, are two different functions; so different, indeed, that it is no exaggeration to say that an electromagnet designed for the one purpose is unfitted for the other. The question of designing electromagnets for either of these purposes will occupy a large part of these pages. The action which an electromagnet exercises on an armature in its neighbourhood may be of several kinds. If the armature is of soft iron, placed nearly parallel to the polar surfaces, the action is one simply of attraction, producing a motion of simple translation, in which it will make no difference which pole is a north pole and which is a south pole. If the armature lies oblique to the line of the poles, there will be a tendency to turn it round as well as to attract it; but, again, if the armature is of soft iron the action will be independent of the polarity of the magnet, that is to say, independent of the direction of the exciting current. If, however, the armature be itself a magnet of steel permanently magnetized, then the direction in which it tends to turn, and the amount and also the sign of the force with which it is attracted, will depend on the polarity of the electromagnet, that is to say, will depend on the direction in which the exciting current circulates. Hence there arises a difference between the operation of a *non-polarized* and that of a *polarized* apparatus, the latter term being applied to those forms in which there is employed a portion—say an armature—to which an initial fixed magnetization has been imparted. Non-polarized apparatus is in all cases independent of the direction of the current. Another class of uses served by electromagnets is the production of rapid vibrations. These are employed in the mechanism of electric trembling bells, in the automatic breaks of induction coils, in electrically-driven tuning-forks such as are employed for chronographic purposes, and in the instruments used in harmonic telegraphy. Special constructions of electromagnets are appropriate to special purposes such as these. The adaptation of electromagnets for the special end of responding to rapidly alternating currents is a closely kindred matter. Lastly, there are certain applications of the electromagnet, notably in the construction of some forms of arc lamp, for which it is specially sought to obtain an equal or approximately equal pull over a definite range of motion. This use necessitates special designs. All these matters will be considered in due course. In the mean time, we must enter upon some of the elementary principles of electromagnetism.

POLARITY.

It is a familiar fact that the polarity of an electromagnet depends upon the sense in which the current is flowing around it. Various rules for remembering the relation of the electric flow and the magnetic force have been given. One of them that is useful is that when one is looking at the north pole of an electromagnet, the current will be flowing around that pole in the sense opposite to that in which the hands of a clock are seen to revolve.

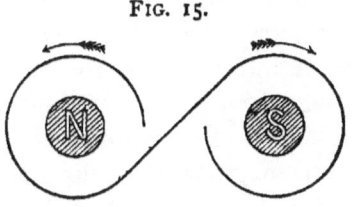

FIG. 15.

CIRCULATION OF CURRENT AROUND A TWO-POLE ELECTROMAGNET.

This necessitates the connexion of the bobbins of a two-pole electromagnet in such a way that the currents shall circulate as indicated in the manner depicted in Fig. 15 Another useful rule, suggested by Maxwell, is illustrated by Fig. 16, namely, that the sense of the circula-

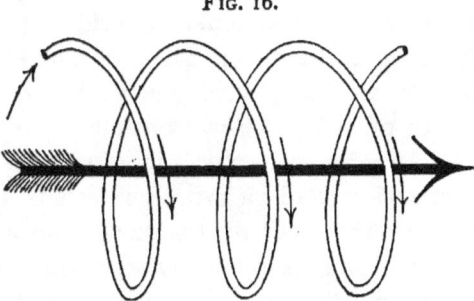

FIG. 16.

DIAGRAM ILLUSTRATING RELATION OF MAGNETIZING CIRCUIT AND RESULTING MAGNETIC FORCE.

tion of the current (whether right or left-handed), and the positive direction of the resulting magnetic force, are related together in the same way as the rotation and the travel of a right-handed screw are associated together. Right-handed rotation of the screw is associated with forward travel.

Right-handed circulation of a current is associated with a magnetic force tending to produce north polarity at the forward end of the core.

POLES AND POLAR SURFACES.

Magnetism, whether resident permanently in a piece of lodestone or of hardened steel, or whether temporarily produced in a bar of soft iron by means of an external circulation of electric currents in a surrounding coil, is something which exists internally in the piece of lodestone or steel or iron, though it may also manifest itself externally. In the case of ordinary lodestones and steel magnets, and in that of those electromagnets that do not form closed magnetic circuits, there are certain external manifestations, most of which are well known. Small pieces of iron or steel will be attracted by and adhere to certain parts of the surface of such magnets. Compass needles brought into the neighbourhood of the same parts are deflected by their influence. Those parts of the surface of a lodestone, or steel magnet, or electromagnet, to which iron filings adhere in tufts, and which act on compass needles, are in common language called the *polar surfaces*, or simply the *poles*, of the magnet. In former days magnetism was regarded as a surface phenomenon, resident (as an invisible fluid or fluids) upon these polar parts of the surface. It is, however, possible to have a magnet, highly magnetized, entirely destitute of poles. For example, it is possible to magnetize a steel ring in such a way (compare Fig. 71, p. 162) that the magnetization shall be entirely internal: the magnetic lines flowing round wholly within the metal, and never coming up to the surface. This is done by overwrapping the ring all round with a magnetizing coil of insulated copper wire, through which an electric current is passed. Of course such a ring magnet is incapable of attracting iron filings, or of perturbing a compass needle. The evidence that such a ring is magnetized internally is two-fold. In the act of magnetizing the ring grows larger to a minute extent. If cut or broken, such a ring forthwith exhibits polar properties at the cut surface. The pole or polar region of a magnet is simply that part of the surface where the internal magnetic lines emerge into the air. Only such parts exhibit any such properties as those to which formerly the virtues of magnetic fluids were attributed. In one sense, therefore, it is true that the external phenomena of poles are accidental. Nevertheless, it was precisely these external phenomena which first drew attention to the subject of magnetism and were first investigated. It is the consideration of them that originated the terms in which magnetic facts are described. The very unit of magnetism that has been adopted as a means of expressing the relative strengths of magnets is based upon the repulsion which two poles are found to exert upon one another when separated by an air space: and the unit of intensity of magnetic force is in turn based on the unit of polar magnetism. It is too late to dream of changing these definitions, which are now universally accepted. Every treatise on magnetism deals copiously with them; it is enough for the present purpose, therefore, to state them for reference.

Magnetic Units.

The international units, now adopted by all electricians, are based upon the absolute system of weights and measures, known as the "C.G.S. system." In this system, which is further explained in Appendix B, the *centimetre* is taken as the unit of length, the *gramme* as the unit of mass, and the *second* as the unit of time. From these three fundamental units are derived all the other physical units. For example, the *dyne* or unit of force in this system, is that force which, if it acts on one gramme for one second, gives to it a velocity of one centimetre per second. The pull of the earth on a mass of one pound (this pull being what is commonly called the pound's weight) is equal (at London) to 444,971 dynes. One dyne is a pull (or push) equal to about $\frac{1}{63}$ of the weight (at London) of one grain. Upon this abstract unit of force is founded the magnetic unit.

The unit quantity of magnetism, or *unit magnetic pole*, is one of such a strength that, when placed at a distance of one centimetre (in air) from a similar pole of equal strength, it repels it with a force of one dyne. The distribution of apparent magnetism at a polar surface is usually stated in terms of the number of units of magnetism (defined as above) per square centimetre, this number being sometimes called the surface-density of the magnetization. It is found that this surface-density cannot, even with the most enormous magnetizing forces, be made to exceed a limiting amount. Ewing finds that the softest iron, subjected temporarily to the highest magnetizing force, will not exhibit more than 1700 units of magnetism per square centimetre. Hard steel will not usually retain permanently a higher magnetization than 500 units per square centimetre of pole surface, or at the rate of 3225 units per square inch. This mode of expressing the facts in surface units is antiquated. The modern mode of regarding the matter is to think of the magnetic lines which flow around the magnetic circuit, emerging from the metal at one polar surface, and re-entering it at the other. Owing to the conventional mode of regarding the matter, according to which one draws one magnetic line per square centimetre to represent a force of one dyne acting on a unit pole, we have to draw, or imagine as drawn, as many lines per square centimetre as there are dynes of force on a unit pole. As a matter of fact, the number of dynes of force that are exerted on a unit pole placed between two surfaces of opposite polarity, close together, is numerically equal to 4π (i.e. = 12·566) times the number of surface units of magnetism per square centimetre on either of the surfaces. Hence arises the curious rule that for every unit of magnetism at the surface one must suppose 4π magnetic lines coming up through the surface. For example, if at any polar surface there are 100 units of magnetism, we must imagine 1257 lines emerging through that surface.

In the case of a bar magnet, we conceive these lines as flowing through the metal, emerging at one end (the north pole) and curving round to re-enter the metal at the other end (the south pole). In Fig. 17, these lines are drawn both internally and externally. Every one knows that if we dip

FIG. 17.

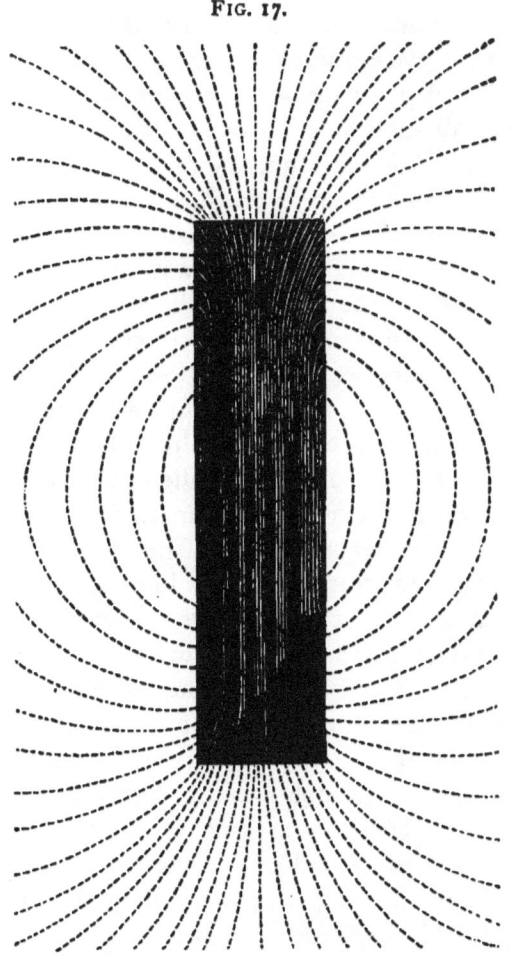

LINES OF FORCE RUNNING THROUGH BAR MAGNET.

such a magnet into iron filings the small bits of iron stick on, more especially at the ends, but not exclusively ; and if you hold it under a piece of paper or cardboard, and sprinkle iron filings on the paper, you obtain curves like those shown on the diagram. They attest the distribution of the magnetic forces in the external space. The magnetism running internally through the body of the iron begins to leak out sideways, and finally all the rest streams out in a great tuft at the end. These magnetic lines pass round to the other end, and there go in again, and the place where the steel is internally most highly magnetized is this place across the middle, where externally no iron filings at all stick to it. Now, we have to think of magnetism from the inside and not the outside. This magnetism extends in lines, coming up to the surface somewhere near the ends of the bar, and the filings stick on wherever the magnetism comes up to the surface. They do not stick on at the middle part of the bar, where the metal is really most completely permeated through and through by the magnetism; there are a larger number of lines per square centimetre of cross-section in the middle region where none come up to the surface.

The space outside the magnet where these magnetic lines are passing from pole to pole is called the *magnetic field* of the magnet. It is obviously more intense, or in other words the lines in it are denser, near to the poles than at some distance away. In this field the lines are so drawn, or imagined, that at any point the direction of the lines at that part of the field shows the direction of the resultant magnetic force at that point. Also, the density of the lines indicates the magnitude of the force. A single free magnet pole placed at any point in the field would tend to move along the magnetic lines, a north pole sailing one way, a south pole sailing the other. If a small pivotted compass needle is set down at any point in the field, it at once turns round and sets itself along the magnetic line at that point. If a needle is set down on a curved magnetic line, and is free not only to turn round but to shift its whole position, it will be pulled sideways and always toward the inner side of the curve. Small unmagnetized needles, and filings of iron, if placed in the magnetic field, are magnetized by the influence of the magnet, and turn round and point along the lines. This explains the familiar experiment of sprinkling iron filings on a sheet of paper or glass over a magnet. The filing-figure thus obtained (Fig. 18) is useful for studying the manner in which the magnetic lines are distributed in the surrounding field.

FIG. 18.

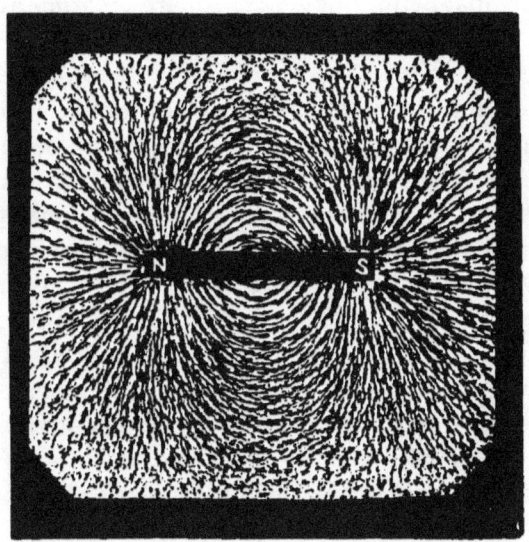

FILING-FIGURE OF THE BAR MAGNET.

Now in studying electromagnets by the light of the modern principle of the magnetic circuit, we have to think of these things as internal, not as mere superficial, phenomena. We have to think of iron and steel as being good conductors of magnetic lines; we have to calculate what cross-section of iron to allow for any desired number of magnetic lines to permeate through it; and we have to calculate the magnetizing power that will be needful to apply in order to force any given number of magnetic lines into the circuit. Symbols and definitions are required in dealing with these matters. In order to express the intensity of the magnetic force at any point in a magnetic field, it is usual, as explained previously, to describe it in terms of the number of magnetic lines per square centimetre that are to be considered as traversing the air at the point in question; the intensity of the force in the field (or *intensity of field*, as it is often called), being denoted by the letter H. The term *the magnetic flux* is often used for the whole number of magnetic lines that flow around the circuit, and denoted by the symbol N. The number of magnetic lines per square centimetre, in any material, is called *the magnetic induction*, or simply *the induction* (also called *the permeation*, or the *internal magnetization*), and for this the symbol B is used. As iron, steel and other magnetic materials are more permeable for magnetic lines than air is, it follows that the same application of magnetizing power that would produce H lines per square centimetre in air at any point in space will produce a greater number of lines per square centimetre if the space in question is occupied by iron instead of air. That is to say, the same magnetizing power that would have produced H lines per square centimetre in air will produce B lines per square centimetre in iron; where B is a number greater than H, according to the permeability of the specimen. Of this more will be said in Chapter III. on the Properties of Iron.

ELEMENTARY PROPOSITIONS IN ELECTROMAGNETICS.

Here it will be convenient to lay down a few elementary propositions dealing with the relation between electric currents and magnetic forces.

1. *Magnetomotive Force*, or Total Magnetizing Power of Electric Current circulating in a Coil.—It is found that when a current flows along in a copper wire that is coiled in several turns around a core, and is thus made to circulate around an interior magnetic circuit, the magnetizing power or tendency of this circulation of electricity is proportional both to the strength of the current so circulating and to the number of turns in the coil. If other things are equal, the total magnetizing power depends on nothing else but these two matters; being independent of the size or

material of the wire, and of its shape, and is the same whether the spirals are close together or wide apart. If S stands for the number of *spirals* in the coil, and *i* be the number of *amperes* of current that are flowing, then S multiplied by *i* will be the number of *ampere-turns* of circulation of current. It is experimentally proved that twenty amperes circulating around five turns exert precisely the same magnetizing power as one ampere circulating one hundred times, or as one hundred amperes circulating once around the core.

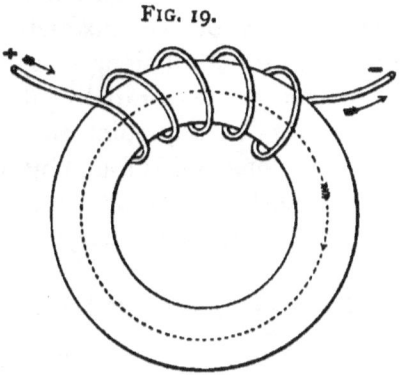

FIG. 19.

MAGNETIZING COIL WOUND AROUND A MAGNETIC CIRCUIT.

In each of these cases the circulation of current is one hundred ampere-turns. To calculate from this the value, in absolute C.G.S. units, of the magnetomotive force, it is requisite to multiply the ampere-turns by $\frac{4}{10}\pi$, or by 1·257. Or, in symbols,

Magnetomotive force = 1·257 × S *i*.

It will be shown in Appendix C that it is possible to avoid the use of this multiplier by taking the ampere-turns themselves as the magnetomotive force.

Some writers call the magnetomotive force the "line-integral of the magnetic forces."

2. *Intensity of Magnetic Force at any point in a long Magnetizing Coil.*—The preceding expression for the total magnetizing power of a coil does not give any information about the variation of the magnetic force at different parts. If in Fig. 19, a closed curve be drawn (the dotted curve) passing through all the spirals, and the question be asked "What is the intensity of the magnetic force at various points on this curve?" it must be replied that the intensity of the

force will vary greatly from point to point, being greatest at the middle of that part of the curve which lies within the spirals. If a uniformly-wound coil were constructed of very great length (say at least one hundred times its own diameter), the intensity of the magnetic force would be very nearly uniformly great all along its axis, until quite close to the ends of the coil, where it rapidly falls off. The expression for the value of H at any point along the axis (save near the ends) of such a long coil is found by considering the magnetomotive force as distributed uniformly along its length; or, in symbols, where l stands for the length of the coil, in centimetres.

$$H = \frac{4\pi}{10}\frac{Si}{l} = 1\cdot 257 \text{ times the ampere-turns per centimetre of length.}$$

Or, if the length is given in inches:

$$H = 0\cdot 495 \text{ times the ampere-turns per inch.}$$

Or, if it is desired to express the intensity of the magnetic force in lines *per square inch* instead of per square centimetre, we shall have:—

$$H_{,,} = 3\cdot 193 \times \text{ the ampere-turns per inch.}$$

In the case where a wire is wound in an annular coil upon an iron ring, so that there are no ends to the coil, H is uniform at all points along the closed curve drawn within the coil, and is calculated as above, taking the mean length along the body of the ring as l. It is obvious that when H is uniform, $H \times l$ gives the total magnetizing power or magnetomotive force.

3. *Intensity of Magnetic Force at centre of a single Ring.*—At the centre of a single ring or circular loop of wire carrying current of i amperes, and of radius r centimetres, the intensity of the magnetic force is calculated by the formula

$$H = \frac{2\pi i}{10\, r} = 0\cdot 6284 \times \frac{\text{amperes}}{\text{radius}}.$$

Some Elementary Propositions.

This is the case of a tangent-galvanometric ring; the number needing to be multiplied by S, if there are S turns in the ring.

4. *Force on Conductor* (carrying current) *in a Magnetic Field.*—Suppose a magnetic field is furnished by a permanent magnet (Fig. 20), and that a conducting wire carrying an electric current is brought into the magnetic field, it is observed to experience a mechanical force in a direction at right angles to its own length, and at right angles to the magnetic lines of the field. In the figure the direction of the flux of magnetic lines is horizontal from right to left between the limbs of the magnet; the direction of the current is horizontal from front to back; and the resulting mechanical force will urge the conductor vertically upwards, as shown by the arrow. Reversing the current would, of course, result in a downward force.

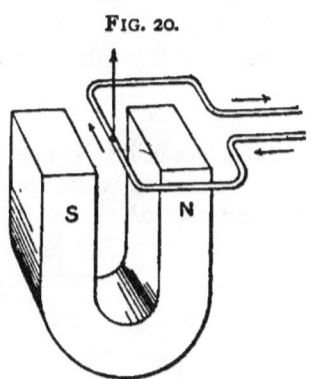

FIG. 20.

ACTION OF MAGNETIC FIELD ON CONDUCTOR CARRYING CURRENT.

The magnitude of this force can be calculated as follows:— Assume the field to be of uniform intensity H, and that the length (centimetres) of conductor lying squarely across the field is l. Then if i is the number of amperes of current, the force (in dynes) will be given by the rule

$$f = Hli \div 10 = H_{,,} l'' i \div 25 \cdot 4,$$

or in grains' weight

$$f = H_{,,} l'' i \div 161.$$

5. *Work done by Conductor* (carrying current) *in moving across Magnetic Field.*—If the conductor moves across a breadth b (centimetres) of the magnetic field, the work done will be expressed (in ergs) as follows:—

$$w = fb = bHli \div 10.$$

But bl is the area of field swept out; and this area multiplied by the number of magnetic lines per square centimetre (H) gives the whole number N of magnetic lines cut in the operation: whence

$$w = Ni \div 10.$$

Proof.—This matter may be independently deduced as follows:— By definition of electric potential, the work done in moving Q units of electricity through difference of potential $V_1 - V_2$ is

$$w = Q (V_1 - V_2).$$

But in cutting N lines of magnetic field in a time of t seconds there is generated an electromotive force $= N \div t$, and this constitutes the difference of electric potential $V_1 - V_2$ and may be substituted for it.

Further, if the current i is expressed in amperes, the quantity Q of electricity expressed in absolute C.G.S. units conveyed through the circuit in t seconds, will be $= it \div 10$.

Substituting the latter value for Q and the former for $V_1 - V_2$, one at once obtains the result—

$$w = Ni \div 10, \text{ as before.}$$

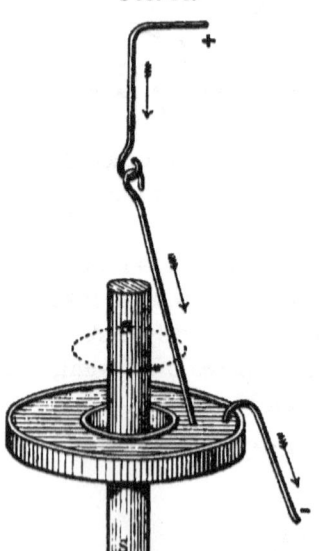

FIG. 21.

ROTATION OF CONDUCTOR OF CURRENT AROUND MAGNETIC POLE.

6. *Rotation of Conductor* (carrying current) *around a Magnetic Pole.*—If a portion of an electric circuit be arranged with movable connexions so that it can slide around a magnet pole, rotation will ensue, the angular force (or torque) being calculated as follows:—

In Fig. 21, a bar magnet stands vertically with an annular mercury cup around its middle. Into this cup dips a wire suspended by a flexible joint or swivel, and conducts a current of i amperes. If the pole-strength of the magnet is m units of magnetism, the whole number of magnetic lines radiating out from that pole will be $4\pi m$; and if the annular cup is close enough to the magnet, the whole of these, practically, will be cut by the conductor in one rotation. Hence, by the preceding article the work done in one revolution will be—

$$w = 4\pi m i \div 10 = 1\cdot 257 \times m i.$$

Hence, dividing by the angle 2π, the torque or turning moment γ will be :—

$$\gamma = 2 m i \div 10 = 0\cdot 2 \times m i.$$

It follows that the pole tends to rotate around the conductor in the reverse sense with an equal torque.

7. *Every electric circuit tends so to alter its configuration as to make the magnetic flux through it a maximum.*—This rule, which in different words was given by Maxwell, is extremely useful to assist one in seeing in what way conductors that are parts of electric circuits will tend to move when situated in any way in a magnetic field. For example, the case presented by Fig. 20, p. 47, can be argued out as follows :—The current in the circuit, as viewed in the cut, is shown by the arrows to be circulating right-handedly. This circulation of current, according to the rule laid down in p. 37, would produce a magnetic flux through the circuit from above to below. But, because of the north pole below it, there is already a magnetic flux of lines through the circuit in the contrary sense, from below upwards. The circuit will therefore tend to move in such a way as to diminish this contrary flux, virtually increasing its own thereby. Another example is afforded by De la Rive's familiar experiment with a floating coil and battery, which is attracted by one pole of a bar magnet, and repelled by the other.

8. *Two electric circuits* (or conductors carrying currents) *are urged by mutual forces to change their configurations so that their mutual magnetic flux may become a maximum.*—This is a generalized statement of the rules often given about the attractions and repulsions of parallel currents and angular currents. One current does not attract or repel another except by virtue of the magnetic fields which they set up in the

surrounding space. When two currents run parallel and in the same direction, each produces a magnetic field around itself, and each tends to move laterally across the other's magnetic field. In the case of two parallel coils like tangent galvanometer coils of equal diameters, lying near together, the mutual force varies inversely as the axial distance between them, and directly as the product of their respective ampere-turns.

Typical Forms of Electromagnets.

We will now consider the classification of forms of electromagnets. I do not pretend to have found a complete classification. There is a very singular book written by Mons. Nicklès, in which he classifies under thirty-seven different heads all conceivable kinds of magnets, bidromic, tridromic, monocnemic, multidromic, and I do not know how many more; but the classification is both unmeaning and unmanageable. It is sufficient for the present purpose simply to pick out those which come under four or five heads, and deal separately with others that do not quite fit under any of these categories.

1. *Bar Electromagnet.*—This consists of a single straight core (whether solid, tubular, or laminated), surrounded by a coil. Fig. 3 (p. 4) depicted Sturgeon's earliest example. Fig. 22 shows a straight magnet with cylindrical core.

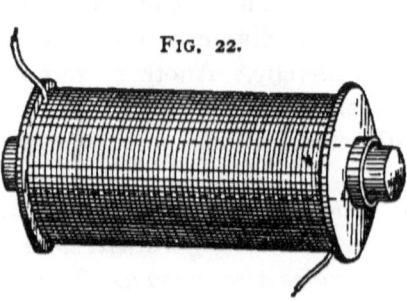

FIG. 22.

BAR ELECTROMAGNET.

2. *Horse-shoe Electromagnet.* — There are two sub-types included in this name. The original electromagnet of Sturgeon (Fig. 1, p. 3) really resembled a horse-shoe in form, being constructed of a single piece of round wrought iron, about half an inch

in diameter and nearly a foot long, bent into an arch. In recent years the other sub-type has prevailed, consisting, as shown in Fig. 23, of two separate iron *cores*, usually cut from circular rod, screwed or riveted into a third piece of wrought iron, the *yoke*. Joule's special form of horse-shoe is shown in Fig. 7, p. 21.

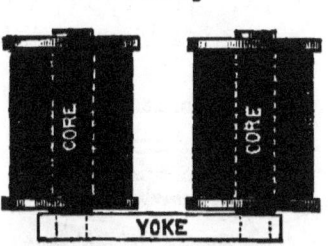

FIG. 23.

TYPICAL TWO-POLE ELECTROMAGNET.

3. *Club-foot Electromagnet.*—Occasionally the preceding form is modified by being furnished with one coil only, the other core being left uncovered. This form is known in France as the *aimant boiteux*, which may be rendered into English as the club-footed electromagnet. The German name is *hinkender Magnet*. Concerning this form, Fig. 24, more

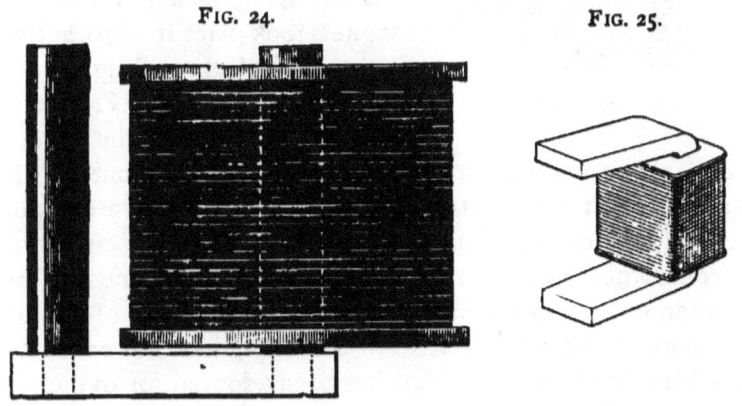

FIG. 24.

CLUB-FOOT ELECTROMAGNET.

FIG. 25.

HORSE-SHOE ELECTROMAGNET WITH ONE COIL ON YOKE.

will be said hereafter. Fig. 25 may be regarded as a kindred variety, being a horse-shoe magnet with one coil upon the yoke, the two limbs being left uncovered.

4. *Iron-clad Electromagnet.*—This form differs from the simple bar magnet in having an iron shell or casing external

to the coils and attached to the core at one end. Such a magnet presents, as depicted in Fig. 26, a central pole at one end surrounded by an outer annular pole of the opposite polarity. The appropriate armature for electromagnets of this type is a circular disk or lid of iron. It is curious how often the use of a tubular jacket to an electromagnet has been reinvented. It dates back to about 1850, and has been variously claimed for Romershausen, for Guillemin, and for Fabre.* It is described in Davis's *Magnetism*, published in Boston in 1855. About sixteen years ago Mr. Faulkner, of Manchester, revived it, under the name of the *Altandae* electromagnet. A discussion upon jacketed electromagnets took place in 1876, at the Society of Telegraph Engineers; and in the same year, Professor Graham Bell used the same form of electromagnet in the receiver of the telephone which he exhibited at the Centennial Exhibition. There are some kindred forms of electromagnet in which the return circuit of the iron comes back outside the coil, either from one end or the other, or from both ends, sometimes in the form of two or more parallel return yokes. All such magnets I propose to call—following the fashion that has been adopted for dynamos—iron-clad electromagnets. There is one used by Mr. Cromwell Varley, in which a straight magnet is placed between a couple of iron caps, which fit over the ends, and virtually bring the poles down close together; the circular rim of one cap being the north pole, and that of the other cap being the

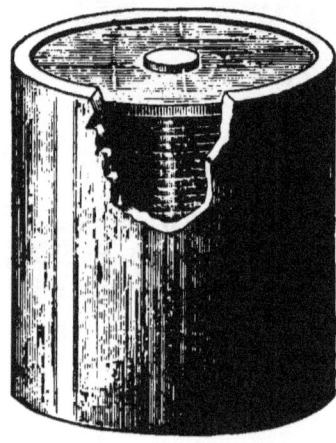

FIG. 26.

IRON-CLAD ELECTROMAGNET.

---

* According to Nicklès (see *Comptes Rendus*, xlv. 1857, p. 253) it was invented by Fabre when working some years previously in Nicklès's laboratory in Paris.

south pole, the two rims being close together. That plan of course produces a great tendency to leak across from one rim to the other all round.

A recent variety of the iron-clad type is depicted in Fig. 27, which has the merit of great simplicity of construction and great holding power. It is used in electromagnetic clutches and brakes.

FIG. 27.

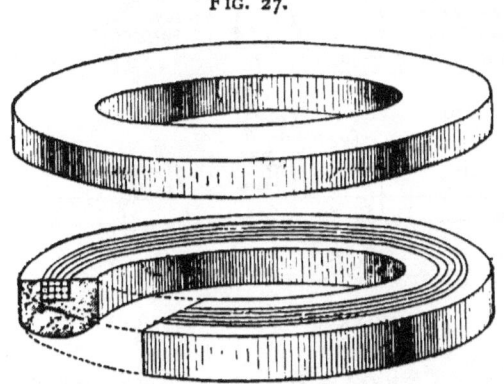

ANNULAR IRON-CLAD ELECTROMAGNET.

Ruhmkorff's Electromagnet.—This form, used for experiments on diamagnetism and on the magnetic rotation of light, was designed in 1846.* Verdet gives the following description of that used by him (Fig. 28) in his researches on magneto-optic rotation. The electromagnet cores are two cylinders of soft iron A B, A' B', each 20 cm. long, and 7·5 cm. diameter, pierced longitudinally to admit a ray of light. They are clamped by soft iron yokes P P' to a soft iron bed-plate R S. For the optical experiments a very nearly uniform magnetic field is procured by screwing on the cores the cylindrical pole-pieces F F', each 14 cm. in diameter and 5 cm. broad. The coils consist each of about 250 metres of copper wire 2·5 mm. in diameter. In another research, Verdet used a vertical horse-shoe electromagnet resembling Fig. 23, p. 51, provided

---

* Ruhmkorff. See *Comptes Rendus*, xxiii. 417 and 538, 1846; and *Ann. de Chim. et de Phys.* (3) xviii. 318, 1846.

with perforated pole-pieces, as Faraday had previously done.

FIG. 28.

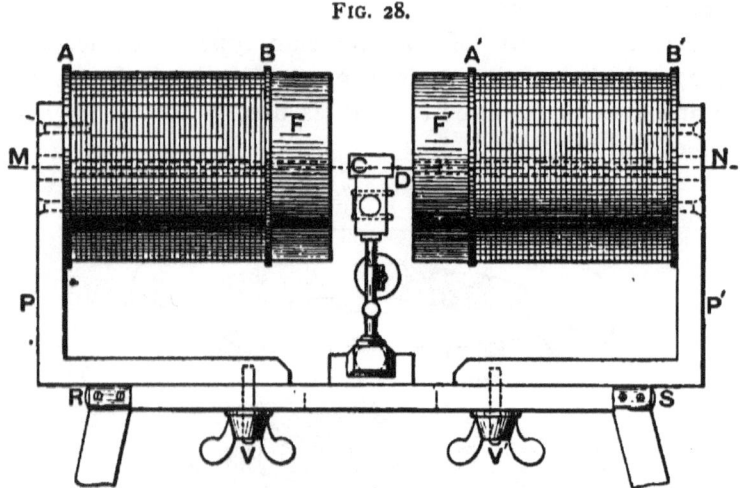

RUHMKORFF'S ELECTROMAGNET.

5. *Coil-and-Plunger.*—A detached iron core is attracted into a hollow coil, or solenoid, of copper wire, when a current of electricity flows round the latter. This is a special form, and will receive extended consideration in Chapter VIII. It will be sufficient here to mention that whereas ordinary electromagnets with fixed cores exert powerful forces on armatures near them, their range of action is usually extremely small; on the other hand, the pull of the coil on the plunger is in general a feebler pull, but one of extended range of action.

6. *Stopped Coil-and-Plunger.*—A whole group of forms have been devised intermediate between the coil-and-plunger form and the ordinary forms with fixed cores. If a coil is arranged with a short fixed core extending part of the way through it, and with a second movable core to be sucked into the coil and finally attracted into close proximity with the fixed core, the action will be of a nature intermediate between the weak long-range pull of the coil-and-plunger and the powerful short-range pull of the ordinary electromagnet.

Sometimes these forms are also iron-clad with an external jacket to better the magnetic circuit.

7. *Electromagnets with Consequent Poles.*—Electromagnets wound with such windings or so connected that the currents circulate in opposite directions over different parts of the core, are peculiar in the fact that they possess " consequent poles." This term is applied to poles that are produced

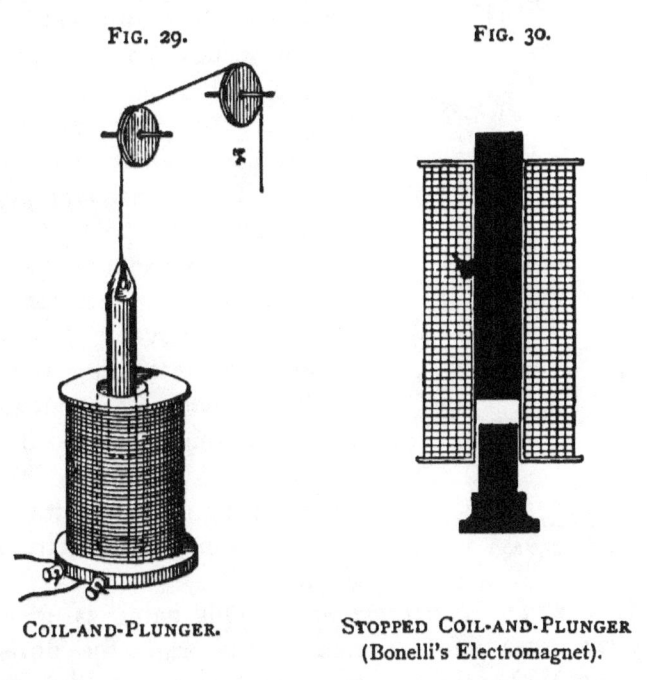

FIG. 29.   FIG. 30.

COIL-AND-PLUNGER.   STOPPED COIL-AND-PLUNGER
(Bonelli's Electromagnet).

between two poles of opposite kind. If, for example, a steel bar is so magnetized as to have a north pole at each end and a south pole at the middle, that middle pole is called a consequent pole; and is regarded as being the magnetic equivalent of two south poles put end to end. Electromagnets with consequent poles are not frequent, except in certain forms of dynamos and of electric motors, wherein they occur sometimes in special forms of field-magnets, sometimes in armatures. An example occurs in the case of the well-known Gramme ring. Suppose a ring core of iron (Fig. 31) is entirely overwound with a copper coil (right-handedly wound in this figure) which is closed on itself. If then a current is introduced into the coil at one side and drawn off at the other, it will have two paths of flow; and half the ring will be subjected to a right-handed circulation of current, while the other half is subjected to a left-handed circulation.

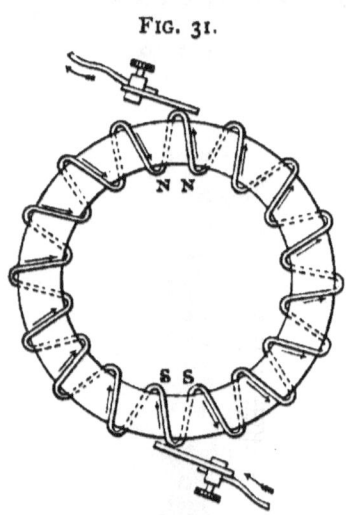

FIG. 31.

RING ELECTROMAGNET WITH CONSEQUENT POLES.

The result will be that each half will act like a magnet having (in this instance) a south pole where the current enters, and a north pole where the current leaves the coil. There will be, therefore, two consequent poles, one double-north pole at the top, one double-south pole at the bottom. The position of these poles is movable; the pole will shift round the ring to any point if the point of connexion between the annular coil and the circuit is shifted. This is indeed the most important feature of this peculiar electromagnetic device.

8. *Circular Electromagnets.*—This name is given to a peculiar class of electromagnets, originally devised by Wilhelm Weber, having cylindrical or pulley-like cores, and coils wound (as in Fig. 159) in grooves around their peripheries. If a simple iron pulley is wound with a magnetizing coil of insulated copper wire, and excited by a current, the whole of one rim will be a north polar surface, and the other rim will be a south polar surface. Such a magnet will attach itself to iron at any part of its periphery. In Fig. 32 a double circular magnet is shown, with a consequent north pole in an intermediate rim, the two outer rims being south poles. It has been proposed to use such electromagnets as a form of driving gear instead of toothed wheels, the requisite adherence of surface being attained magnetically. In another variety,

due to Nicklès, called *paracircular* electromagnets, the coil, instead of being wound on the wheel, is placed outside it, encircling only that segment of it which is to make a magnetic contact.

FIG. 32.

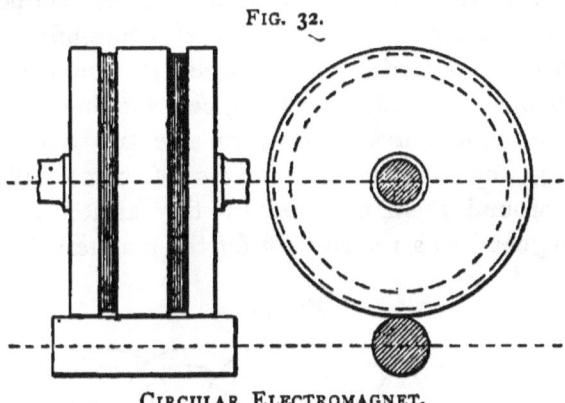

CIRCULAR ELECTROMAGNET.

9. *Miscellaneous Forms.*—Of these there are an innumerable variety, many of which are noticed chiefly in the latter part of this book. Two examples may be given here. Fig. 33

FIG. 33.

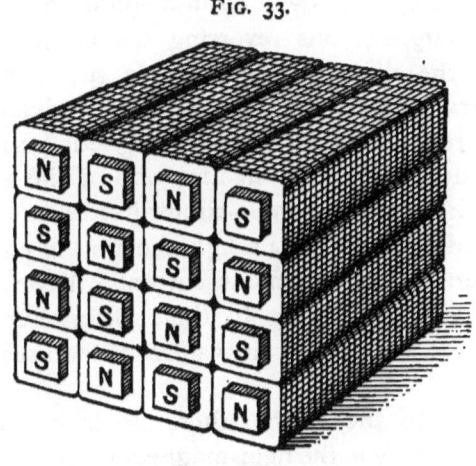

FAGGOT OF ELECTROMAGNETS.

illustrates a faggot of bar electromagnets, each wound with a separate coil of copper wire, and arranged with alternate polarities. This form was intended to provide a powerful electromagnet for the purpose of adherence to an iron slab. As a matter of fact a more powerful effect would have been obtained from the same quantity of iron and copper had a different disposition been adopted, and a simplified magnetic circuit been arranged: for example, it would have been better to have carried all the copper windings in one coil around the eight cores which form the two middle layers, leaving the top and bottom layers of cores without any winding around them, to serve (in the fashion of iron-clad electromagnets) as a return-path for the magnetic lines.

FIG. 34.

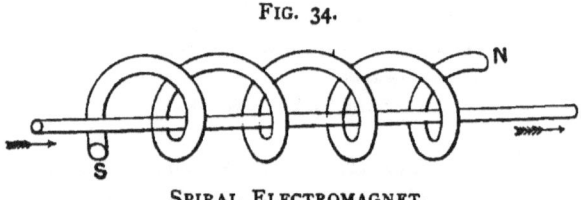

SPIRAL ELECTROMAGNET.

Fig. 34 depicts a spiral electromagnet, in which the iron core, greatly elongated, is wound in a spiral around a straight conductor of copper, thus reversing the usual arrangement. There is not the slightest advantage in such a modification. On the contrary, there are the serious disadvantages that, owing to the relatively considerable magnetic permeability of air, the magnetism created in any one of the turns of the spiral will leak across and short-circuit itself to a considerable degree, instead of going on through the iron to emerge at the pole; and further, an enormous current will be required to energize the magnet, owing to there being but one single conductor in the electric circuit instead of the usual numerous convolutions.

In addition to these forms there are many others, some multipolar, used only in the field-magnets of dynamo machines and of electric motors. For further information about these the reader is referred to treatises on the subject of dynamo-electric machinery, and also to Chapter XII., p. 350.

MATERIALS OF CONSTRUCTION.

Three classes of materials enter into the construction of electromagnets :—Magnetic materials for cores ; electric conducting materials for coils ; insulating materials for preventing leakage of electric current.

I. *Magnetic Materials for Cores.*—Iron and steel are the only materials that need be discussed, though nickel, cobalt, and lodestone (magnetic oxide of iron), also belong to the class of magnetic materials. The magnetic properties of iron and steel are so all-important that Chapter III. is specially devoted to their consideration. Their mechanical properties are so well known as to need no notice here.

II. *Electric Conducting Materials for Coils.*—One material only need here be discussed—namely copper, being the metal universally employed. Other metals have been tried in place of copper wires : for example, iron, German-silver, and silver. Iron was suggested with the idea that, being also a magnetic substance, it would act in double capacity, carrying electricity and permeated simultaneously by the magnetic lines. This application is, however, to be condemned : firstly, because iron wires offer much more electrical resistance than copper wires of the same size, and heat more, and are wasteful of electrical power ; secondly, because the iron wires, being wound transversely with insulating material between them, are badly situated for carrying the longitudinal magnetic lines. Silver has been suggested instead of copper, as being a better electrical conductor. In recent years, however, ways have been found of refining copper, notably by electro-metallurgical processes, and producing it of a quality that rivals silver, and in some cases surpasses it in electric conducting power. "High conductivity" copper is now an established commercial article; and none should be tolerated in the construction of electromagnets that has a lower conductivity than 98 per cent. of that of pure copper. German silver is unfitted for electromagnet coils by reason of its poor conductivity, being only about 8 per cent. of that of pure copper; the only circumstance in its favour being its constancy at all temperatures in this low conducting power. Rules about the winding of copper wire coils, guiding the constructor in the gauge and quantity required, will be given in due course in Chapter VI. In the mean time it may be pointed out that the constructor is not limited to the use of simple round wires, though these are usually employed in the coils of all small electromagnets. For larger coils requiring thicker conductors for carrying many amperes of current, a large sectional area of conductor is needed. Round wires of large sectional area become inadvisable in such cases for two reasons—they are difficult to bend, and they leave great intersticial spaces in the winding. In such cases it is better to employ stranded wires made up of seven smaller wires twisted together and overspun with suitable insulating material, or else to use wire of retangular section either drawn through rectangular dies or built up of narrow copper ribbons bound together by a serving of tape. All the manufacturers of copper wire for electrical purposes now regularly supply both stranded wires and wires built up of strips to any required gauge, these having become things required every day in the construction of dynamos and other electrical machines.

III. *Insulating Materials for preventing Leakage of Electric Current.*—Insulating materials are required to prevent the electric current from leaking from one part of the coil to another, and also to prevent it from leaking from the coil into the core. The coil also requires to be held together mechanically; this is usually accomplished by winding it on a bobbin or former. If the bobbin is of wood, ebonite, or other non-metallic substance, it itself aids as an insulator in lessening the chance of leakage from coil to core. If the bobbin is of metal, then care must be taken to prevent the electric current from leaking from the coil into the metal of the bobbin. Theoretically, any one point of leakage from coil to core or from coil to bobbin will not matter, as the current will not flow, even along a leakage path, from the coil into the

Insulating Materials.

neighbouring metal unless it can find also a way out back into the circuit. But the very fact that such a possible leakage path exists increases the probability of another leak being established at some other point, resulting in a breakdown. Complete insulation of the coil is therefore most desirable in itself.

The tendency for a leak or short-circuit to occur between any two conductors depends on the difference of their electric pressures. The greater their difference in electric pressure, the more likely is the insulating material between them to break down and establish a leak. There is generally very little difference in electric pressure between one turn of wire and the turn that lies next to it in the same layer, but there may be a considerable difference in electric pressure between the wire in one layer and the wire that lies over it in the next layer. Hence, it is more important to insulate well between layer and layer, than between wire and wire in the same layer. For the very same reasons it is still more important to insulate well between coil and bobbin, because the wire of all the layers comes close up to the cheeks of the bobbin. A numerical example will help here. Suppose an electromagnet is to be used across a pair of electric mains which are supplied with current from a dynamo at a pressure of 100 volts; that is to say, there is a difference of pressure of 100 volts between the mains. Suppose this electromagnet to possess a coil of wire consisting of twenty layers with fifty turns in each layer. The layers are supposed to be wound in the usual way, beginning at one end and returning back, so that the second layer ends just over the spot where the first layer began. Then if there is 100 volts difference of pressure applied to the coil to drive the current through the convolutions of its coil, there will be between a point at the beginning of one layer and a point at the end of the next (which lies on the top of it) just one-tenth of all the coil, and therefore one-tenth of all the pressure—*i.e.*, 10 volts. But between these two points in the windings there will be 100 turns of wire; hence the difference of pressure between wire and wire will be only $\frac{1}{10}$ volt. Yet, since the pressure between one end of the coil and the other end is 100 volts, and as both ends come close to the bobbin, the tendency to spring a leak into the metal of the bobbin will be far greater than the tendency to leak from layer to layer. It may be approximately taken that the electric tension on the insulating material—or tendency to pierce it with a spark—is proportional to the square of the volts of electric pressure. Hence, in the case given, the insulating material between coil and bobbin has to stand a tension one hundred times as great as that between layer and layer, and one million times as great as that between wire and wire.

Wire Insulation.—For large electromagnets, to be used on ordinary circuits with pressures not exceeding 500 volts, it is sufficient to use as wire insulation a double covering of cotton, which is afterwards well soaked with shellac varnish and dried for some hours at steam-heat. For small electromagnets, such as are used in telegraphic and telephonic work and in the best electric-bell work, silk-covered wires are preferable, the coils being afterwards baked and immersed in a bath of melted paraffin wax. Coatings of gutta-percha or indiarubber, or of tarred hemp or ozokerited tape, are not recommended for the wires that are to be wound on the coils of electromagnets. Stranded wires of large size and rectangular conductors made of strip should be taped with a strong cotton tape, and served with shellac varnish after being wound on the bobbins or magnets.

Layer Insulation.—The insulation between layer and layer, *particularly between the first turn of one layer and the last of the next*, must be good. In small electromagnets no particular precautions are needed beyond care in winding; but, in the case of larger electromagnets, if wound with many turns of fine wire to be used on high-pressure circuits, it is well to lay between the layers a wrapping of thin Willesden paper or of thin vulcanized fibre, or even of thin canvas or cotton cloth varnished with shellac varnish.

Core and Bobbin Insulation.—In those cases where coils

are to be wound direct on the cores without any bobbin between, the core should be itself well insulated by being painted with some good tough and non-conducting paint, such as bath-japan or Aspinall's enamel, or covered with varnished canvas, or with a well-fitting tube of vulcanized fibre, or with several thicknesses of well-varnished paper of tough quality. Metal bobbins should be served in the same way. The faces of the cheeks of bobbins should be most thoroughly and carefully protected. In cases where coils are wound direct on cores, metal cheeks or "magnet heads" are often used to hold up the coils at the ends. Such cheeks, whether attached to the core or to a separate bobbin, ought in all large electromagnets to be separated from the coils by the interposition of a sheet of vulcanized fibre or of dermatine or Willesden paper, or, failing these, of oil-cloth or tough paper, varnished or enamelled. Only by such precautions can break-downs be avoided. Trouble often arises from the end of the wire which comes up from the innermost layer of the coil to the outside. This leading-out end often breaks off in a most annoying way, necessitating rewinding; or if not actually broken it works loose and endangers the insulating layers between itself and other parts of coil or bobbin. It is well in many cases, before winding the coil, to provide as a leading-out end a specially strong wire, or piece of stranded cable, or strip of copper; and such should always be extra well insulated.

Fireproof Insulation.—In certain cases of rare occurrence it is needful to provide insulation that will not break down even if the coils become red hot. Asbestos, though itself a poor insulator and bulky, is then the only possible material for covering wires; asbestos sheet and mica may both be used for layer-insulation. An insulating paint containing asbestos is also to be had. A naked copper wire with a rather thicker thread of spun asbestos laid between may serve as coil, with asbestos cloth between successive layers. Stoneware washers and bushes may be inserted for bringing out the leading-out wires.

High-pressure Insulation.—When electromagnets are to be constructed for use on extra high-pressure circuits exceeding 1,000 volts, it is absolutely essential to secure the most perfect insulation between layers and between coil and bobbin. Varnished paper, canvas, and vulcanized fibre all break down. Layers of thin mica secured in position, and sheets of good ebonite, are almost the only materials of any avail. Some constructors, instead of winding the coils in layers, wind them between cloisons of ebonite fixed at intervals along an ebonite tube surrounding the core—the construction habitually used in the building of induction coils. Ozokerited paper in several successive layers, consolidated by pressure while hot (as in the Ferranti mains) seems to be the only other material worth naming in this regard.

CHAPTER III.

THE PROPERTIES OF IRON.

A KNOWLEDGE of the magnetic properties of iron of different kinds is absolutely fundamental to the theory and design of electromagnets. No excuse is therefore necessary for treating this matter with some fulness. In all modern treatises of magnetism the usual terms are defined and explained, and some of them have been explained in Chapter II. Magnetism, which was formerly treated of as though it were something distributed over the end surfaces of magnets, is now known to be a phenomenon of internal structure; and the appropriate mode of considering it is to treat the magnetic materials—iron and the like—as being capable of acting as good conductors of the magnetic lines; in other words, as possessing magnetic permeability. The precise notion now attached to this word is that of a numerical co-efficient. Suppose a magnetic force —due, let us say, to the circulation of an electric current in a surrounding coil—were to act on a space occupied by air, there would result a certain number of magnetic lines in that space. In fact, the intensity of the magnetic force, symbolized by the letter H, is often expressed by saying that it would produce H magnetic lines per square centimetre in air. Now, owing to the superior magnetic power of iron, if the space subjected to this magnetic force were filled with iron instead of air, there would be produced a larger number of magnetic lines per square centimetre. This larger number in the iron expresses the degree of magnetization in the iron; it is symbolized* by the letter B. The ratio of B to H expresses the permeability of the material. The usual symbol for permeability is the Greek letter μ. So we may say that B is equal to μ times H. For example, a certain specimen of iron, when subjected to a magnetic force capable of creating, in air, 50 magnetic lines to the square centimetre, was found to be permeated by no fewer than 16,062 magnetic lines per square centimetre. Dividing the latter figure by the former, gives as the value of the permeability at this stage of the magnetization 321, or the permeability of the iron is 321 times that of air. The permeability of such non-magnetic materials as silk, cotton, and other insulators, also of brass, copper, and all the non-magnetic metals, is taken as 1, being practically the same as that of the air.

This mode of expressing the facts is, however, complicated by the fact of the tendency in all kinds of iron to magnetic saturation. In all kinds of iron the magnetizability of the material becomes diminished as the actual magnetization is pushed further. In other words, when a piece of iron has been magnetized up to a certain degree, it becomes, from that degree onward, less permeable to further magnetization, and though actual saturation is never reached, there is a practical limit beyond which the magnetization cannot well be pushed. Joule was one of the first to establish this tendency toward magnetic saturation. Modern researches have shown numeri-

---

* The following are the various ways of expressing the three definitions :—
   B—The internal magnetization.
      The magnetic induction.
      The induction.
      The intensity of the induction.
      The permeation.
      The number of lines per square centimetre in the material.
   H—The magnetizing force at a point.
      The magnetic force at a point.
      The intensity of the magnetic force.
      The number of lines per square centimetre that there would be in air.
   μ—The magnetic permeability.
      The permeability.
      The specific conductivity for magnetic lines.
      The magnetic multiplying power of the material.

cally how the permeability diminishes as the magnetization is pushed to higher stages. The practical limit of the magnetization, B, in good wrought iron is about 20,000 magnetic lines to the square centimetre, or about 125,000 lines to the square inch; and, in cast iron the practical saturation limit is nearly 12,000 lines per square centimetre, or about 70,000 lines per square inch.

In designing electromagnets, before calculations can be made as to the size of a piece of iron required for the core of a magnet for any particular purpose, it is necessary to know the magnetic properties of that piece of iron; for it is obvious that if the iron be of inferior magnetic permeability, a larger piece of it will be required in order to produce the same magnetic effect as might be produced with a smaller piece of higher permeability. Or again, the piece having inferior permeability will require to have more copper wire wound on it; for in order to bring up its magnetization to the required point, it must be subjected to higher magnetizing forces than would be necessary if a piece of higher permeability had been selected.

A convenient mode of studying the magnetic facts respecting any particular brand of iron is to plot on a diagram the curve of magnetization—i.e. the curve in which the values, plotted horizontally, represent the magnetic force, H, and the values plotted vertically those that correspond to the respective magnetization, B. In Fig. 35, which is modified from the researches of Professor Ewing,* are given five curves, relating to soft iron, hardened iron, annealed steel, hard-drawn steel, and glass-hard steel. It will be noticed that all these curves have the same general form. For small values of H the values of B are small, and as H is increased B increases also. Further, the curve rises very suddenly, at least with all the softer sorts of iron, and then bends over and becomes nearly horizontal. When the magnetization is in the stage below the bend of the curve, the iron is said to be far from the state of saturation. But when the magnetization has been

* *Phil. Trans.*, 1885.

pushed beyond the bend of the curve, the iron is said to be in the stage approaching saturation; because at this stage of magnetization it requires a large increase in the magnetizing force to produce even a very small increase in the magnetization. It will be noted that for soft wrought iron the stage of

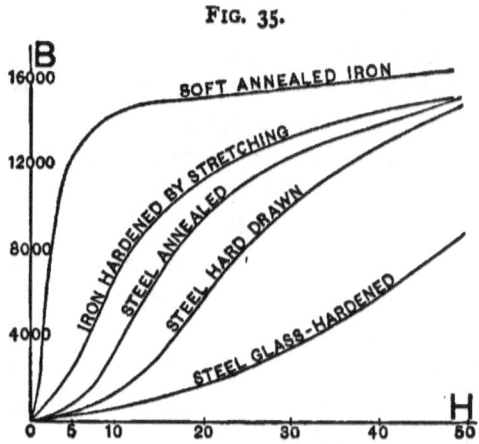

FIG. 35.

CURVES OF MAGNETIZATION OF DIFFERENT MAGNETIC MATERIALS.

approaching saturation sets in when B has attained the value of about 16,000 lines per square centimetre, or when H has been raised to the value of about 50. As we shall see, it is not economical to push B beyond this limit; or, in other words, it does not pay to use stronger magnetic forces than those of about H = 50.

METHODS OF MEASURING PERMEABILITY.

There are four sorts of experimental methods of measuring permeability.

1. *Magnetometric Methods.*—These are due to Müller, and consist in surrounding a bar of the iron in question by a magnetizing coil, and observing the deflexion its magnetization produces in a magnetometer.

2. *Balance Methods.*—These methods are a variety of the preceding, a compensating magnet being employed to balance

the effect produced by the magnetized iron on the magnetometric needle. Von Felitzsch used this method, and it has received a more definite application in the magnetic balance of Professor Hughes. The author has had a large number of observations made by students of the Technical College, by its means, upon sundry samples of iron and steel. None of these methods are, however, to be compared with those that follow.

3. *Inductive Methods.*—There are several varieties of these, but all depend on the generation of a transient induction-current in an exploring coil which surrounds the specimen of iron, the integral current being proportional to the number of magnetic lines introduced into, or withdrawn from, the circuit of the exploring coil. Three varieties may be mentioned.

(*A.*) *Ring Method.*—In this method, due to Kirchhoff, the iron under examination is made up into a ring, which is wound with a primary or exciting coil, and with a secondary or exploring coil. Determinations on this plan have been made by Stowletow, Rowland, Bosanquet, and Ewing; also by Hopkinson. Rowland's arrangement of the experiment is

FIG. 36.

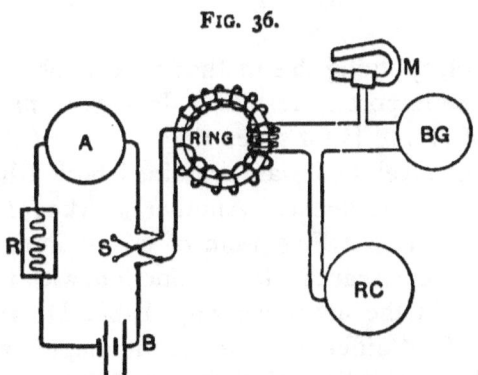

RING METHOD OF MEASURING PERMEABILITY.
(Rowland's Arrangement.)

shown in Fig. 36, in which B is the exciting battery, S the switch for turning on or reversing the current, R an adjustable resistance, A an amperemeter, and B G the ballistic galvanometer, the first swing of which measures the integral induced current. R C is an earth-inductor or reversing coil wherewith to calibrate the readings of the galvanometer; and above is an arrangement of a coil and a magnet to assist in bringing the swinging needle to rest between the observations. The exciting coil and the exploring coil are both wound upon the ring; the former is distinguished by being drawn with a thicker line. The usual mode of procedure is to begin with a feeble exciting current, which is suddenly reversed, and then reversed back. The current is then increased, reversed, and re-reversed; and so on, until the strongest available points are reached. The values of the magnetizing force H are calculated from the observed value of the current by the following rule:—If the strength of the current, as measured by the amperemeter, be i, the number of spires of the exciting coil S, and the length in centimetres of the coil (i.e. the mean circumference of the ring) be l, then H is given by the formula—

$$H = \frac{4\pi}{10} \times \frac{Si}{l} = 1{\cdot}2566 \times \frac{Si}{l}.$$

Bosanquet, applying this method to a number of iron rings, obtained some important results. In Fig. 37 are plotted out the values of H and B for seven rings. One of these, marked J, was of cast steel, and was examined both when soft and afterwards when hardened. Another, marked I, was of the best Lowmoor iron. Five were of crown iron, of different sizes. They were marked for distinction with the letters G, E, F, H, K. In the accompanying Table II. are set down the values of B at different stages of the magnetization.

I have the means here of illustrating the induction method of measuring permeability. Here is an iron ring, having a cross section of almost exactly one square centimetre. It is wound with an exciting coil supplied with current by two accumulator cells; over it is also wound an exploring coil of 100 turns connected in circuit (as in Rowland's arrangement)

TABLE II.—VALUES OF B IN FIVE CROWN IRON RINGS.

Name.	G.	E.	F.	H.	K.
Mean Diam. Bar thickness.	21·5 cm. 2·535	10·035 cm. 1·298	22·1 cm. 1·292	10·735 cm. 0·7137	22·725 cm. 0·7544
Magnetizing Force.					
0·2	126	73	62	82	85
0·5	377	270	224	208	214
1	1,449	1,293	840	675	885
2	4,564	3,952	3,533	2,777	2,417
5	9,900	9,147	8,293	8,479	8,884
10	13,023	13,357	12,540	11,376	11,388
20	14,911	14,653	14,710	14,066	13,273
50	16,217	15,704	16,062	15,174	13,890
100	17,148	16,677	17,900	16,134	14,837

with a ballistic galvanometer which reflects a spot of light upon yonder screen. In the circuit of the galvanometer is also included a reversing earth-coil. As a matter of fact this earth-coil is of such a size, and wound with so many convolutions of wire, that when it is turned over, the amount of cutting of magnetic lines is equal to 840,000, or is the same as if 840,000 magnetic lines had been cut once. By adjusting the resistance of the galvanometer circuit, it is arranged that the first swing due to the induced current, when I suddenly turn over the earth-coil, is 8·4 scale divisions. Then, seeing that our exploring coil has 100 turns, it follows that when in our subsequent experiment with the ring we get an induced current from it, each division of the scale over which the spot swings will mean 1000 lines in the iron. I turn on my exciting current. See: it swings about 11 divisions. On breaking the circuit it swings nearly 11 divisions the other way. That means that the magnetizing force carries the magnetization of the iron up to 11,000 lines; or, as the cross section is about 1 sq. cm., B 11,000. Now, how much is H? The exciting coil has 180 windings, and the exciting current through the amperemeter is just 1 ampere. The total excitation is just 180 "ampere-turns." We must, according to our rule given above, multiply this by 1·2566 and divide by the mean circumferential length of the coil,

FIG. 37.

BOSANQUET'S DATA OF MAGNETIC PROPERTIES OF IRON AND STEEL RINGS.

which is about 32 cm. This makes H = 7. So if B = 11,000 and H = 7, the permeability (which is the ratio of them) is about 1570. It is a rough and hasty experiment, but it illustrates the method.

Bosanquet's experiments settle the debated question whether the outer layers of an iron core shield the inner layers from the influence of magnetizing forces. Were this the case, the rings made from thin bar iron should exhibit higher values of B than do the thicker rings. This is not so; for the thickest ring, G, shows throughout the highest magnetizations.

(*B.*) *Bar Method.*—This method consists in employing a long bar of iron instead of a ring. It is covered from end to end with the exciting coil, but the exploring coil consists of but a few turns of wire situated just over the middle part of the bar. Rowland, Bosanquet, and Ewing have all employed this variety of method ; and Ewing specially used bars the length of which was more than 100 times their diameter, in order to get rid of errors arising from end effects.

(*C.*) *Divided Bar Method.*—This method, due to Dr. Hopkinson,[*] is illustrated by Fig. 38.

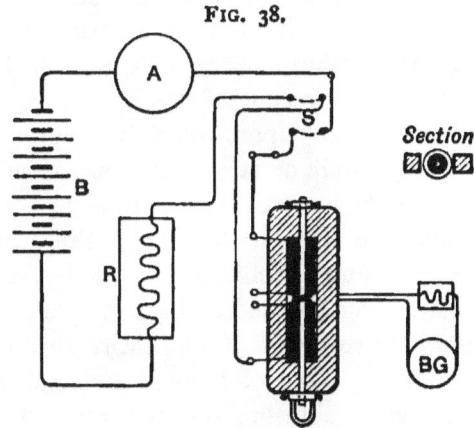

FIG. 38.

HOPKINSON'S DIVIDED BAR METHOD OF MEASURING MAGNETIC PERMEABILITY.

The apparatus consists of a block of annealed wrought iron about 18 inches long, 6½ wide, and 2 deep, out of the middle of which is cut a rectangular space to receive the magnetizing coils.

The test samples of iron consist of two rods, each 12·65 mm. in diameter, turned carefully true, and slide in through holes bored in the ends of the iron blocks. These two rods meet in the middle, their ends being faced true so as to make a good contact. One of them is secured firmly, and the other has a handle fixed to it, by means of which it can be withdrawn. The two large magnetizing coils do not meet, a space being left between them. Into this space is introduced the little exploring coil, wound upon an ivory bobbin,

[*] *Phil. Trans.*, 1885, p. 456.

through the eye of which passes the end of the movable rod. The exploring coil is connected to the ballistic galvanometer, B G, and is attached to an india-rubber spring (not shown in the figure), which, when the rod is suddenly pulled back, causes it to leap entirely out of the magnetic field. The exploring coil had 350 turns of fine wire; the two magnetizing coils had 2008 effective turns. The magnetizing current, generated by a battery, B, of eight Grove cells, was regulated by a variable liquid resistance, R, and by a shunt resistance. A reversing switch and an amperemeter, A, were included in the magnetizing circuit. By means of this apparatus the sample rods to be experimented upon could be submitted to any magnetizing forces, small or large, and the actual magnetic condition could be examined at any time by breaking the circuit and simultaneously withdrawing the movable rod. This apparatus, therefore, permitted the observation separately of a series of increasing (or decreasing) magnetizations without any intermediate reversals of the entire current. Thirty-five samples of various irons of known chemical composition were examined by Hopkinson, the two most important for present purposes being an annealed wrought iron and a grey cast iron, such as are used by Messrs. Mather and Platt in the construction of dynamo machines. Hopkinson embodied his results in curves, from which it is possible to construct, for purposes of reference, numerical tables of sufficient accuracy to serve for future calculations. The curves of these two samples of iron are reproduced in Fig. 39, but with one simple modification. British engineers who, unfortunately, are condemned by local circumstances to use inch measures instead of the international metric system, prefer to have the magnetic facts also stated in terms of square inch units instead of square cm. units. This change has been made in Fig. 39, and the symbols B_{u} and H_{u} are chosen to indicate the numbers of magnetic lines to the square inch in iron and in air respectively. The permeability, or multiplying power of the iron, is the same, of course, in either measure. In Table III. are given the corresponding data in square inch measure; and in Table IV. the data in square cm. measure for the same specimens of iron.

It will be noted that Hopkinson's curves are double, there being one curve for the ascending magnetizations, and a

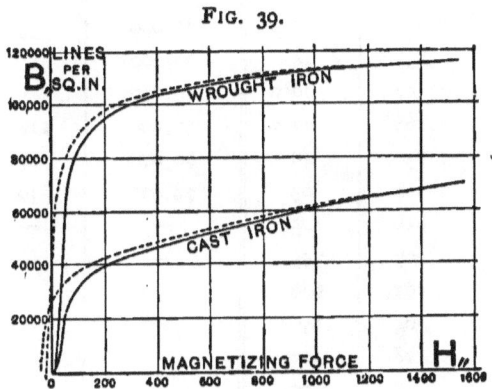

CURVES OF MAGNETIZATION OF IRON.

separate one, a little above the former, for descending magnetizations. This is a point of a little importance in designing electromagnets. Iron, and particularly hard sorts of iron, and steel, after having been subjected to a high degree of magnetizing force, and subsequently to a lesser magnetizing force, are found to retain a higher degree of magnetization than if the lower magnetizing force had been simply applied. For example, reference to Fig. 37 shows that the wrought iron, where subjected to a magnetizing force gradually rising from zero to $H_{\shortparallel}$ = 200, exhibits a magnetization of $B_{\shortparallel}$ = 95,000; but after $H_{\shortparallel}$ has been carried up to over 1000, and then reduced again to 200, $B_{\shortparallel}$ does not come down again to 95,000, but only to 98,000. Any sample of iron which showed great retentive qualities, or in which the descending curve differs widely from the ascending curve, would be unsuitable for constructing electromagnets, for it is important that there should be as little residual magnetism as possible in the cores. It will be noted that the curves for cast iron

TABLE III.—SQUARE INCH UNITS.

Annealed Wrought Iron.			Grey Cast Iron.		
B_{u}	μ	H_{u}	B_{u}	μ	H_{u}
30,000	2,926	10·2	25,000	833	30
40,000	2,857	14	30,000	445	53·5
50,000	2,392	20·9	40,000	245	163
60,000	2,166	27·7	50,000	112	447
70,000	1,750	40	60,000	64	940
80,000	1,368	63	70,000	40	1750
90,000	856	105	—	—	—
100,000	407	245	—	—	—
110,000	161	686	—	—	—
120,000	64	1850	—	—	—
130,000	28	4500	—	—	—
140,000	18	7630	—	—	—

TABLE IV.—SQUARE CENTIMETRE UNITS.

Annealed Wrought Iron.			Grey Cast Iron.		
B	μ	H	B	μ	H
5,000	3,000	1·66	4,000	800	5
9,000	2,250	4	5,000	500	10
10,000	2,000	5	6,000	279	21·5
11,000	1,692	6·5	7,000	133	42·
12,000	1,412	8·5	8,000	100	80
13,000	1,083	12	9,000	71	127
14,000	823	17	10,000	53	188
15,000	526	28·5	11,000	37	292
16,000	320	50	—	—	—
17,000	161	105	—	—	—
18,000	90	200	—	—	—
19,000	54	350	—	—	—
20,000	30	666	—	—	—

Curves of Magnetic Properties.

Fig. 40.

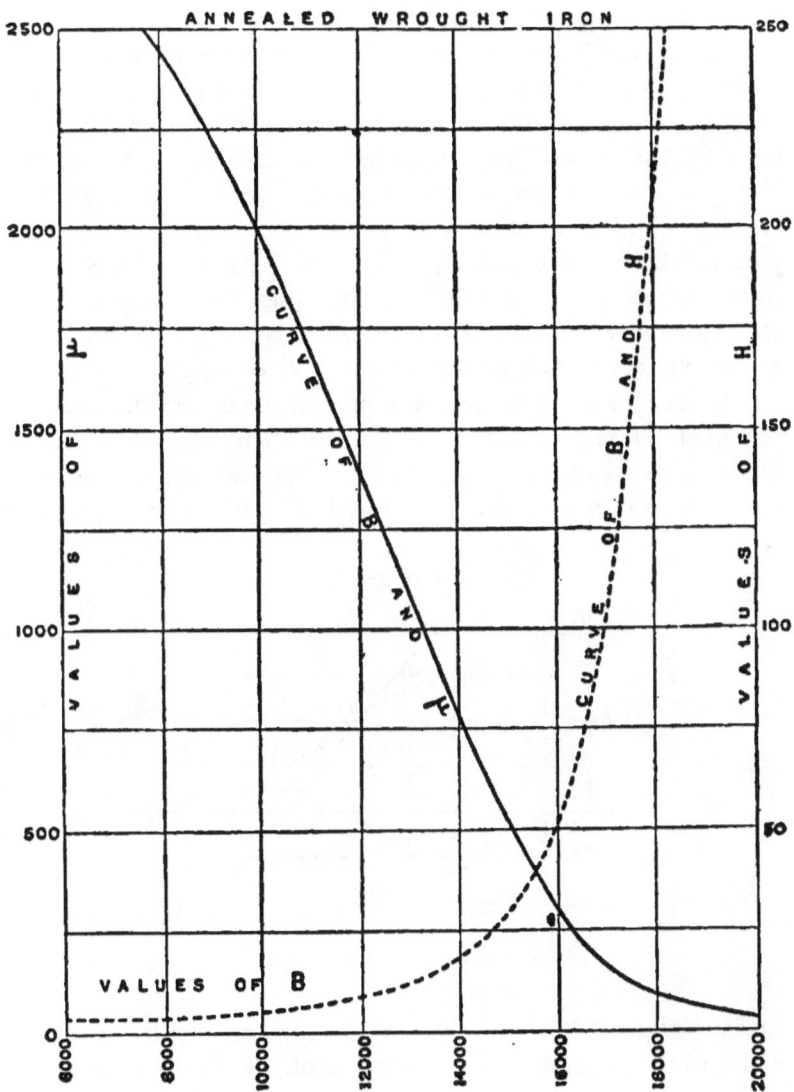

Curves of Magnetic Properties of Annealed Wrought Iron, Plotted from Table IV.

show more of this residual effect than do those for wrought iron. The numerical data in Tables III. and IV. are averages of the ascending and descending values.

As an example of the use of the Tables, we may take the following:—How strong must the magnetizing force be in order to produce in wrought iron a magnetization of 110,000 lines to the square inch? Reference to Table III., or to Fig. 39, shows that a magnetizing field of 664 will be required, and that at this stage of the magnetization the permeability of the iron is only 166. As there are 6·45 square cm. to the square inch, 110,000 lines to the square inch corresponds very nearly to 17,000 lines to the square cm., and $H_{,,} = 664$ corresponds very nearly to $H = 100$.

A very useful alternative mode of studying the results obtained by experiment is to construct curves, such as those of Fig. 41, in which the values of the permeability are plotted out vertically in correspondence with the values of B plotted

FIG. 41.

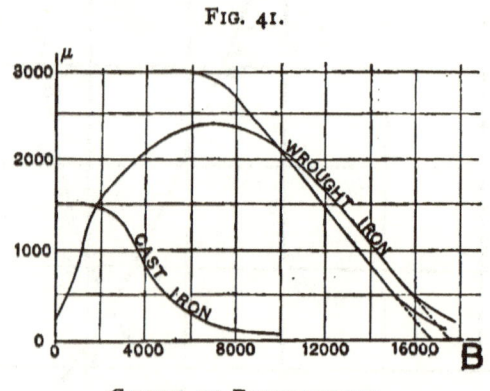

CURVES OF PERMEABILITY.

horizontally. Two of these relate respectively to Hopkinson's data for cast iron and for wrought iron, as given in Table IV. The third curve corresponds to Bidwell's data, Table V. It will be noticed that in the case of Hopkinson's specimen of annealed wrought iron, between the points where $B = 7000$ and $B = 16,000$, the mean values of μ lie almost on a straight line, and might be approximately calculated from the equation :—

$$= (17{,}000 - B) \div 3\cdot 5.$$

The data given in Table IV. are plotted out to a larger scale in Fig. 40 to facilitate the calculation of values of μ and of H from any given value of B. For example, if it is desired to know how great the permeability (for iron of this kind) when B is forced up to 12,000 lines to the square cm., reference to the curve of B and μ shows that at that stage of magnetization μ will be about 1400.

To facilitate the operation of making observations on new brands of iron, Mr. J. Swinburne has recently* devised a method of experimenting which obviates the use of a ballistic galvanometer. The reader is referred to the original paper for further details.

4. *Traction Methods.* — Another group of methods of measuring permeability is based upon the law of magnetic traction. Of these there are several varieties.

(*D.*) *Divided Ring Method.*—Mr. Shelford Bidwell has kindly lent me the apparatus with which he carried out this method. It consists of a ring of very soft charcoal iron rod 6·4 mm. in thickness, the external diameter being 8 cm., sawn into two half rings, and then each half carefully wound over with an exciting coil of insulated copper wire of 1929 convolutions in total. The two halves fit neatly together; and in this position it constitutes practically a continuous ring. When an exciting current is passed round the coils, both halves become magnetized and attract one another; the force required to pull them asunder is then measured. According to the law of traction, which will occupy us in the second lecture, the tractive force (over a given area of contact) is proportional to the square of the number of magnetic lines that pass from one surface to the other through the contact joint. Hence the force of traction may be used to determine B; and on calculating H as before, we can determine the

* See *The Electrician*, vol. xxv. p. 648, October 10, 1890.

permeability. The following Table V. gives a summary of Mr. Bidwell's results:—

TABLE V.—SQUARE CENTIMETRE MEASURE.

Soft Charcoal Iron.		
B	μ	H
7,390	1899·1	3·9
11,550	1121·4	10·3
15,460	386·4	40
17,330	150·7	115
18,470	88·8	208
19,330	45·3	427
19,820	33·9	585

(*E.*) *Divided Rod Method.*—In this method, also used by Mr. Bidwell, an iron rod, hooked at both ends, was divided across the middle, and placed within a vertical surrounding magnetizing coil. One hook was hung up to an overhead support; to the lower hook was hung a scale-pan. Currents of gradually-increasing strength were sent around the magnetizing coil from a battery of cells, and note was taken of the greatest weight which could in each case be placed in the scale-pan without tearing asunder the ends of the rods.

(*F.*) *Permeameter Method.*—This is a method which the author has himself devised for the purpose of testing specimens of iron. It is essentially a workshop method, as distinguished from a laboratory method. It requires no ballistic galvanometer, and the iron does not need to be forged into a ring or wound with a coil. For carrying it out a simple instrument is needed, which he ventures to denominate as a *permeameter.* Outwardly, it has a general resemblance to Dr. Hopkinson's apparatus, and consists (Fig. 42) of a rectangular piece of soft wrought iron, slotted out to receive a magnetizing coil, down the axis of which passes a brass tube.

The block is 12 in. long, 6½ in. wide, and 3 in. in thickness. At one end the block is bored to receive the sample of iron that is to be tested. This consists simply of a thin rod about a foot long, one end of which must be carefully surfaced up. When it is placed inside the magnetizing coil, and the exciting current is turned on, the rod sticks tightly at its lower end to the surface of the iron block; and the force required to detach it (or, rather, the square root of that force) is a measure of the permeation of the magnetic lines through its end face. In the first permeameter which I constructed the magnetizing coil is 13·64 cm. in length, and has 371 turns of wire. One ampere of exciting current consequently produces a magnetizing force of $H = 34$. The wire is thick enough to carry 30 amperes, so that it is easy to reach a magnetizing force of 1000. The current I now turn on is 25 amperes. The two rods here are of "charcoal iron" and "best iron" respectively; they are of quarter-inch square stuff. Here is a spring balance, graduated carefully and provided with an automatic catch so that its index stops at the highest reading. The tractive force of the charcoal iron is about 12½ lb., while that of the "best" iron is only 7½ lb. B is about 19,000 in the charcoal iron, and H being 850, μ is about 22·3.

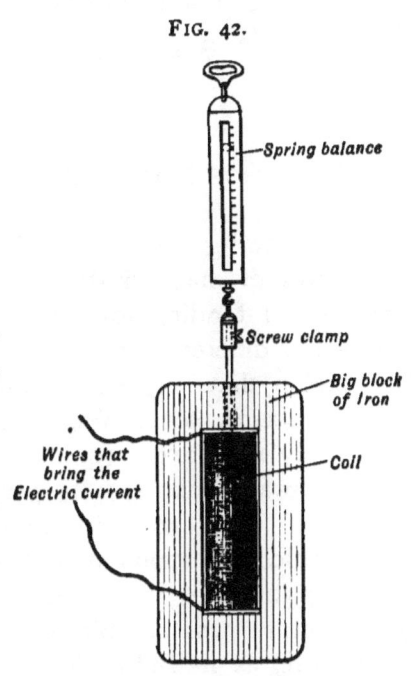

FIG. 42.

THE PERMEAMETER.

The law of traction which is used in calculating B will occupy us much in the next chapter, but meantime I content myself in stating it here for use with the permeameter. The formula for calculating B when the core is thus detached by a pull of P pounds (the coil being left behind), the area of contact being A square inches, is as follows:—

$$B = 1317 \times \sqrt{P \div A} + H.$$

Limits of Magnetization and Permeability.

In reviewing the results obtained, it will be noted that the curves of magnetization all possess the same general features, all tending toward a practical maximum, which, however, is different for different materials. Joule expressed the opinion that *no force of current could give an attraction equal to 200 lb. per sq. in.*, the greatest he actually attained being only 175 lb. per sq. in. Rowland was of opinion that the limit was about 177 lb. per sq. in. for an ordinary good quality of iron, even with infinitely great exciting power. This would correspond roughly to a limiting value of B of about 17,500 lines to the sq. cm. This value has, however, been often surpassed. Bidwell obtained 19,820, or possibly a trifle more, as in Bidwell's calculation the value of H has been needlessly discounted. Hopkinson gives 18,250 for wrought iron, and 19,840 for mild Whitworth steel. Kapp gives 16,740 for wrought iron, 20,460 for charcoal iron in sheet, and 23,250 for charcoal iron in wire. Bosanquet found the highest value in the middle bit of a long bar to run up in one specimen to 21,428, in another to 29,388, in a third to 27,688. Ewing, working with extraordinary magnetic power, forced up the value of B in Lowmoor iron to 31,560 (when μ came down to 3), and subsequently to 45,350. This last figure corresponds to a traction exceeding 1000 lb. to the square inch.

In the following table are given some of Ewing's figures relating to the magnetization of Swedish iron in very strong magnetic fields:—

TABLE VI.—SWEDISH IRON (Square Centimetre Measure).

H	B	μ
1,490	22,650	15·20
3,600	24,650	6·85
6,070	27,130	4·47
8,600	30,270	3·52
18,310	38,960	2·13
19,450	40,820	2·10
19,880	41,140	2·07

Cast iron falls far below these figures. Hopkinson, using a magnetizing force of 240, found the values of B to be 10,783 in grey cast iron, 12,408 in malleable cast iron, and 10,546 in mottled cast iron. Ewing, with a magnetizing force nearly fifty times as great, forced up the value of B in cast iron to 31,760. Mitis metal, which is a sort of cast wrought iron, being a wrought iron rendered fluid by addition of a small percentage of aluminium, is, as I have found, more magnetizable than cast iron, and not far inferior to wrought iron. It should form an excellent material for the cores of electromagnets for many purposes where a cheap manufacture is wanted.

It was at one time supposed that the values of B would show a limiting value at about 20,000, for example, in wrought iron. Ewing's figures, obtained with enormous magnetizing forces, show that this is not so; but, on the other hand, they show that B − H does tend to a limit. In other words, that part of B which is directly due to the presence of the iron, does tend to a true saturation limit. This maximum appears to be about 21,360 in wrought iron, and 15,580 in cast iron.

EFFECT OF AIR-GAP IN MAGNETIC CIRCUIT.

All the preceding results refer exclusively to that which goes on in the iron itself, the curves of magnetization referring to the magnetic materials only. They tell us (in terms of H) the magnetic force required to drive B lines of magnetization through a single cubic cm. of the material. If we are to deal with an actual piece of iron that is more than 1 sq. cm. in cross section, and more than 1 cm. in length, all that is necessary to represent the facts (so long as we are dealing with magnetization that is entirely internal to the iron) is to change the scale of the curves. For example, suppose we are dealing with an iron ring made of a piece of square bar annealed wrought iron (of the same brand as Hopkinson used), the size of the iron being 2 cm. in the side, and its (mean) length 80 cm.; we shall now have to plot out (instead of B and H) N, the whole flux of magnetic lines within the section of the iron, and Hl, the line-integral of the magnetizing force around the length of the iron circuit.

Taking the curve of B and H in Fig. 40, the scales will then have to be changed as follows:—Since the area is 4 sq. cm., N at any stage of the magnetization will be equal to four times B at that stage. Hence the point on the horizontal scale called B = 16,000, will now have to be called N = 64,000. And since l, the length of the bar, is 80 cm., the same point that now stands for H = 50 on the vertical scale (on the right side), which is the corresponding value of the magnetic force, will have to be called Hl = 4000. With these changes of scales the curve will then serve to represent the magnetic behaviour of the whole ring; it will tell us how much integral magnetizing force we must exert (by means of a current in a coil) in order to drive up N, the total flux of magnetic lines, to any desired amount. If we know Hl, it is easy to calculate the requisite ampere-turns, because (as shown on p. 46) the ampere-turns multiplied by 1·257 are equal to the line-integral of the magnetizing force.

But if there is an air-gap in the magnetic circuit, or if there is a gap filled with any non-magnetic material, seeing that all these things possess a permeability that is equal to that of air (*i.e.* = 1), it is evident that to force the same number of magnetic lines across a layer of such inferior permeability will necessitate an increase in the amount of magnetizing power that must be applied.

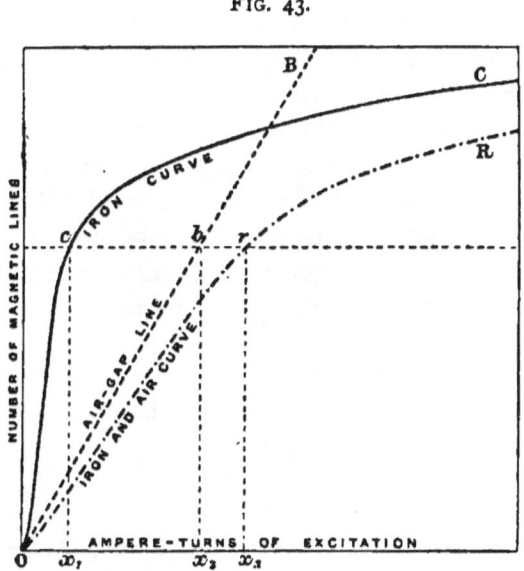

FIG. 43.

CURVE OF MAGNETIZATION OF MAGNETIC CIRCUIT WITH AIR-GAP.

This is made plainer by reference to Fig. 43, in which the curve O c C represents the relation between the number of magnetic lines in an iron bar and the number of ampere-turns of excitation (H $l \div$ 1·257) needed to force these magnetic lines through the iron. For example, to reach the height c, the excitation has to be of the value represented by the length O x_1. On the same diagram the line O b B represents the relation between the flux of magnetic lines across the air-gap and the ampere-turns required to force these lines across. If the gap were 1 sq. cm. in section, and 1 cm. long, 0·795 ampere-turns of current would produce field H = B = 1. In this case the gap is supposed to be of larger area and shorter than 1 cm., the line sloping up at such a slope that the length O x_2 represents the ampere-turns requisite to bring up the magnetic flux to b, which is at the same height on the scale as a. It is then easy to put the two things together, for the total amount of excitation required to force these magnetic lines through air and iron will (neglecting leakage) be the sum of the separate amounts. The point x_3 is chosen so that $O\,x_3$ is equal to the sum of $O\,x_1$ and $O\,x_2$, or that the distance of point r from the vertical axis is equal to the sum of the respective distances of c and b. If the same thing is done for a large number of corresponding points, the resultant curve $O\,r\,R$ may be constructed from the two separate curves. It will be seen then that, in general, the presence of a gap in the magnetic circuit has the effect of causing the magnetic curve to rake over, the initial slope being determined by the air-gap.

The student should compare some interesting experiments made in Paris by M. Leduc,[*] who, however, falls into an error respecting tubular cores.

Effect of Straight Cores of Various Lengths.

From the foregoing remarks it must be evident that when a short iron core is placed in a magnetizing coil, since the only return path for its magnetic lines is through air, an external magnetizing force of much greater intensity must be applied to it than to a long piece, or to a ring, to bring up its magnetization to an equal degree. Fig. 44, which is taken from the researches of Prof. Ewing,[†] relates to an annealed soft iron wire, the length of which, at first equal to 200 times its own diameter, was reduced to 100 and then to 50 times its own diameter. The curves drawn with continuous lines relate to observations made with increasing magnetic forces; those in dotted lines to observations made with forces decreasing back to zero. The three sloping lines, O A, O B,

---

[*] *La Lumière Electrique*, xxviii. p. 520, 1888.
[†] *The Electrician*, xxiv. p. 591, Apr. 18, 1890.

Effect of Length of Rod.

OC, may be taken to show that part of the applied magnetic force which was used in driving the magnetic lines through air. For example, taking the case of the wire 50 diameters

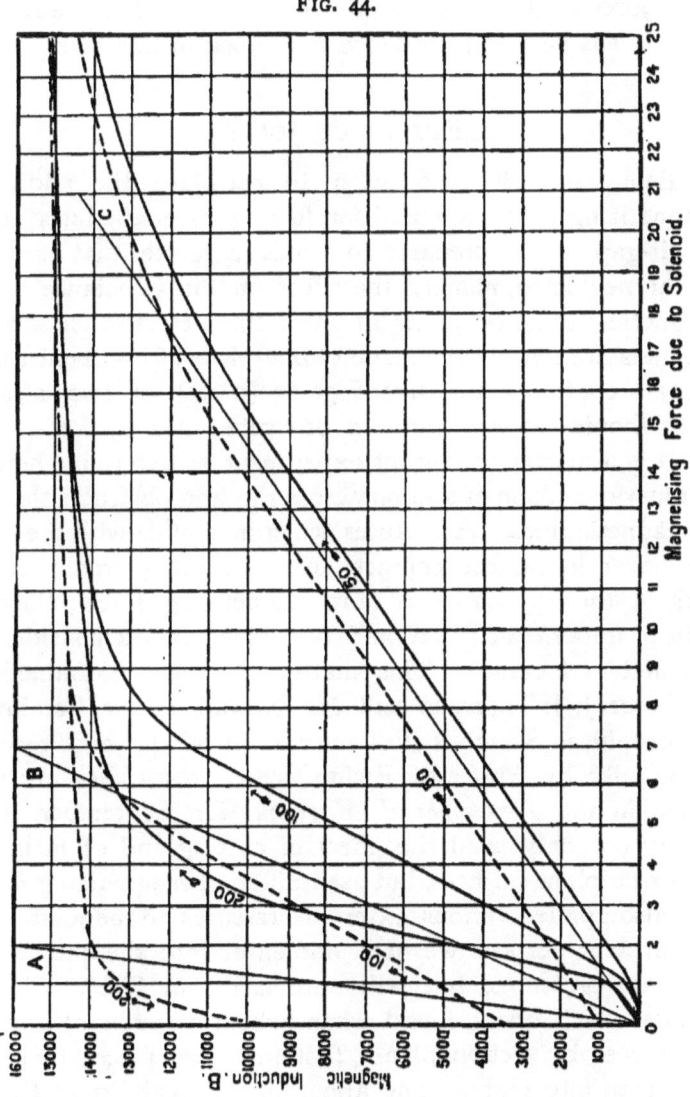

Fig. 44.

MAGNETIZATION OF SOFT IRON RODS OF VARIOUS LENGTHS.

long. To raise B in it to 9000 required a magnetizing force of H = 14. But of this force of 14, 13 parts were really required to drive these lines through the air return path, as shown by the crossing of the line O C across the level of B = 9000 at the point where H = 13; so that really only H = 1 was needed to produce B = 9000 in this wire.

EFFECT OF JOINTS.

Being now in a position to calculate the additional magnetizing power required for forcing magnetic lines across an air-gap, we are prepared to discuss a matter that has been so far neglected, namely, the effect on the reluctance of the magnetic circuit of joints in the iron. Horse-shoe electromagnets are not always made of one piece of iron bent round. They are often made, like Fig. 23 (p. 51), of two straight cores shouldered and screwed, or riveted into a yoke.

It is a matter purely for experiment to determine how far a transverse plane of section across the iron obstructs the flow of magnetic lines. Armatures, when in contact with the cores, are never in perfect contact, otherwise they would cohere without the application of any magnetizing force; they are only in imperfect contact, and the joint offers a considerable magnetic reluctance. This matter has been examined by Professor J. J. Thomson and Mr. Newall, in the Cambridge Philosophical Society's *Proceedings*, in 1887; and recently more fully by Professor Ewing, whose researches are published in the *Philosophical Magazine* for September, 1888. Ewing not only tried the effect of cutting and of facing up with true plane surfaces, but used different magnetizing forces, and also applied various external pressures to the joint. For our present purpose we need not enter into the questions of external pressures, but will summarize in Table VII. the results which Ewing found when his bar of wrought iron was cut across by section planes, first into two pieces, then into four, then into eight. The apparent permeability of the bar was reduced at every cut.

TABLE VII.—EFFECT OF JOINTS IN WROUGHT-IRON BAR (not compressed).

H	B				Mean thickness of equivalent air-space for one cut.		Thickness of iron of equivalent reluctance per cut.	
	Solid.	Cut in Two.	In Four.	In Eight.	Centimetres.	Inches.	Centimetres	Inches.
7·5	8,500	6,900	4,809	2,600	0·0036	0·0014	4	1·57
15	13,400	11,550	8,900	5,550	0·0030	0·0012	2·53	1
30	15,350	14,550	12,940	9,800	0·0020	0·0008	1·10	0·433
50	16,400	15,950	15,000	13,300	0·0013	0·0005	0·43	0·169
70	17,100	16,840	16,120	15,200	0·0009	0·0004	0·22	0·087

Suppose we are working with the magnetization of our iron pushed to about 16,000 lines to the sq. cm. (*i. e.* about 150 lb. per square inch) traction, requiring a magnetizing force of about H = 50; then, referring to Table VII., we see that each joint across the iron offers as much reluctance as would an air-gap 0·0005 of an inch in thickness, or adds as much reluctance as if an additional layer of iron about $\frac{1}{6}$th of an inch thick had been added. With small magnetizing forces the effect of having a cut across the iron with a good surface on it is about the same as though you had introduced a layer of air $\frac{1}{800}$th of an inch thick, or as though you had added to the iron circuit about 1 inch of extra length. With large magnetizing forces, however, this disappears, probably because of the attraction of the two surfaces across that cut. The stress in the magnetic circuit, with high magnetic forces running up to 15,000 or 20,000 lines to the sq. cm., will of itself put on a pressure of 130 to 230 lbs. to the square inch, and so these resistances are considerably reduced; they come down in fact to about $\frac{1}{20}$th of their initial value. When Ewing especially applied compressing forces, which were as large as 3200 lb. to the sq. in., which would of themselves ordinarily, in a continuous piece of iron, have diminished the magnetizability, he found the diminution of the magnetizability of iron itself was nearly compensated for by the better conduction of the cut surface. The old surface, cut and compressed in that way, closes up as it were magnetically—does not act like a cut at all; but at the same time you lose just as much as you gain, because the iron itself becomes less magnetizable.

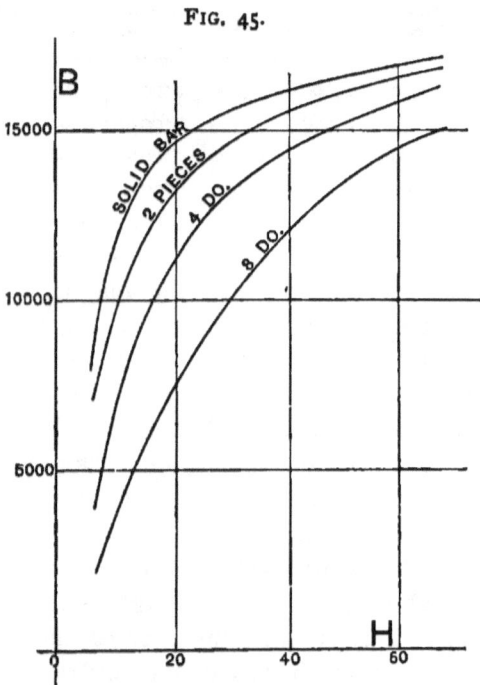

FIG. 45.

EWING'S CURVES FOR EFFECT OF JOINTS.

The above results of Ewing's are further represented by the curves of magnetization drawn in Fig. 45. When the faces of a cut were carefully surfaced up to true planes, the disadvantageous effect of the cut was reduced considerably, and under the application of a heavy external pressure almost vanished.

The influence of compression was notable. When a compression of 3210 lb. per sq. in. (= 226 kilogrammes per sq. cm.) was applied to the iron bar, the joint showed, under magnetic forces, a reluctance which decreased as the magnetic force was increased. The following table gives the values of H and of B in the solid bar and the bar after being cut, together with the mean thickness of the equivalent air-space.

TABLE VIII.—EFFECT OF COMPRESSION ON JOINTS.

H	B under compression of 3210 lb. per sq. inch.		Thickness of mean equivalent air-space. (Millimetres.)
	Solid bar.	Bar cut in eight.	
7·5	7,500	3,600	0·020
10	10,000	4,900	0·019
20	13,900	8,300	0·018
30	15,200	10,700	0·017
50	16,500	13,750	0·011
70	17,200	15,700	0·007

When various loads were tried, the effect of increasing the load, in a weak magnetic field, was practically to close up well-faced joints, as the following table shows :—

TABLE IX.—EFFECT OF VARIOUS LOADS ON JOINTS.

Load : kilos. per sq. centimetre.	B (when $H = 5$).		Thickness of equivalent air-space. (Millimetres.)
	Before cutting.	After cutting and facing.	
0	5,600	4,700	0·022
56·5	5,400	4,670	0·020
131	4,700	4,200	0·017
169·5	4,050	3,800	0·010
226	3,650	3,650	0

EFFECTS OF STRESS.

A piece of iron when placed under stress is somewhat changed in its magnetic properties. If a longitudinal pull is applied to iron whilst it is being magnetized, it is found at first to increase its permeability, whilst a longitudinal push, tending to compress it, decreases its permeability. This is very well shown by the figures given in the second column of the last table, wherein it appears that a compression of 226 kilos. per sq. cm., or nearly $1\frac{1}{2}$ ton per sq. in., brought down the value of B in a wrought-iron bar from 5600 to 3650; or diminished the permeability from 1120 to 730. Stress also impairs the softness of iron. A piece of annealed iron wire hardened by previous stretching, behaves more like a piece of steel, as may be seen by reference to Ewing's curves, Fig. 35, p. 70. Twisting stresses also affect the magnetic quality. The reader should consult Ewing's papers on magnetism.

Another important matter is that all such actions as hammering, rolling, twisting, and the like, impair the magnetic quality of annealed soft iron. Pieces of annealed wrought iron which have never been touched by a tool, provided they do not constitute actually closed magnetic circuits, show hardly any trace of residual magnetization, even after the application of magnetic forces. But the touch of the file will at once spoil it. Sturgeon pointed out the great importance of this point. In the specification for tenders for instruments for the British Postal Telegraphs, it is laid down as a condition to be observed by the constructor, that the cores must not be filed after being annealed. The continual hammering of the armature of an electromagnet against the poles may in time produce a similar effect.

Effects of Vibration.

The effects of vibration on magnetism are to diminish all residual actions, and to cause the specimen more rapidly to assume the mean state corresponding to the magnetic force present. If a specimen of soft iron is examined while under rapid vibration, it is found that in it there is scarcely any difference between the ascending and descending curves of magnetization. A single tap on a wire of soft iron will at once destroy any residual magnetism in it.

EFFECTS OF HEAT.

When iron is warmed, its magnetic properties undergo singular changes. Rise of temperature produces different effects at different stages of the magnetization, and the effects differ in different materials. In soft iron, in weak magnetic fields, the effect of raising the temperature is to produce an increase of permeability, which goes on until the specimen is at a full red heat, about 760° C., when it reaches the enormous value of 10,000, after which point it suddenly falls, and when the temperature of 780° C. (about) is reached, the iron ceases to be a magnetic body, its permeability at that temperature, and at all higher temperatures, not differing sensibly from that of air or vacuum. But if placed in a very strong magnetic field, the action of raising the temperature produces a diminution of permeability, at first slight, then more rapid until the temperature of 780° C. is reached, when again all magnetism disappears. In steel, the effect is curiously different. In both soft and hard steels, the effect, in a very weak magnetic

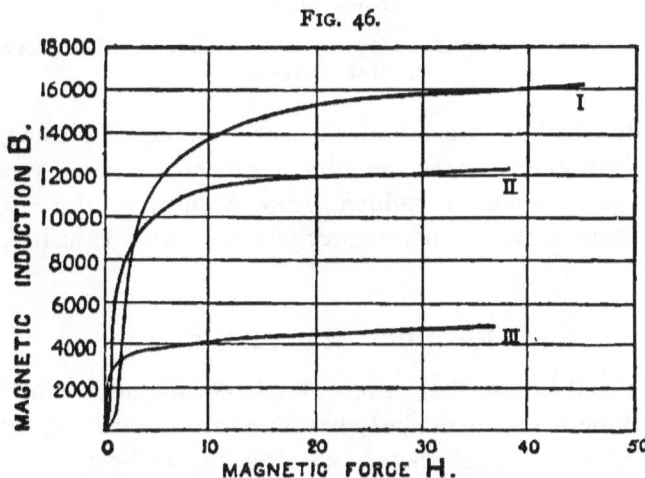

FIG. 46.

MAGNETIZATION OF MILD STEEL AT VARIOUS TEMPERATURES.

field of about $H = 0.2$, of raising the temperature is to increase the permeability, until it is heated to a point close up to 700° C., when it suddenly drops to zero. In a field of about H = 2, its magnetization at the lower temperatures is greater, and the final drop sets in at a temperature considerably below 700°. In a strong field of H = 40, the permeability drops steadily as the temperature rises. At high temperatures, too, all residual effects are smaller.

Fig. 46, shows the effects of temperature in modifying the magnetic curve of steel, the three curves given relating respectively to temperatures of 12° (C.), 620°, and 715°. Fig. 47, which, like Fig. 46, is taken from Hopkinson's researches,

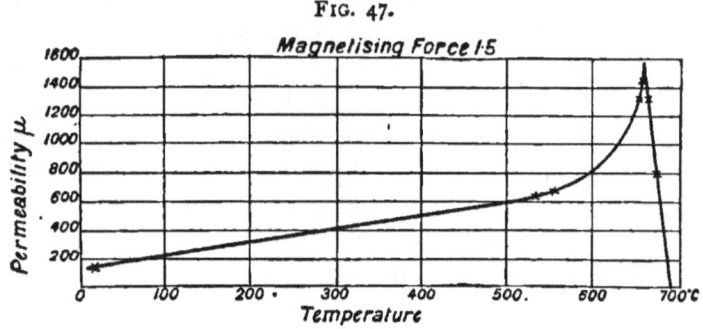

FIG. 47.

RELATION BETWEEN PERMEABILITY (IN WEAK FIELD) AND TEMPERATURE IN HARD STEEL.

shows how in hard steel, under a weak magnetizing force, the magnetization increases as the temperature rises, until a certain point when a sudden drop occurs, to the critical temperature at which the magnetization entirely vanishes.

RESIDUAL MAGNETISM.

It is well known that several kinds of magnetic materials —lodestone, steel, particularly hardened steel, and hard sorts of iron—exhibit residual magnetism after having been subjected to magnetic forces. It is also known that closed circuits of soft iron—even of the very softest—will exhibit a considerable amount of residual magnetism so long as the circuit which they constitute is unbroken. A very simple illustration of this is afforded by any electromagnet possessing in its core and well-fitting armature a compact magnetic circuit. If it is excited by passing a current, which is then quietly turned off, the armature usually does not drop off, and may even require considerable force to detach it; but when once so detached will not again adhere, the residual magnetization not being permanent In like manner a steel horse-shoe magnet, if magnetized powerfully while its keeper is across its poles, may become "supersaturated"; that is to say, magnetized to a higher degree of magnetization than it can retain in permanence, a portion of this residual magnetization disappearing the first time the keeper is removed. All these residual phenomena are part of a wide subject of magnetic after-effects. Owing to causes presently to be discussed, magnetic forces, if sufficiently powerful, produce effects on the molecules of a magnetizable body which remain after the cause has passed away, with the result that if the causes change in a continuous manner the effects also change in a continuous manner, but suffer a retardation in phase, the cause lagging after the effect. This must not be confused with an alleged time-lag of magnetism to which many things have been supposed to be due which were really due to quite other things. The present considerations relate to retardations in phase rather than in time, and occur no matter whether the operations themselves are conducted quickly or slowly.

Reference to Fig. 39, p. 75, will show that when the magnetizing force H is gradually increased from zero to a high value, and is then gradually decreased to zero, the resulting internal magnetization B first increases to a maximum, and then decreases, but does not come back to zero. The curve descending from the maximum does not coincide with the ascending curve. In fact, when the magnetizing force has been entirely removed there remained (in this specimen) a residual magnetization of about 47,000 lines to the sq. in., or about 7300 lines per sq. cm. It has been proposed to give the name of the *remanence* to the number of lines per sq. cm. that thus remain as the residual value of B. To remove this *remanence*, a negative magnetizing force must be applied.

Suppose enough magnetizing force has been used, the curve will descend and cut the horizontal axis at a point to the left of the origin; and with greater negative magnetizing forces, the specimen will begin to be magnetized with magnetic lines running through it in the reversed direction. The particular value of the negative magnetizing force which is needed to bring the remanent magnetization to zero has been termed by Hopkinson *the coercive force.* In the specimen of wrought iron in question the coercive force (in C.G.S. measure) is about 2, or in sq. in. measure about 13. The force thus required to deprive any specimen of its remanent magnetization may be taken as a measure of the tendency of iron of this particular quality to retain permanent magnetism. Hard kinds of iron and steel always show more coercive force than soft kinds of iron. For example, whilst that of soft wrought iron is about 2, that of hard steel may be as much as 50. Some further data about hard steels are given in Chapter XVI., on Permanent Magnets.

Hysteresis.

Professor Ewing, who has particularly studied the residual effects exhibited by various qualities of iron and steel, has given the name of *hysteresis* to this tendency of the effects to lag, in phase, behind the causes that produce them. The appropriate mode of studying hysteresis is to subject the specimen to a complete cycle (or to a number of successive cycles) of magnetizing forces. For example, let the magnetizing force begin at zero, and increase to a high value (say to $H = 200$) and then decrease back to zero, then reverse and increase to a high negative value, and finally return to zero. Such a cycle is given in Fig. 48, which is taken from Ewing's researches, and relates to a series of experiments made with a piece of annealed steel pianoforte wire. The curve begins in the centre of the diagram, and as H is increased positively, the curve rises at first concavely to the right, then turns over, and when $H = 90$, B has risen to a little over 14,000. When

Hysteresis.

H is then reduced back to zero the curve turns back on itself, but does not fall as fast as it previously rose, for when H is reduced to 20, B has gone down only to 12,000, and when H = 0 the remanence is about 10,500. If at this point H had been again increased to 90, B would have run up again to 14,000, as shown by the thin line. If, however, the magnetizing force is reversed, the curve descends to the left, and cuts the horizontal axis at − 24, which is therefore the value of the coercive force. On increasing the reversed magnetizing force to H = − 90, the reversed magnetization increases to

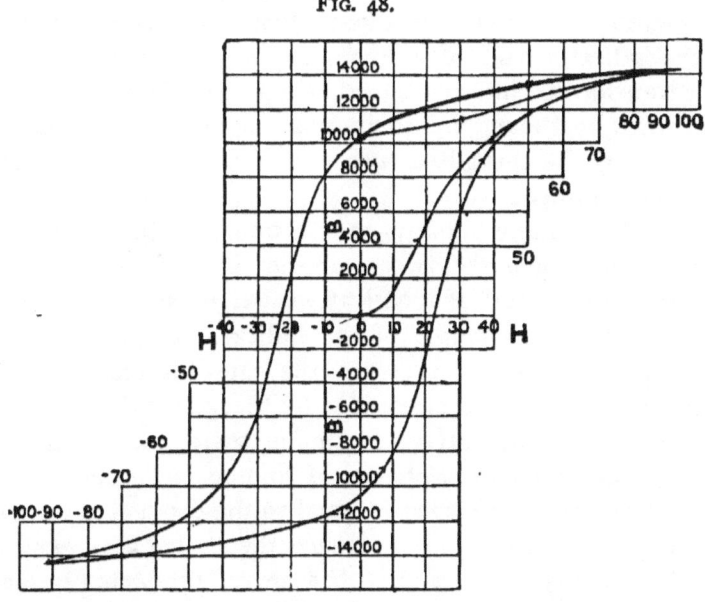

FIG. 48.

CYCLE OF MAGNETIC OPERATIONS ON ANNEALED STEEL WIRE.

the value B = − 14,000, or a little more. Then when these reversed magnetizing forces are reduced to zero, the curve returns towards the right, crossing the vertical axis at B = − 10,500 (the negative remanence); and on re-reversing the magnetizing force it is found that when H = + 24, the magnetization is once more zero. After this point increasing

H causes the magnetization to run up very rapidly, not quite following its former track, but coming up as before to the apex, when H is raised to the same maximum of 90.

Cycles of Magnetization.

Such cycles of magnetization as that which has just been described, if carried out on any specimen of iron or steel, always yield curves that exhibit, like Fig. 48, an enclosed area. This fact has been shown by Warburg[*] and by Ewing[†] to possess a special significance, for the area inclosed is a measure of the work wasted in carrying the iron through a complete cycle of magnetizations. Just as the area traced out on the indicator-card of a steam engine is a measure of the heat transformed into useful work in the cycle of operations performed by the engine, so in this magnetic cycle the area enclosed by the curve is a measure of the work transformed into (useless) heat.

To study more closely the meaning of an area on a diagram in which the two quantities plotted out are magnetizing force and magnetization, we must glance for a moment at the principle of the indicator diagram. In the indicator diagram the pressure (the cause) is recorded vertically and the volume swept out by the piston (the effect) is traced horizontally. If when the pressure has the average value p, the elementary change of volume is dv, the element of work thereby performed is equal to the product of the two; or, in symbols, $dw = p\,dv$. Hence the entire area enclosed, which is simply the sum of such a set of products represents the sum of all the elements of work done in the cycle; or in symbols,

$$w = \int p\,dv.$$

In the same way there are in the magnetic cycle the two variables, the magnetizing force H (the cause), and the mag-

---

[*] *Wied. Ann.*, xiii. 1881, p. 141.

[†] *Proc. Roy. Soc.*, xxxi. 1881, p. 22; xxxiv. 1884, p. 39; and xxxv. 1885, p. 1; and *Phil. Trans.*, 1885, pt. ii., 523.

netization B (the effect). If while the magnetizing force has the average value H the magnetization increases by an amount d B, the elementary work done is proportional to the product of the two; or, in a whole cycle, the area enclosed (which is the sum of the elementary areas corresponding to such products) is proportional to the whole work so done; or, in symbols

$$w = \int H \, d B.$$

In Fig. 49, which relates to the specimen of annealed wrought iron examined by Hopkinson (see Figs. 39 and 40), the values of H_u are plotted vertically, and those of B_u

FIG. 49.

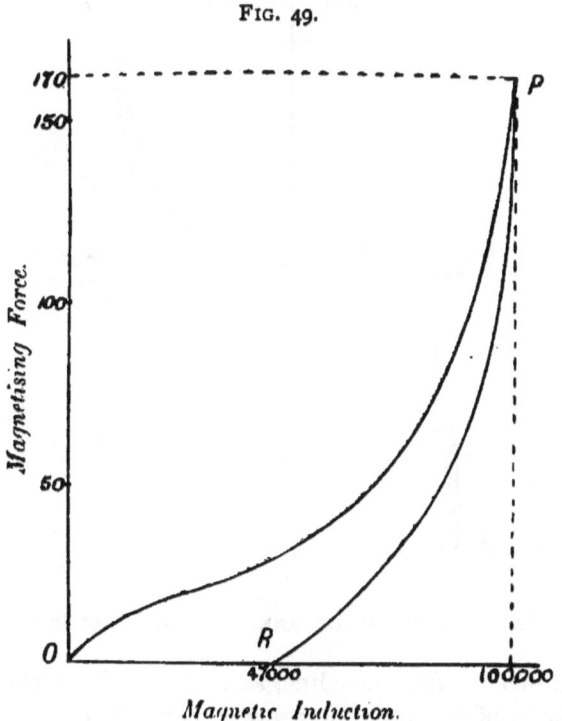

WORK DONE IN INCREASING MAGNETIZATION, AND RESTORED IN DECREASING.

horizontally, this position being chosen to correspond to the p and v of the indicator diagram. The curve that starts from

O represents the set of observations made with magnetic forces, gradually increasing to about $H_{,,} = 170$, when the resulting number of magnetic lines per square inch in the iron is about 100,000. The second curve represents the result of then decreasing the magnetizing forces back to zero, when there remains behind a residual magnetization of no fewer than 47,000 lines per square inch.

Applying the principle of the indicator diagram to these curves, it will be seen that the area below the first of these

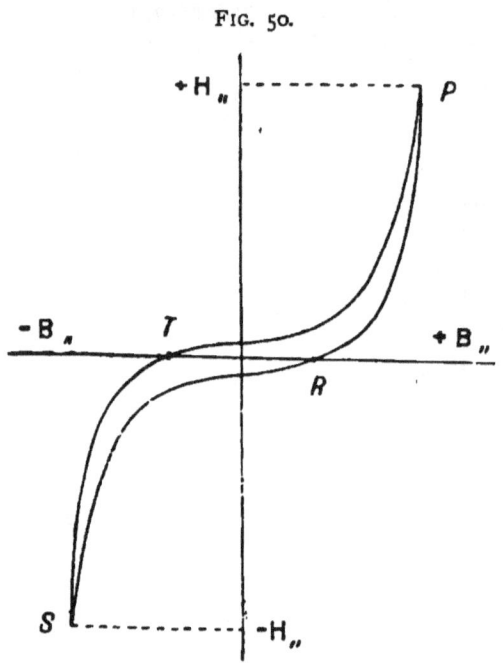

FIG. 50.

MAGNETIC CYCLE FOR ANNEALED WROUGHT IRON.

curves, down to the zero line, and bounded on the right by the ordinate of the point P, represents the integral product of magnetic force into magnetization, and is consequently proportional to the work done (per cubic inch of the iron) in bringing about the magnetic state of things implied by the position on the diagram of point P. Similarly, the area under the descending curve represents the work restored in the demagnetizing of the specimen down to the state of things represented by the point R, when the magnetizing force has been removed. The narrow area enclosed between the curves represents the work not so restored, and for the expenditure of which the residual magnetism is all that there is to show. If the magnetizing forces are carried through a complete cycle of alternations from $+H_{\prime\prime}$ to $-H_{\prime\prime}$ and back, the resulting magnetization goes through a corresponding cycle, with the result that a narrow closed area is completed between the curves in the diagram (Fig. 50). This figure relates to the wrought iron specimen, and should be compared with that given for soft steel (Fig. 48) in which, however, the plotting is given in C.G.S. measure and with B vertically.

For the sake of comparison, a curve for wrought iron and one for steel are given side by side in Fig. 51. In all these

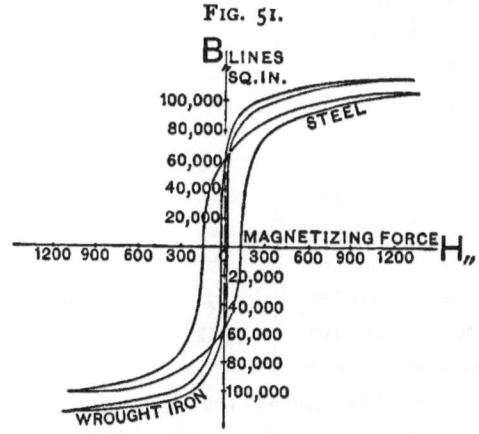

HYSTERESIS IN WROUGHT IRON AND IN STEEL.

cases the closed area represents the work which has been wasted or dissipated in subjecting the iron to these alternate magnetizing forces. In very soft iron, where the ascending and descending curves are close together, the enclosed area is small; and as a matter of fact, very little energy is dissipated in a cycle of magnetic operations. On the other hand, with hard iron, and particularly with steel, there is a great width between the curves, and there is great waste of energy. Hysteresis may be regarded as a sort of internal or molecular magnetic friction, by reason of which alternate magnetizations cause the iron to grow hot. Hence the importance of understanding this curious effect, in view of the construction of electromagnets that are to be used with rapidly alternating currents. The following figures of Table X. give the number of watts (1 watt = $\frac{1}{746}$ of a horse-power) wasted by hysteresis in well-laminated soft wrought iron when subjected to a succession of rapid cycles of magnetization.

TABLE X.—WASTE OF POWER BY HYSTERESIS.

B	$B_{\prime\prime}$	Watts wasted per cubic foot at 10 cycles per second.	Watts wasted per cubic foot at 100 cycles per second.
4,000	25,800	40	400
5,000	32,250	47·5	575
6,000	38,700	75	750
7,000	45,150	92·5	925
8,000	51,600	111	1,110
10,000	64,500	156	1,560
12,000	77,400	206	2,060
14,000	90,300	202	2,620
16,000	103,200	324	3,240
17,000	109,650	394	3,940
18,000	116,100	487	4,870

It will be noted that the waste of energy increases as the magnetization is pushed higher and higher in a dispropor-

* The proofs of these matters are as follows. In a magnetic field of strength H it will require H units of work to move a unit of magnetism along a length of 1 centimetre against the magnetizing forces. Hence, since there are 4 π magnetic lines to each unit of magnetism, the work done in one complete cycle on a single cubic centimetre of the iron will be equal to $\frac{1}{4\pi} \int H \, dB$. If H and B are in C.G.S. units, the work will be given in ergs per cubic centimetre. Hence if this number is multiplied by the number of cycles per second and divided by 10^7, the result will express the number of watts of power wasted.

tionate degree, the waste when B is 18,000 being six times that when B is 6,000. There is some experimental evidence to show that in very rapid cycles of magnetization the loss by hysteresis is less than in slow cycles.

Hopkinson has made the remark that the area $\int H\,dB$ is approximately equal to a rectangle, the height of which is double the remanence, and the breadth of which is double the coercive force.

Ewing has given the following values of the energy wasted in a magnetic cycle of strong magnetization on various brands of iron and steel:—

TABLE XI.—WASTE OF ENERGY BY HYSTERESIS.

Brand experimented upon.	Ergs per cubic centimetre lost in one complete cycle of magnetization.
Very soft annealed iron	9,300
Less soft ,, ,,	16,300
Hard drawn iron wire	60,000
Annealed steel wire	70,500
Glass hard steel wire	76,000
Pianoforte steel wire (ordinary state)	116,000
,, ,, ,, (annealed)	94,000
,, ,, ,, (glass hard)	117,000

These figures are surpassed by some of the brands examined by Hopkinson, who found that oil-hardened tungsten steel, the sort chosen for making permanent magnets because of its great coercive force, wasted no less than 216,864 ergs per cubic centimetre per cycle.

Cycle of Operations in the Stroke of an Electromagnet.—Passing from the properties of the material to those of a definite electro-magnetic apparatus, the cycle of relations between magnetizing force and magnetism may be still traced out by means of diagrams. Fig. 52 relates to a certain horseshoe electromagnet. The curve O P represents the ascending curve of the electromagnet, when its armature was about $\frac{1}{25}$

of an inch away. Under these circumstances, owing to the reluctance to magnetization of the air-gaps, a given magnetizing power (ampere-turns of circulation of current) will produce less magnetization than if the magnetic circuit were closed. The second curve Q R is the descending curve of the same electromagnet when the armature is close up to the core, closing the magnetic circuit. Now, suppose such an electromagnet to be used to do the work of drawing up its armature. We may consider the successive operations of producing a complete stroke, exactly as the engineer discusses the cycle

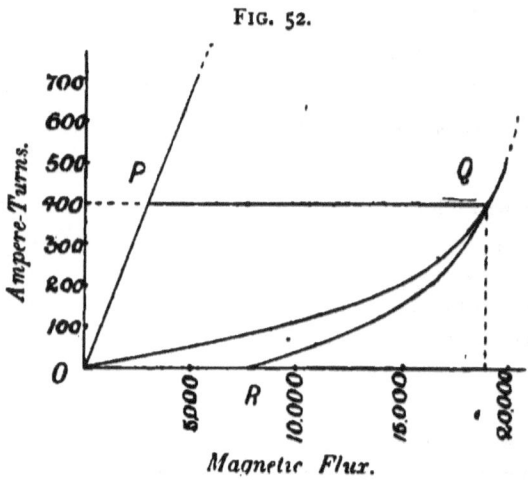

FIG. 52.

CYCLE OF OPERATIONS.

of operations in the stroke of the piston in the cylinder. Starting at O, with the armature away from the core, the magnetizing current is turned on, and the magnetization is carried to the point corresponding to P. Suppose, now, the armature is allowed to move up to the core. The magnetization is increased, because of the more perfect condition of the magnetic circuit. During the movement of the armature mechanical work is performed, and as the magnetizing power is kept all the time at a constant value, the line P Q represents (mechanically) this operation. If, now, the magnetizing current is cut off, the magnetization follows the right-hand descending curve to R. During this operation a certain amount of energy is restored magnetically. If, now, the armature is permitted to fly back to its initial position, thus opening the gaps in the magnetic circuit, the residual magnetization will disappear almost entirely (provided the iron is soft), and the energy it represents is wasted as heat. The area O P Q R O represents the whole quantity of energy magnetically applied to the system. Part of this, O Q R O, is wasted by hysteresis; the other part, O P Q O, is utilized in doing mechanical work. As a matter of fact, these four operations do not occur separately in a stroke; for the armature usually begins to move before the magnetism has grown up to the point P, and it begins to fly back before the residual magnetism has entirely disappeared. The actual cycle is not made up of two straight lines and two magnetic curves, any more than the actual diagram on the indicator card of a steam engine consists of the two isothermals and two adiabatics of the ideal case called Carnot's cycle.

As a matter of fact the motion of the armature itself sets up in the copper wire an induced electromotive force, which tends to diminish the magnetizing current during the time that the motion lasts. The efficiency of the arrangement as an engine depends indeed upon the magnitude of such counter-electromotive forces, and upon the degree to which they cut down the working current. But this diagram does not represent the electrical losses; it exhibits solely the magnetic quantities concerned. Were a greater amount of space available, one might enlarge upon the mode in which the energy is transferred electrically to the magnetic system; how the approach of the armature, by increasing the magnetic induction as it completes the magnetic circuit, sets up counter-electromotive forces, the electric energy utilized being proportional to the integral product of the current into the counter-electromotive force so generated. This is too wide a field for discussion in this chapter; it involves the whole theory of the efficiency of electric motors, for which the reader is referred to treatises on dynamo-electric machines or on the electric transmission of power.

It might also be pointed out that this diagram does not exhibit the mechanical value of the force at any part of the stroke. During the part of the stroke corresponding to P Q, in which the armature is approaching the core, there is an ever-increasing pull. On the other hand, during the return stroke, the magnetism is undergoing the transition from R to O, and work is being done upon the armature instead of by it. There is an analogy, however, here also, to Carnot's cycle diagram. That diagram does not exhibit the thermal quantities concerned in the different parts of the cycle. Heat is taken in along one of the isothermals and rejected along the other, but the length or position of those isothermals does not exhibit the quantities thus supplied to the engine or expelled from it. Heat values are not exhibited upon the Carnot diagram of mechanical work; neither are mechanical values exhibited upon my diagram of magnetic work.

In no case can an electromagnet do any work, save by such a change of the configuration of the system as shall alter the magnetization of the magnetic circuit, the tendency of the configuration being always so to change as to make the magnetic flux a maximum.

Magnetic Creeping.

Another kind of after-effect was discovered by Ewing, and named by him "viscous hysteresis." This is the name given to the gradual creeping-up of the magnetization when a magnetic force is applied with absolute steadiness to a piece of iron. This gradual creeping-up may go on for half-an-hour or more, and amount to several per cent. of the total magnetization. This is a true, but slow, magnetic lag, and must not be confounded either with the lag of phase discussed already under the name hysteresis, or with the apparent lag due to the retardation of the magnetizing current resulting from self-induction, or with the apparent lag observable in unlaminated iron cores due to eddy-currents circulating in the mass of the iron itself.

To Demagnetize Iron.

In order to deprive a specimen of iron of all traces of remanent magnetism, it must be subjected to a series of alternate magnetizing forces of decreasing amplitude. The reason of this is that, to remove the remanence left after the operation of any given magnetizing force requires the application of a reversed magnetizing force of lesser amount than that originally applied. One mode of doing this to specimens of iron is to place them within a tubular coil through which an alternating electric current of sufficient strength is being sent, and then to withdraw the piece of iron slowly from the coil.

Watches which have become magnetized by being accidentally brought too near to a powerful electromagnet may be demagnetized in this way. A simpler method, though not always equally effective, is to hang the watch by a cord, and set it spinning (by twisting the cord rapidly) while it is near the pole of a powerful electromagnet; and, while spinning, withdraw it to a distance.

Self-Demagnetizing Effects of Poles. Properties of Short Pieces of Iron.

It is sometimes said that the poles of a magnet tend to produce a self-demagnetizing force. It is certain that rings and other closed magnetic circuits exhibit a much greater amount of sub-permanent magnetism than do pieces of iron that are not closed on themselves. (See pp. 105 and 234.) The matter may be put in a clearer light by the following considerations. First, retentiveness for magnetism is a property found only in solid substances, never in any liquid or gas through which magnetic lines have been passing, nor has such a property been observed in any of the metals usually classed as non-magnetic. Secondly, to create a magnetic flux in a magnetic circuit, whether in air or iron, requires the expenditure of energy; and so long as the flux persists it represents a store of potential energy, just as truly as a bent spring does.* Thirdly, potential energy always tends to run down to a minimum. Hence, when a magnetized system is left to itself, and the impressed magnetizing forces are removed, there is an immediate tendency for all magnetism to disappear unless there is such a mutual action between the parts of the system as tends to keep them in the state in which they have been left. In liquids and gases there is no such mutual action; they instantly demagnetize. In magnetic solids, such as iron, there is, however, such a mutual action between the parts,† with the result that when the magnetic flux lies wholly within iron (as in the case of rings and of horse-shoe electromagnets with their keepers actually in contact) there is a very great sub-permanent remanence. On opening a gap in the iron circuit, the magnetic flux must now traverse a layer of air, which not only is relatively very impermeable, but tends of itself to become demagnetized. The same residual (sub-permanent) force which, if acting in iron alone, was sufficient to maintain a considerable flux in the iron circuit, is unable to maintain this flux across the air gap, and at once the flux diminishes in that part of the magnetic circuit, precisely as does the flux through the polar parts of a permanent magnet when the keeper is taken off. (See experiment, p. 212). But in the soft iron ring each part acts as keeper to the rest, and if at one part the keeper-action is thus destroyed, the magnetism of the other parts at once falls off. Hence a gap in the magnetic circuit tends to demagnetize.

* The analogy is even closer than at first sight appears. If the spring is perfectly elastic, *all* the energy spent in bending it is given back when it is allowed to return. But if it is imperfectly elastic, and on being allowed to fly back, does not return perfectly to its original form but shows a residual strain, then of course only part of the energy is given back. Part is lost, having been wasted in heat, in producing that part of the effect which is remanent. This is perfectly analogous to the loss of energy by hysteresis in the magnetizing of a piece of iron.

† See below, Ewing's *Theory of Magnetism*, p. 110.

The old way of looking at the facts was to say that the magnetism on the surface of the poles tended to push itself away from the polar surface laterally toward the neutral part of the magnet.

If the reader will now return to Fig. 44, p. 87, which relates to the behaviour of long and short iron cylindrical cores, he will be better able to understand why it is that for the short cores the ascending and descending curves of magnetization are much nearer together than they are for the long cores. He will also be able to see a reason why short cylinders and spheres of iron show scarcely any residual magnetization at all, have no magnetic memory. Without going into elaborate theoretical considerations concerning the properties of ellipsoids,* and merely reflecting that such short pieces, when removed from external magnetizing forces and surrounded by air, form very small parts of an entire magnetic circuit, it must be evident that there is reason enough why they should be practically self-demagnetizing.

This matter has a practical bearing; for it is most important in the use of electromagnets that their armatures should not stick on when the current has been broken. To destroy the sub-permanent magnetization that makes its appearance when the circuit is quite closed there have been several devices suggested. Hecquet† interposes a thin sheet of copper or of paper between the cylindrical cores and the yoke. This layer of non-magnetic material, interposed in the magnetic circuit, of course, increases slightly its reluctance, and diminishes slightly the magnetization. Trotter has found a similar device necessary in the field-magnets of certain dynamo-machines.

Another and more common device is to interpose between the armature and the pole of the electromagnet a distance-piece of some non-magnetic material. In some cases a pin or stud of brass is inserted in the polar face, with its head

---

* See Ewing, in *The Electrician*, vol. xxiv. p. 340, for a very clear mathematical account of the properties of ellipsoids.

† *Les Mondes*, xxxviii. 733, 1875.

slightly raised above the surface. In other cases a slip of paper or card is interposed, or cemented to the face of the armature or of the pole.

Ewing's Theory of Induced Magnetization.

Professor Ewing has recently propounded a theory of induced magnetism to account for the facts observed in magnetic circuits. Accepting the general idea propounded by Poisson and by Weber that a magnet must be regarded as an assemblage of elementary magnets, all previously magnetized, and that the act of magnetization merely consists in bringing them into allignment, he has shown that it is not needful, as Maxwell and others have supposed, to postulate the existence of internal frictional forces to resist the motions of the molecular magnets. Neither is it necessary to suppose that in the demagnetized state the molecules group themselves is closed rings or chains, as Hughes suggested.

Ewing has, in fact, shown that the whole of the facts are explicable on the supposition that the elementary molecular magnets are subject to mutual magnetic forces. This he has illustrated by constructing a model made of a large number of pivoted magnetic needles which are arranged at definite distances apart, and take up positions which, in the absence of external magnetic forces, are such as will simply produce a mutual neutralization of all external action, the positions of the individual needles of the mass being miscellaneous, not directed, on the whole, in any one direction more than in another. On such an assembly of compass needles an externally applied magnetic force, if small, produces a small and strictly proportional effect; and when removed leaves the needles precisely as they were before. This corresponds with the fact that *small* magnetizing forces produce no remanent magnetization in any material of any shape. On increasing the external magnetizing force, however, there comes a point when, for individual groups of needles, any further increase produces a sudden upsetting of equilibrium, and one or more needles suddenly swing round and take up a new position. This instability, and sudden increase in the number of needles pointing in one direction, corresponds to the instability and sudden increase in permeability exhibited, particularly by soft iron, which manifests itself in the abrupt rising of the ascending curve of magnetization when the magnetizing force reaches a certain value. If, after some of the needles have thus taken up new positions of equilibrium, the external magnetizing force be removed, the needles do not now return to their original positions; they retain a residual set, and, as a matter of fact, can as a whole, act as a magnet.

This corresponds to the remanent magnetism always observed after the application of a large enough magnetizing force. Like the remanent magnetism, it can be almost completely and instantly destroyed by mechanical shocks. Then, if a very large external magnetizing force is applied, the whole of the needles will turn round and point in nearly one direction, their own mutual actions being comparatively feeble and not able to turn them much aside; yet a still greater force will be able to produce a very slight additional directive action. All these things correspond so closely to the behaviour of magnetic bodies as to make this theory a most helpful one. The loss of energy by hysteresis is represented in the model, by the energy lost by the needles in beating against the air as they swing suddenly round to take up new positions, and oscillate until brought to rest by irreversible actions such as friction or eddy-currents. As for the further striking analogies presented by the model in the effects of stress, temperature, and the like, the reader is referred to Prof. Ewing's original paper.*

* *Proc. Roy. Soc.*, June 19, 1890; also *The Electrician*, xxv. pp. 514, 541, and 550.

CHAPTER IV.

PRINCIPLE OF THE MAGNETIC CIRCUIT—THE LAW OF MAGNETIC TRACTION—DESIGN OF ELECTROMAGNETS FOR MAXIMUM TRACTION.

IN this chapter we have to discuss the law of the magnetic circuit in its application to the electromagnet, and in particular to dwell upon some experimental results which have been obtained from time to time by different authorities as to the relation between the construction of the various parts of an electromagnet, and the effect of that construction on its performance. We have to deal not only with the size, section, length, and material of the iron cores, and of the armatures of iron, but to speak in particular about the way in which the shaping of the core and of the armature affects the performance of the electromagnet in acting on its armature, whether in contact or at a distance. But before we enter on the last more difficult part of the subject, we will deal solely and exclusively with the law of force of the magnet upon its armature when the two are in contact with one another; in other words, with *the law of traction*.

PRINCIPLE OF THE MAGNETIC CIRCUIT.

Some account is given in the Preface to this book of the history of the principle of the magnetic circuit, showing how the idea had gradually grown up, perforce, from a consideration of the facts. The law of the magnetic circuit was first thrown into shape in 1873 by Professor Rowland, of Baltimore. He pointed out that if you consider any simple case, and find (as electricians do for the electric circuit) an expression for the magnetizing force which tends to drive the magnetism round the circuit, and divide that by the resistance to magnetization reckoned also all round the circuit, the quotient of those two gives you the total amount of flow or flux of magnetism. That is to say, one may calculate the quantity of magnetism that passes in that way round the magnetic circuit in exactly the same way as one calculates the strength of the electric current by the law of Ohm. Rowland, indeed, went a great deal further than this, for he applied this very calculation to the experiments made by Joule more than thirty years before, and from those experiments deduced the degree of magnetization to which Joule had driven the iron of his magnets, and by inference obtained the amount of current that he had been causing to circulate. Now this law requires to be written out in a form that can be used for future calculation. To put it in words without any symbols, we must first reckon out from the number of turns of wire in the coil, and the number of amperes of current which circulates in them, the whole *magnetomotive force*—the whole of that which tends to drive magnetism along the piece of iron—for it is in fact, proportional to the strength of the current, and the number of times it circulates. Next we must ascertain the resistance which the magnetic circuit offers to the passage of the magnetic lines. This was Joule's own expression, which was afterwards adopted by Rowland; and, for short, so as to avoid having four words, we might simply call it the *magnetic resistance*. Mr. Oliver Heaviside has suggested as an advisable alternative term, magnetic *reluctance*, in order that we may not confuse the resistance to magnetism in the magnetic circuit with the resistance to the flow of current in an electric circuit. However, we need not quarrel about terms; magnetic reluctance is sufficiently expressive. Then having found these two, the quotient of them gives us a number representing the quantity or number of magnetic lines which flow round the circuit. If we adopt a term which is used on the Continent, we may call it simply *the magnetic flux*, the flux of magnetism being the analogue of the flow of electricity in the

I

electric law. The law of the magnetic circuit may then be stated as follows :—

$$\text{Magnetic flux} = \frac{\text{magnetomotive force}}{\text{reluctance}}.$$

However, it is more convenient to deal with these matters in symbols, and therefore the symbols which I use, and have long been using, ought now to be explained. For the number of spirals in a winding I use the letter S; for the strength of current, or number of amperes, the letter i; for the length of a bar, or core, I am going to use the letter l; for the area of cross-section, the letter A; for the permeability of the iron which we discussed in the last Chapter, the Greek symbol μ; and for the total magnetic flux, the number of magnetic lines, I use the letter N. Then our law becomes as follows :—

$$\text{Magnetomotive force} \quad \frac{4\pi S i}{10};$$

$$\text{Magnetic reluctance} \quad \Sigma \frac{l}{A\mu};$$

$$\text{Magnetic flux} \quad N = \frac{\frac{4\pi S i}{10}}{\Sigma \frac{l}{A\mu}}$$

If we take the number of spirals and multiply by the number of amperes of current, so as to get the whole amount of circulation of electric current expressed in so many ampere-turns, and multiply by 4π, and divide by 10, in order to get the proper unit (that is to say, multiply it by 1·257), that gives us the magnetomotive force. For magnetic reluctance, calculate out the reluctance exactly as you would the resistance of an electric conductor to the flow of electricity, or the resistance of a conductor of heat to the flow of heat; it will be proportional to the length, inversely proportional to the cross-section, and inversely proportional to the conductivity, or, in the present case, to the magnetic permeability. Now if the circuit is a simple one, we may simply write down here the length, and divide it by the area of the cross-section and the permeability, and so find the value of the reluctance. But if the circuit be not a simple one, if you have not a simple ring of iron of equal section all round, it is necessary to consider the circuit in pieces as you would an electric curcuit, ascertaining separately the reluctance of the separate parts, and adding all together. As there may be a number of such terms to be added together, I have prefixed the expression for the magnetic reluctance by the sign Σ of summation. But it does not by any means follow, because we can write a thing down as simply as that, that the calculation of it will be a very simple matter.

In the case of magnetic lines we are quite unable to do as one does with electric currents to insulate the flow. An electric current can be confined (provided we do not put it in at 10,000 volts pressure, or anything much bigger than that) to a copper conductor by an adequate layer of adequately strong—and I use the word "strong" both in a mechanical and electrical sense—of adequately strong insulating material. There are materials whose conductivity for electricity as compared with copper may be regarded perhaps as millions of millions of times less; that is to say, they are practically perfect insulators. There are no such things for magnetism. The most highly insulating substance we know of for magnetism is certainly not 10,000 times less permeable to magnetism than the most highly magnetizable substance we know of, namely, iron in its best condition; and when one deals with electromagnets where curved portions of iron are surrounded with copper, or with air, or other electrically insulating material, one is dealing with substances whose permeability, instead of being infinitely small, compared with that of iron, is quite considerable.

We have to deal mainly with iron when it has been well magnetized. Its permeability, compared with air, is then from 1000 to 100 roughly; that is to say, the permeability of air compared with the iron is not less than from $\frac{1}{100}$th to $\frac{1}{1000}$th part. That means that it is quite possible to have a very considerable leakage of magnetic lines from iron into air occurring to complicate one's calculations, and prevent an accurate estimate being made of the true magnetic reluctance of any part of the circuit. Suppose, however, that we have got over all these difficulties, and made our calculations of the magnetic reluctance; then dividing the magnetomotive force by the reluctance gives us the whole number of magnetic lines.

There, then, is in its elementary form the law of the magnetic circuit stated exactly as Ohm's law is stated for electric circuits. But as a general rule one requires this magnetic law for certain applications, in which the problem is not to calculate from those two quantities what the total of magnetic lines will be. In most of the cases a rule is wanted for the purpose of calculating back. You want to know how to build a magnet so as to give you the requisite number of magnetic lines. You start by assuming that you need to have so many magnetic lines, and you require to know what magnetic reluctance there will be, and how much magnetomotive force will be needed. Well, that is a matter precisely analogous to those which every electrician comes across. He does not always want to use Ohm's law in the way in which it is commonly stated, to calculate the current from the electromotive force and the resistance; he often wants to calculate what is the electromotive force which will send a given current through a known resistance. And so do we. Our main consideration will be here directed to the question how many ampere-turns of current circulation must be provided in order to drive the required quantity of magnetism through any given magnetic reluctance. Therefore, we will state our law a little differently. What we want to calculate out is the number of ampere-turns required. When once we have got that, it is easy to say what the copper wire must consist of; what sort of wire, and how much of it. Turning then to our algebraic rule, we must transform it, so as to get all the other things beside the ampere-turns, to the other side of the equation. So we write the formula :—

$$Si = \frac{N \cdot \Sigma \frac{l}{A\mu}}{1 \cdot 257}.$$

We shall have then the ampere-turns equal to the number of magnetic lines we are going to force round the circuit, multiplied by the sum of the magnetic reluctances, divided by 1·257. Now this number, 1·257, is the constant that comes in when the length, l, is expressed in centimetres, the area in square centimetres, and the permeability in the usual numbers.

Many persons, unfortunately—I say so advisedly because of the waste of brain labour that they have been compelled to go through—prefer to work in inches and pounds and feet. They have, in fact, had to learn tables instead of acquiring them naturally without any learning. If the lengths be specified in inches, and areas in square inches, then the constant is a little different. The constant in that case, for inches and square inch measures, is 0·3132, so that the formula becomes :—

$$Si = N \times \Sigma \frac{l'''}{A''\mu} \times 0 \cdot 3132.$$

Here it is convenient to leave the law of the magnetic circuit and come back to it from time to time as we require. With the guidance provided by this law, the various points that come under review can be arranged and explained one after another, so that there does not now remain—if one applied this law with judgment—a single fact about electro-magnets which is either anomalous or paradoxical. Paradoxical some things may seem in form, but they all reduce to what is perfectly rational when one has a guiding principle of this kind to tell one how much magnetization one will get under given circumstances, or to tell one how mnch magnetizing power one requires in order to get a given quantity of magnetization. The word "magnetization" is used here in the popular sense, not in the narrow mathematical sense in which it has sometimes been used (i. e. for the magnetic moment per unit cube of the material). It is used here simply to express the fact that the iron or air, or whatever it may be, has been subjected to the process which results in there being magnetic lines of force induced through it.

THE LAW OF TRACTION.

Now let us apply this law of magnetic circuit in the first place to the traction, that is to say, the lifting power of electromagnets. The law of traction was assumed in the previous chapter, and made the basis of a method of measuring the amount of permeability. The law of magnetic traction was stated once for all by Maxwell, in his great treatise, and it is as follows :—

$$P \text{ (dynes)} = \frac{B^2 A}{8 \pi},$$

where A is the area in square centimetres. This becomes

$$P \text{ (grammes)} = \frac{B^2 A}{8 \pi \times 981}.$$

That is, the pull in grammes per square centimetre is equal to the square of the magnetic induction, B (being the number of magnetic lines to the square centimetre), divided by 8π, and divided also by 981. To bring grammes into pounds you divide by 453·6; so that the formula then becomes :—

$$P \text{ (lb.)} = \frac{B^2 A}{11,183,000};$$

or if square inch measures are used :—

$$P \text{ (lb.)} = \frac{B_{''}^2 A''}{72,134,000}.$$

To save future trouble we will now calculate out from the law of traction the following table ; in which the traction in grammes per square centimetre or in pounds per square inch is set down opposite the corresponding value of B.

TABLE XII.—MAGNETIZATION AND MAGNETIC TRACTION.

B lines per sq. cm.	B$_{\prime\prime}$ lines per sq. in.	Dynes per sq. centim.	Grammes per sq. centim.	Kilogrs. per sq. centim.	Pounds per sq. inch.
1,000	6,450	39,790	40·56	·0456	·577
2,000	12,900	159,200	162·3	·1623	2·308
3,000	19,350	358,100	365·1	·3651	5·190
4,000	25,800	636,600	648·9	·6489	9·228
5,000	32,250	994,700	1,014	1·014	14·39
6,000	38,700	1,432,000	1,460	1·460	20·75
7,000	45,150	1,950,000	1,987	1·987	28·26
8,000	51,600	2,547,000	2,596	2·596	36·95
9,000	58,050	3,223,000	3,286	3·286	46·72
10,000	64,500	3,979,000	4,056	4·056	57·68
11,000	70,950	4,815,000	4,907	4·907	69·77
12,000	77,400	5,730,000	5,841	5·841	83·07
13,000	83,850	6,725,000	6,855	6·855	97·47
14,000	90,300	7,800,000	7,550	7·550	113·1
15,000	96,750	8,953,000	9,124	9·124	129·7
16,000	103,200	10,170,000	10,390	10·39	147·7
17,000	109,650	11,500,000	11,720	11·72	166·6
18,000	116,100	12,890,000	13,140	13·14	186·8
19,000	122,550	14,360,000	14,630	14·63	208·1
20,000	129,000	15,920,000	16,230	16·23	230·8

This simple statement of the law of traction assumes that the distribution of the magnetic lines is uniform all over the area we are considering; and that, unfortunately, is not always the case. When the distribution is not uniform, then the mean value of the squares becomes greater than the square of the mean value, and consequently the pull of the magnet at its end face may, under certain circumstances, become greater than the calculation would lead you to expect—greater than the average of B would lead you to suppose. If the distribution is not uniform over the area of contract, then the accurate expression for the tractive force (in dynes) will be

$$= \frac{1}{8\pi} \int B^2 \, dA,$$

the integration being taken over the whole area of contact.

This law of traction has been verified by experiment. The most conclusive investigations were made in 1886[*] by Mr. R. H. M. Bosanquet, of Oxford, whose apparatus is depicted in Fig. 53. He took two cores of iron, well faced, and surrounded them both by magnetizing coils, fastened the upper one rigidly, and suspended the other one, on a lever with a counterpoise weight. To the lower end of this core he hung a scale-pan, and measured the traction of one upon the other when a known current was circulating a known number of times round the coil. At the same time he placed an exploring coil round the joint, that exploring coil being connected, in the manner described in Chapter III., on p. 70, with a ballistic galvanometer, so that at the moment when the two surfaces parted company, or at the moment when the magnetization was released by stopping the magnetizing current the galvanometer indication enabled him to say exactly how many magnetic lines went through that exploring coil. So that, knowing the area, you could calculate the number per square centimetre, and you could therefore compare B^2 with the pull per square centimetre obtained directly on the scale-pan. Bosanquet found that even when the surfaces were not absolutely perfectly faced the correspondence was very close indeed, not varying by more than 1 or 2

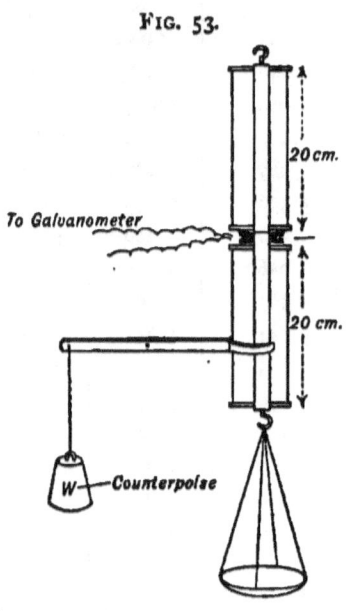

FIG. 53.

BOSANQUET'S VERIFICATION OF THE LAW OF TRACTION.

---

[*] *Phil. Mag.*, Dec. 1886; see also *The Electrician*, xviii., Dec. 3, 1886.

per cent., except with small magnetizing forces, say forces less than five C.G.S. units.

Knowing how irregular the behaviour of iron is when the magnetizing forces are so small as this, one is not astonished to find a lack of proportionality. The correspondence was, however, sufficiently exact to say that the experiments verified the law of traction, that the pull is proportional to the square of the magnetic induction through the area, and integrated over the area.

Design of Electromagnets for Traction.

Now the law of traction being in that way established, one at once begins to get some light upon the subject of the design of electromagnets. Indeed, without going into any mathematics, Joule had foreseen this when he in some instinctive sort of way seemed to consider that the proper way to regard an electromagnet for the purpose of traction was to think how many square inches of contact surface it had. He found that he could magnetize iron up until it pulled with a force of 175 lb. to the square inch, and he doubted whether a traction as great as 200 lb. per square inch could be obtained.

In the following Table, Joule's results (see Table I., p. 23) are re-calculated, and the values of B deduced :—

Table XIII.—Joule's Results Re-calculated.

Description of Electromagnet.		Section.		Load.		Pounds per sq. in.	Kilos per sq. cm.	B	Ratio of load to weight.
		sq. in.	sq. cm.	lb.	kilos.				
Joule's own electro-magnets	No. 1	10	64·5	209	947	104·5	7·35	13,600	139
	No. 2	0·196	1·26	49	22	125	8·75	14,700	324
	No. 3	0·0436	6·28	12	5·4	137·5	9·75	15,410	1286
	No. 4	0·0012	0·0077	0·202	0·09	81	5·7	11,830	2384
Nesbit's..	..	..	4·5	29·1	142·8	647	158·5	11·2	16,550	28
Henry's..	..	..	3·94	25·3	750	346	95	6·7	12,820	36
Sturgeon's	..	..	0·196	1·26	53	22·6	127·5	8·95	14,850	114

I will now return to the data on Table XII., and will ask you to compare the last column with the first. Here are the various values of B, that is to say, the amounts of magnetization you get into the iron. You cannot conveniently crowd more than 20,000 magnetic lines through the sq. cm. of the best iron, and, as a reference to the curves of magnetization shows, it is not expedient in the practical design of electromagnets to attempt, except in extraordinary cases, to crowd more than about 16,000 magnetic lines into the sq. cm. The simple reason is this, that if you are working up the magnetic force, say from 0 up to 50, a magnetizing force of 50 applied to good wrought iron will give you only 16,000 lines to the sq. cm., and the permeability by that time has fallen to about 320. If you try to force the magnetization any further, you find that you have to pay for it so heavily. If you want to force another 1000 lines through the sq. cm., to go from 16,000 to 17,000, you have to add on an enormous magnetizing force; you have to double the whole force from that point to get another 1000 lines added. Obviously it would be much better to take a larger piece of iron and not to magnetize it too highly—to take a piece a quarter as large again, and to magnetize that less forcibly. It does not therefore pay to go much above 16,000 lines through a sq. cm.—that is to say, expressing it in terms of the law of traction, and the pounds per square inch, it does not pay to design your electromagnet so that it shall have to carry more than about 150 lb. to the square inch. This shall be our practical rule: let us at once take an example. If you want to design an electromagnet to carry a load of one ton, divide the ton, or 2240 lb., by 150, and that gives the requisite number of square inches of wrought iron, namely, 14·92, or say 15. Of course one would work with a horse-shoe shaped magnet, or something equivalent—something with a return circuit—and calculate out the requisite cross-section, so that the total area exposed might be sufficient to carry the given load at 150 lb. to the square inch. And, as a horse-shoe magnet has two poles, the cross-section of the bar of which it

Designing Electromagnets. 123

is made must be 7½ square inches. If of round iron, it must be about 3½ inches in diameter; if of square iron, it must be 2¾ inches each way.

That settles the size of the iron, but not the length. Now the length of the iron, if one only considers the law of the magnetic circuit, ought to be as short as it' can possibly be made. Reflect for what purpose we are designing. The design of an electromagnet is to be considered, as every design ought to be, with a view to the ultimate purpose to be served by that which you are designing. The present purpose is the actual sticking on of the magnet to a heavy weight, not acting on another magnet at a distance, not pulling at an armature separated from it by a thick layer of air; we are dealing with traction in contact.

The question is—How long a piece of iron shall we need to bend over? The answer is—Take length enough, and no more than enough, to permit of room for winding on the necessary quantity of wire to carry the current which will give the requisite magnetizing power. But this latter we do not yet know; it has to be calculated out by the law of the magnetic circuit. That is to say, we must calculate the magnetic flux, and the magnetic reluctance as best we can; then from these calculate the ampere-turns of current; and from this calculate the needful quantity of copper wire, so arriving finally at the proper length of the iron core. It is obvious, the cross-section being given, and the value of B being prescribed, that settles the whole number of magnetic lines, N, that will go through the section. It is self-evident that length adds to the magnetic reluctance, and, therefore, the longer the length is, the greater have to be the number of ampere-turns of circulation of the current; while the less the length is, the smaller need be the number of ampere-turns of circulation.

Therefore, you should design the electromagnet as stumpy as possible, that is to say, make it a stumpy arch, even as Joule did when he came across the same problem, and arrived, by a sort of scientific instinct, at the right solution.

You should have no greater length of iron than is necessary in order to get the windings on. Then you see we cannot absolutely calculate the length of the iron until we have an idea about the winding, and we must settle, therefore, provisionally, about the windings. Take a simple ideal case. Suppose we had an indefinitely long, straight iron rod, and we wound that from end to end with a magnetizing coil. How thick a coil, how many ampere-turns of circulation per inch length, will you require in order to magnetize up to any particular degree? It is a matter of very simple calculation. You can calculate exactly what the magnetic reluctance of an inch length of the core will be. For example, if you are going to magnetize up to 16,000 lines per sq. cm., the permeability will be 320. You can take the area anything you like, and consider the length of one inch; you can therefore calculate the magnetic reluctance per inch of conductor, and then you can at once say how many ampere-turns per inch would be necessary in order to give the desired indication of 16,000 magnetic lines to the sq. cm.

Then, knowing the properties of copper wire, and how it heats up when there is a current; and knowing also how much heat you can get rid of per square inch of surface, it is a very simple matter to calculate what minimum thickness of copper the fire insurance companies would allow you to use. They would not allow you to have too thin a copper wire, because if you provide an insufficient thickness of copper, you still must drive your amperes through it to get a sufficient number of ampere-turns per inch of length; and if you drive those amperes through copper winding of an insufficient thickness the copper wire will over-heat, and your insurance policy will be revoked.

You therefore are compelled, by the practical consideration of not over-heating, to provide a certain thickness of copper wire winding. I have made a rough calculation for certain cases, and I find that for such small electromagnets as one may ordinarily deal with, it is not necessary in any practical case to use a copper wire winding the total thickness of which is greater than about half an inch; and, as a matter of fact, if you use as much thickness as half an inch, you need not then wind the coil all along, for if you will use copper wire winding, no matter what the size, whether thin or thick, so that the total thickness of copper outside the iron is half an inch, you can, without over-heating, using good wrought iron, make one inch of winding do for 20 inches length of iron. That is to say, you do not really want more than $\frac{1}{40}$ of an inch of thickness of copper outside the iron to magnetize up to the prescribed degree of saturation that indefinitely long piece of which we are thinking, without over-heating the outside surface in such a way as to violate the insurance rules. Take it roughly, if you wind to a thickness of half an inch, the inch length of copper will magnetize 20 inches length of iron up to the point where B equals 16,000. If, then, we have a bar bent into a sort of horse-shoe in order to make it stick on to a perfectly-fitting armature also of equal section and quality, we really do not want more than one inch along the inner curve for every 20 inches of iron.

An extremely stumpy magnet, such as I have sketched in Fig. 54, will therefore do, if one can only get the iron sufficiently homogeneous throughout. If instead of crowding the wire near the polar parts, we could wind entirely all round the curved part, though the layer of copper winding would be half an inch thick inside the arch, it would be much less outside. Such a magnet, provided the armature fitted with perfect accuracy to the polar surfaces, and provided a battery were arranged to send the requisite number of amperes of current through the coils, would pull with a force of one ton, the iron being but $3\frac{1}{2}$ inches in diameter. For my own part, in this case I should prefer not to use round iron, one of square or rectangular section being more convenient; but the round iron would take less copper in winding, as each turn would be of minimum length if the section were circular.

Now, this sort of calculation requires to be greatly modified directly one begins to deal with any other case. A stumpy short magnetic circuit with great cross-section is clearly the right thing for the greatest traction. You will get the given magnetization and traction with the least amount of magnetizing force when you have the area as great as possible, and the length as small as possible.

FIG. 54.

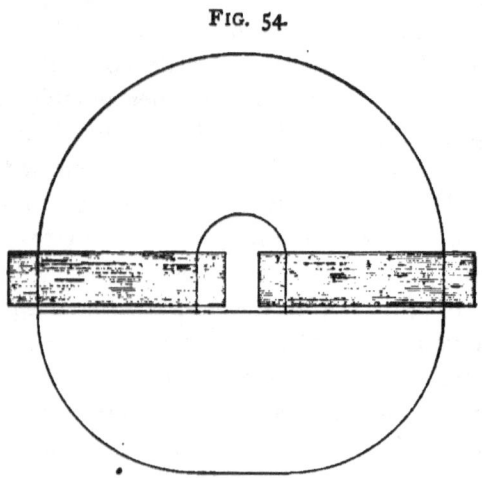

STUMPY ELECTROMAGNET.

You will kindly note that I have given you as yet no proofs for the practical rules that I have been using: they must come later. Also, I have said nothing about the size of the wire, whether thick or thin; that does not in the least matter, for the ampere-turns of magnetizing power can be made up in any desired way. Suppose we want on any magnet one hundred ampere-turns of magnetizing power, and we choose to employ a thin wire that will only carry half an ampere, then we must wind 200 turns of that thin wire. Or, suppose we choose to wind it with a thick wire that will carry ten amperes, then we shall want only ten turns of that wire. The same weight of copper, heated up by the corresponding current to an equal degree of temperature, will have equal magnetizing power when wound on the same core. But the rules about winding the copper will be considered later.

WEIGHT AND TRACTION OF MAGNETS.

Now if you look in the text-books that have been written on magnetism for information about the so-called lifting power or portative force of magnets—in other words, their traction—you will find that from the time of Bernoulli downwards, the law of portative force has claimed the attention of experimenters, who, one after another, have tried to give the law of portative force in terms of the weight of the magnets; usually dealing with permanent magnets, not electromagnets. D. Bernoulli gave * a rule of the following kind, which is commonly known as Häcker's rule—

$$P = a \sqrt[3]{W};$$

where W is the weight of the magnet, P the greatest load it will sustain, and a a constant depending on the units of weight chosen, on the quality of the steel, and on its goodness of magnetization. If the weights are in kilogrammes then a is found, for the best steels, to vary from 18 to 24 in magnets of horse-shoe shape. This expression is equivalent to saying that the power which a magnet can exert—he was dealing with steel magnets, there were no electromagnets in Bernoulli's time—is equal to some constant multiplied by the three-halfth root of the weight of the magnet itself. The rule is accurate only if you are dealing with a number of magnets all of the same geometrical form; all horse-shoes, let us say, of the same general shape, made from the same sort of steel, similarly magnetized. In former years I pondered much on Häcker's rule, wondering how on earth the three-halfth root of the weight could have anything to do with the magnetic pull; and having cudgelled my brains for a considerable time, I saw that there was really a very simple meaning in it.

What I arrived at † was this. If you are dealing with a given material, say hard steel, the weight is proportional to

---

* *Acta Helvetica*, iii. p. 233, 1758.
† *Philosophical Magazine*, July 1888.

the volume, and the cube root of the volume is something proportional to the length, and the square of the cube root forms something proportional to the square of the length, that is to say, to something of the nature of a surface. What surface? Of course the polar surface.

This complex rule, when thus analyzed, turns out to be merely a mathematician's expression of the fact that the pull for a given material magnetized in a given way is proportional to the area of the polar surface; a law which in its simple form Joule seems to have arrived at naturally, and which in this extraordinarily academic form was arrived at by comparing the weight of magnets with the weight which they would lift.

You will find it stated in many books that a good magnet will lift twenty times its own weight. There never was a more fallacious rule written. It is perfectly true that a good steel horse-shoe magnet weighing 1 kilogramme ought to be able to pull with a pull of 20 kilogrammes on a properly-shaped armature. But it does not follow that a magnet which weighs 2 kilogrammes will be able to pull with a force of 40 kilogrammes. It ought not to, because a magnet that weighs 2 kilogrammes has not poles twice as big if it is the same shape. In order to have poles twice as big you must remember that three-halfth root coming in. If you take a magnet that weighs eight times as much, it will have twice the linear dimensions and four times the surface; and with four times the surface in a magnet of the same form, similarly magnetized, you will have four times the pull. With a magnet eight times as heavy you will have only four times the pull. The pull, when other things are equal, goes by surface, and not by weight, and therefore it is ridiculous to give a rule saying how many times its own weight a magnet will pull.

It is also narrated as a very extraordinary thing that Sir Isaac Newton had a magnet—a loadstone—which he wore in a signet ring, which would lift 234 times its own weight. Professor G. Forbes, in his *Lectures on Electricity* describes a small iron-clad electromagnet, weighing about 3 oz., which would sustain a load 600 times its own weight. I have had an electromagnet which would lift 2500 times its own weight, but then it was a very small one, and did not weigh with its copper coil more than a grain and a half. When you come to small things, of course the surface is large proportionally to the weight; the smaller you go, the larger becomes that disproportion. This all shows that the old law of traction in that form was practically valueless, and did not guide you to anything at all, whereas the law of traction as stated by Maxwell, and explained further by the law of the magnetic circuit, proves a most useful rule.

CALCULATION OF EXCITING POWER NEEDED.

From this digression let us return to the law of the magnetic circuit in order to calculate the exciting power required to produce the magnetism. I have given in Chapter II., when speaking of permeability, the following rule for calculating the magnetic induction B:—Take the pull in pounds, and the area of cross-section in square inches; divide one by the other, and take the square root of the quotient; then multiplying by 1317 gives B; or multiplying by 8494 gives $B_{\prime\prime}$. We have therefore a means of stepping from the pull per square inch to $B_{\prime\prime}$, or from $B_{\prime\prime}$ to the pull per square inch. Now the other rule of the magnetic circuit also enables us to get from the ampere-turns down to $B_{\prime\prime}$, for on p. 117 we have the following expression for the ampere-turns:—

$$S i = N \times \Sigma \frac{l''}{A''\mu} \times 0 \cdot 3132,$$

and N, the whole number of magnetic lines in the magnetic circuit, is equal to $B_{\prime\prime}$ multiplied by A'', or

$$N = B_{\prime\prime} A''.$$

From these we can deduce a simple direct expression, provided we assume the quality of iron as before, and also assume that there is no magnetic leakage, and that the area of cross-section is the same all round the circuit, in the armature as well as in the magnet core. So that l'' is simply the mean total path of the magnetic lines all round the closed magnetic circuit. We may then write

$$S\,i = \frac{B_{\prime\prime}\,l''}{\mu} \times 0\cdot 3132;$$

whence

$$B_{\prime\prime} = \frac{\mu \times S\,i}{l'' \times 0\cdot 3132}.$$

But, by the law of traction, as stated above,

$$B_{\prime\prime} = 8494\sqrt{\frac{P\;(\text{lb.})}{A\;(\text{sq. in.})}}.$$

Equating together these two values of $B_{\prime\prime}$, and solving, we get for the requisite number of ampere-turns of circulation of exciting currents :—

$$S\,i = 2661 \times \frac{l''}{\mu} \times \sqrt{\frac{P\;(\text{lb.})}{A\;(\text{sq. in.})}}.$$

This, put into words, amounts to the following rule for calculating the amount of exciting power that is required for an electromagnet pulling at its armature, in the case where there is a closed magnetic circuit with no leakage of magnetic lines :—Take the square root of the pounds per square inch; multiply this by the mean total length (in inches) all round the iron circuit; divide by the permeability (which must be calculated from the pounds per square inch by help of Table XII. and Table II.); and finally multiply by 2661 : the number so obtained will be the number of ampere-turns.

One goes then at once from the pull per square inch to the number of ampere-turns required to produce that pull in a magnet of given length and of the prescribed quality. In the case where the pull is specified in kilogrammes, the area of section in sq. cm., and the length in cm., the formula becomes

$$S\,i = 3951 \cdot \frac{l}{\mu} \sqrt{\frac{P}{A}}$$

As an example, take a magnet core of round annealed wrought iron, half an inch in diameter, 8 inches long, bent to horse-shoe shape. As an armature, another piece, 4 inches long, bent to meet the former. Let us agree to magnetize the iron up to the pitch of pulling with 112 lb. to the square inch. Reference to Table XII. shows that B_{μ} will be about 90,000 and Table II. shows that in that case μ will be about 907. From these data calculate what load the magnet will carry, and how many ampere-turns of circulation of current will be needed.

Ans.—Load (on two poles) = 43·97 lb.
Ampere-turns needed = 372·5

N.B.—In this calculation it is assumed that the contact surface between armature and magnet is perfect. It never is; the joint increases the reluctance of the magnetic circuit, and there will be some leakage. It has been shown in Chapter III., p. 91, how to estimate these effects; it will be shown in Chapter VI. how to allow for them in the calculations.

Effect of Diminishing Polar Surface.

Here let me go to a matter which has been one of the paradoxes of the past. In spite of Joule, and of the laws of traction, showing that the pull is proportional to the area, you have this anomaly, first pointed out by Moll,[*] that if you take a bar-magnet having flat-ended poles, and measure the pull which its pole can exert on a perfectly flat armature, and then deliberately spoil the truth of the contact surface, rounding it off, so making the surface gently convex, the convex pole, which only touches at a portion of its area instead of over the whole, will be found to exert a bigger pull than the perfectly flat one. It has been shown by various experi-

[*] *Edin. Journ. Sci.*, iii. p. 240, 1830; and *Pogg. Ann.*, xxiv., p. 632, 1833.

menters, particularly by Nicklès, that if you want to increase the pull of a magnet with armatures, you may reduce the polar surface.

Old steel magnets were frequently purposely made with a rounded contact surface. There are plenty of examples. Suppose you take a straight round core, or one leg of a horse-shoe which answers equally, and take a flat-ended rod of iron of same diameter as an armature; stick it on endwise, and measure the pull when a given amount of ampere-turns of current is circulating round. Then, having measured the pull, remove it and file it a little, so as to reduce it at the edges, or take a slightly narrower piece of iron, so that it will actually be exerting its power over a smaller area; you will get a greater pull. What is the explanation of this extraordinary fact? A fact it is, as can readily be shown. Here, Fig. 55, is a small electromagnet which we can place with its poles upwards. This was very carefully made, the iron poles very nicely faced, and on coming to try them it was found they were nearly equal, but one pole, A, was a little stronger than the other. We have, therefore, rounded the other pole, B, a little, and here I will take a piece of iron, C, which has itself been slightly rounded at one end, though it is flat at the other. I now turn on the current to the electromagnet, and I take a spring-balance with which to measure the pull at either of the two poles. When I put the flat end of C to the flat pole A, so that there is an excellent contact, I find the pull about 2½ lb. Now try the round end of C on the flat pole A; the pull is about 3 lb. The flat end of C on the round pole B is also about 3 lb., perhaps a little more. But if now I put together two surfaces that are both rounded, I get almost

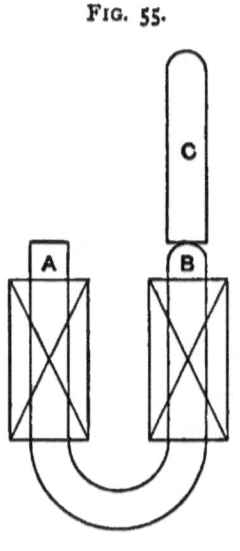

FIG. 55.

EXPERIMENT ON ROUNDING ENDS.

Effect of Rounding Pole.

exactly the same pull as at first with the two flat surfaces. I have made many experiments on this, and so have others. Take the following case:—There is hung up a horse-shoe magnet, one pole being slightly convex and the other absolutely flattened, and there is put on at the bottom a square bar armature, over which is slipped a hook to which weights can be hung. Which end of the armature do you think will be detached first?

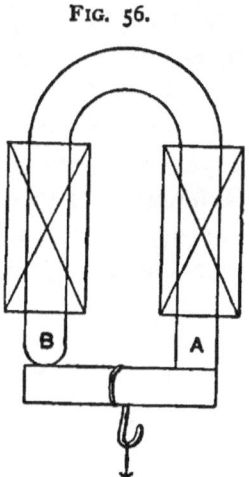

Fig. 56.

EXPERIMENT OF DETACHING ARMATURE.

If you were going simply by the square inches, you would say this square end will stick on tighter; it has more gripping surface. But, as a matter of fact, the other sticks tighter. Why? We are dealing here with a magnetic circuit. There is a certain total magnetic reluctance all round it, and the whole number of magnetic lines generated in the circuit depends on two things—on the magnetizing force, and on the reluctance all round; and, saving a little leakage, it is the same number of magnetic lines which come through at B as go through at A. But here, owing to the fact that there is at B a better contact at the middle than at the edges of the pole, the lines are crowded into a smaller space, and therefore at that particular place $B_{\prime\prime}$ the number of lines per square inch runs up higher, and when you square the larger number, its square becomes still larger in proportion. In comparing the square of smaller B with the square of greater $B_{\prime\prime}$ the square of the smaller $B_{\prime\prime}$ over the larger area turns out to be less than the square of the larger $B_{\prime\prime}$ integrated over the smaller area. It is the law of the square coming in.

You must not jump to the conclusion from this that there would be any benefit in rounding both poles. By rounding a pole you impair the magnetic circuit at that point, and the *other* pole holds on *less* tightly as a result.

As an example, take the case of a magnet pole formed on the end of a piece of round iron 1·15 inch in diameter. The flat pole will have 1·05 inch area. Suppose the magnetizing forces are such as to make B_u = 90,300, then, by Table XII., the whole pull will be 118·75 lb., and the actual number of lines through the contact surface will be N = 94,815. Now suppose the pole be reduced by rounding off the edge till the effective contact area is reduced to 0·9 square inch. If all these lines were crowded through that area, that would give a rate of 105,630 per square inch. Suppose, however, that the additional reluctance and the leakage reduced the number by 2 per cent., there would still be 103,500 per square inch. Reference to Table XII. shows that this gives a pull of 147·7 lb. per square inch, which, multiplied by the reduced area 0·9, gives a total pull of 132·9 lb., which is larger than the original pull.

Let me show you yet another experiment. This is the same electromagnet (Fig. 56), which has one flat pole and one rounded pole. Here is an armature, also bent, having one flat and one rounded pole. If I put flat to flat, and round to round, and pull at the middle, the flat to flat detaches first; but if we take round to flat, and flat to round, we shall probably find they are about equally good—it is hard to say which holds the stronger. On the whole, the rounded armature on the flat pole sticks less tightly than the flat armature on the rounded pole.

Contrast between Flat and Pointed Poles.

We are now in a position to understand the bearing of some researches made about forty years ago by Dr. Julius Dub, which, like a great many other good things, lie buried in the back volumes of Poggendorff's *Annalen*.* Some account of them is also given in Dr. Dub's now obsolete book entitled 'Elektromagnetismus.'

The first of Dub's experiments to which I will refer relates

---

* See *Pogg. Ann.*, lxxiv., p. 465; lxxx., p. 497; xc., p. 248; cv., p. 49.

to the difference in behaviour between electromagnets with flat and those with pointed pole ends. He formed two cylindrical cores, each six inches long, from the same rod of soft iron, one inch in diameter. Either of these could be slipped into an appropriate magnetizing coil. One of them had the end left flat, the other had its end pointed, or, rather, it was coned down until the flat end was left only $\frac{1}{2}$ inch in diameter, possessing therefore only one-fourth of the amount of contact surface which the other core possessed. As an armature there was used another piece of the same soft iron rod, twelve inches long. The pull of the electromagnet on the armature at different distances was carefully measured, with the following results :—

Distance apart in inches.	Pull on Flat Pole (lb.).	Pull on Pointed Pole (lb.).
0	3·3	5·2
0·0055	1·1	1·8
0·0110	0·9	0·75
0·0165	0·71	0·50
0·022	0·60	0·42
0·044	0·38	0·20
0·088	0·19	0·09

These results are plotted out in the curves in Fig. 57. It will be seen that in contact, and at very short distances, the reduced pole gave the greater pull. At about ten mils distance there was equality, but at all distances greater than ten mils the flat pole had the advantage. At small distances the concentration of magnetic lines gave, in accordance with the law of traction, the advantage to the reduced pole. But this advantage was, at the greater distances, more than outweighed by the fact that with the greater widths of air-gap the use of the pole with larger face reduced the magnetic reluctance of the gap and promoted a larger flow of magnetic lines into the end of the armature.

Fig. 57.

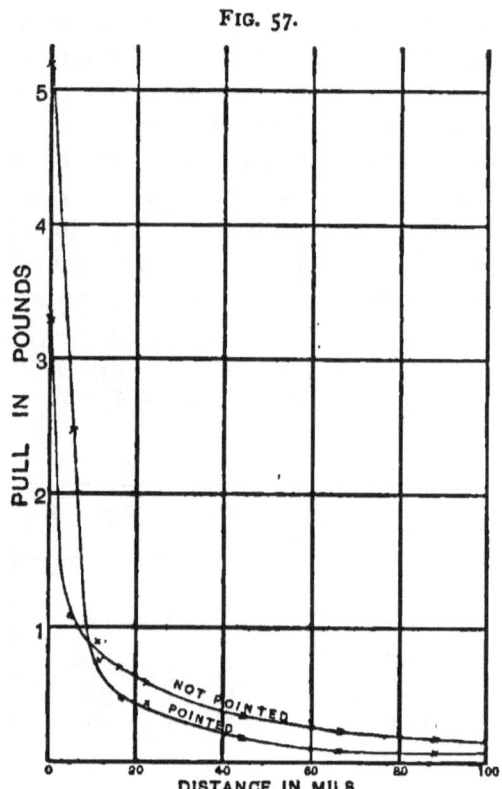

CONTRASTED EFFECT OF FLAT AND POINTED POLES.

EXPLORATION OF SURFACE DISTRIBUTION OF MAGNETISM.

The law of traction can again be applied to test the so-called distribution of free magnetism on the surface. Consider what is meant by the phrase. On p. 41, there was given in Fig. 17 a sketch of the way in which the lines of magnetization traverse the interior of a bar magnet and emerge at the surface. Wherever such lines emerge at a surface, there iron filings will stick on. The old way of stating the fact of such emergence was to say that at these parts of the surface there was free magnetism, the distribution of which over the surface was a matter of much speculation, of great mathematical toil, and of many experiments. The distribution of the magnetic lines at their emergence on the surface we may explore by the method of traction. We can thereby arrive at a kind of measure of the amount of surface density of the free magnetism. I do not like to have to use these ancient terms, because they suggest the ancient notion that magnetism was a fluid, or rather two fluids, one of which was plastered on at one end of the magnet and the other at the other, just as you might put red paint or blue paint over the ends. I only use that term because it is already more or less familiar.

One of the ways of experimentally exploring the so-called distribution of free magnetism must be mentioned here, because it is often used in this class of experiments. It consists in measuring, at various points of the surface, the force required to detach a *proof-piece* consisting of a small sphere, ellipsoid, or rod of iron. This method, due originally to Plücker,* and used by him, and later by Vom Kolke,† by Tyndall,‡ by Lamont,§ and by Jamin,‖ is known in France as the *méthode du clou*, from the circumstance that an iron nail may be used. Plücker himself used little prolate spheroids, 14 mm. long and 8 mm. in diameter, of various sorts of iron and steel; attaching them by a thread to the beam of a balance. Vom Kolke used as an exploring rod a piece of soft iron wire weighing 1·7 gramme, 2·6 cm long, 0·45 cm. thick, pointed at the end. In another set of experiments he used a small polished iron globe 50·3 cm. in diameter.

Tyndall used three smooth soft iron spheres, 0·95 inch, 0·48 inch, and 0·29 inch in diameter. He came to the conclusion that the pull required to detach was proportional simply to the strength of the magnetism. Lamont used a

* *Pogg. Ann.*, lxxxvi., 1852, p. 11.
† *Pogg. Ann.*, lxxxi., 1850, p. 321; and *Wied. Ann.*, iii., 1878, p. 437.
‡ *Pogg. Ann.*, lxxxiii., 1851, p. 1; and *Phil. Mag.*, April 1851. See also Tyndall's *Diamagnetism*, p. 321.
§ *Abhandl. d. Münchener Akad.*, vi., 479; and Lamont's *Magnetismus*, p. 325.
‖ *Journal de Physique*, v., 1876, p. 41; and vii., 1878, p. 38.

short rod of iron rounded at the ends. Jamin assumed as a standard an iron rod 15 cm. long and 0·1 cm. in diameter, but employed for greater convenience short iron wires ending in a small iron ball. He held that the pull required to detach was proportional to the square of the magnetism on the surface at the point examined. The method is discussed by Lamont* and also by Chrystal.† The latter makes the remark that it is not very easy to see, in the complexity of effects depending on contact, induction, and the like, what is the quantity that is being measured. He adds that long-shaped bodies are preferable; for when bodies of high magnetic permeability and of nearly spherical shape are employed, differences in form produce far more effect on the experimental results than the susceptibility of the material does.

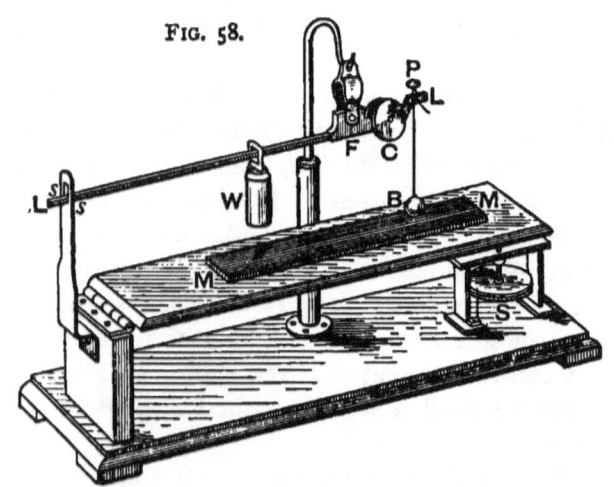

FIG. 58.

AYRTON'S APPARATUS FOR MEASURING SURFACE DISTRIBUTION OF PERMAMENT MAGNETISM.

The little piece of apparatus shown in Fig. 58 was arranged by my friend and predecessor, Professor Ayrton, for the purpose of teaching his students at the Finsbury College.‡ Here is a bar magnet M M of steel, marked in centimetres from end to end; over the top of it there is

* *Op. cit.*, p. 325–8.
† *Encycl. Britannica*, Art. MAGNETISM, p. 242.
‡ See Ayrton's *Practical Electricity*, Fig. 5A, p. 24.

a little steelyard, consisting of a weight W, sliding along an arm L L. At the end of that steelyard there is suspended a small bullet B of iron. If we bring that bullet into contact with the bar magnet anywhere near the end, and equilibrate the pull by sliding the counterpoise along the steelyard arm, we shall obtain the definite pull required to detach that piece of iron. The pull will be proportional, by Maxwell's rule, to the square of the number of magnetic lines coming up from the bar into it. Shift the magnet on a whole centimetre, and attach the bullet a little further on; now equilibrate it, and we shall find it will require a rather smaller force to detach it. Try it again, at points along from the end to the middle. The greatest force required to detach it will be found at the extreme corner, and a little less a little way on, and so on until we find at the middle the bullet does not stick on at all, simply because there are here no magnetic lines leaking. The method is not perfect, because it obviously depends on the magnetic properties of the little bullet, and whether much or little saturated with magnetism. Moreover, the presence of the bullet perturbs the very thing that is to be measured. Leakage into air is one thing; leakage into air perturbed by the presence of the little bullet of iron, which invites leakage into itself, is another thing. It is an imperfect experiment at the best, but a very instructive one.

A preferable method of experimenting consists in using, as suggested by Rowland, a very small coil of insulated copper wire connected to a sensitive ballistic galvanometer. This coil, termed a *magnetic proof-plane*, is laid on the surface of the magnet, and then suddenly removed. The throw of the galvanometer measures the intensity of the normal flux of magnetic lines at the point in question.

Effect of Fixing Masses of Iron upon Ends of Cores.

Here is a paradoxical experiment. I have here a bar electromagnet, which we will connect to the wires that bring the exciting current. Opposite one end of the iron core, and about 18 inches distant, is a small compass needle, with a feather attached to it as a visible indicator, so that, when we turn on the current, the electromagnet will act on the needle and you will see the feather turn round. It is acting there at a certain distance. The magnetizing force is mainly spent, not to drive magnetism round a circuit of iron, but to force it through the air, flowing from one end of the iron core out into the air, passing by the compass needle, and streaming round again, invisible, into the other end of the iron core. It ought to increase the flow if we can in any way aid the magnetic lines to flow through the air. How can I aid this flow? By putting on something at the other end to help the magnetic lines to get back home. Here is a flat piece of iron; putting it on here at the hinder end of the core ought to help the flow of magnetic lines. You see that the feather makes a rather larger excursion. Taking away the piece of iron diminishes the effect. So also in experiments on tractive power, it can be proved that the adding of a mass of iron at the far end of a straight electromagnet greatly increases the pulling power at the end that you are working with; while, on the other hand, putting the same piece of iron on the front end as a pole-piece greatly diminishes the pull.

Here, clamped to the table, is a bar electromagnet excited by the current; and here is a small piece of iron attached to a spring balance, by means of which I can measure the pull required to detach it. With the current which I am employing, the pull is about $2\frac{1}{2}$ lb. I now place upon the front end of the core this block of wrought iron; it is itself strongly held on, but the pull which it itself exerts on the small piece of iron is small. Less than half a pound suffices to detach it. I now remove the iron block from the front end of the core, and place it upon the hinder end; and now I find that the force required to detach the small piece of iron from the front end is about $3\frac{1}{2}$ lb., instead of $2\frac{1}{2}$ lb. The front end exerts a bigger pull when there is a mass of iron attached to the hinder end. Why? The whole iron core, including its front end, becomes more highly magnetized, because there is now a better way for the magnetic lines to emerge at the other end and come round to this. In short, we have diminished the magnetic reluctance of the air part of the magnetic circuit, and the flow of magnetic lines in the whole magnetic circuit is thereby improved. So it was also when the mass of iron was placed across the front end of the core; but the magnetic lines streamed away backwards from its edges, and few were left in front to act upon the small bit of iron. So the law of magnetic circuit action explains this anomalous behaviour. Facts like these have been well known for a long time to those who have studied electromagnets.

In Sturgeon's book there is a remark that bar magnets pull better if they are armed with a mass of iron at the distant end, though Sturgeon did not see what we now know to be the explanation of it. The device of fastening a mass of iron to one end of an electromagnet in order to increase the magnetic power of the other end was patented by Siemens in 1862.

The next experiments to be described relate to the employment of polar extensions or pole-pieces attached to the core. These experiments, which are due to Dr. Julius Dub, are so curious, so unexpected, unless you know the reasons why, that I invite your especial attention to them. If an engineer had to make a firm joint between two pieces of metal, and he feared that a mere attachment of one to the other was not adequately strong, his first and most natural impulse would be to enlarge the parts that come together—to give one as it were a broader footing against the other. And that is precisely what an engineer, if uninstructed in the true principles of magnetism, would do in order to make an electromagnet stick more tightly on to its armature. He would enlarge the ends of one or both; he would add pole-pieces to give the armature a better foothold. Nothing, as you will see, could be more disastrous.

Dub employed in these experiments a straight electromagnet having a cylindrical soft iron core, 1 inch in diameter, 12 inches long; and as armature a piece of the same iron,

6 inches long. Both were flat-ended. Then six pieces of soft iron were prepared of various sizes, to serve as pole-pieces. They could be screwed on at will, either to the end of the magnet core or to that of the armature. To distinguish them we will call them by the letters A, B, C, &c. Their dimensions were as follows, the inches being presumably Bavarian inches:—

Piece.	Diameter.	Length.
	inches.	inches.
A	2	1
B	1⅞	1¼
C	1⅜	2
D	2	½
E	1½	1
F	1	2

Of the results obtained with these pieces we will select eight. They are those illustrated by the eight collected sketches in Fig. 59. The pull required to detach was measured, also the attraction exerted at a certain distance apart. It will be noted that, in every case, putting on a pole-piece to the end of the magnet diminished both the pull in contact and the attraction at a distance; it simply promoted leakage and dissipation of the magnetic lines.

Experiment.	On Magnet.	On Armature.	Traction.	Attraction.
I.	none	none	48	22
II.	D	none	30	10
III.	E	none	32	11·5
IV.	C	none	35	13·5
V.	D	A	20	7·5
VI.	none	B	50	25
VII.	none	D	43	25
VIII.	none	C	50	18

Experiments with Pole-Pieces. 143

The worst case of all was that in which there were pole-pieces both on the magnet and on the armature. In the last three cases the pull was increased, but here the enlarged piece

FIG. 59.

DUB'S EXPERIMENTS WITH POLE-PIECES.

was attached to the armature, so that it helped those magnetic lines which came up into it to flow back laterally to the bottom end of the electromagnet, while thus reducing the magnetic reluctance of the return path through the air, and so increasing the total number of magnetic lines, did not spread unduly those that issued up from the end of the core.

The next of Dub's results relate to the effect of adding these pole-pieces to an electromagnet 12 inches long, which was being employed, broadside-on, to deflect a distant compass needle (Fig. 60).

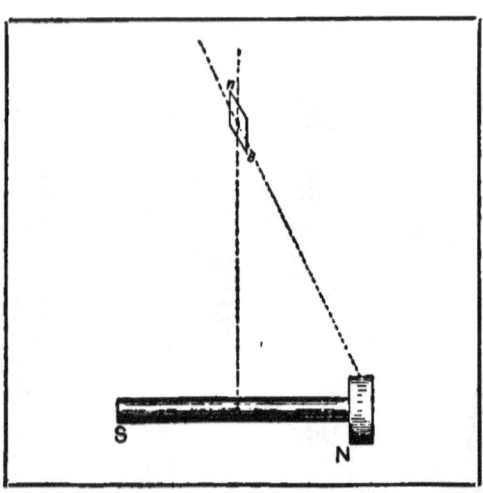

FIG. 60.

DUB'S DEFLEXION EXPERIMENT.

Pole-piece used.	Deflexion (degrees).
none	34·5
A	42
B	41·5
C	40·5
D	41
E	39
F	38

In another set of experiments of the same order, a permanent magnet of steel, having poles $n\ s$, was slung horizontally by a bifilar suspension, to give it a strong tendency to set in a particular direction. At a short distance laterally was fixed the same bar electromagnet, and the same pole pieces were again employed. The results of attaching the pole-pieces at the near end are not very conclusive; they slightly increased the deflexion. But in the absence of information as to the distance between the steel magnet and the electromagnet, it is difficult to assign proper values to all the causes at work. The results were:—

Pole-piece used.	Deflexion (degrees).
none	8·5
A	9·2
B	9·5
C	10
D	8·8

When, however, the pole-pieces were attached to the distant end of the electromagnet, where their effect would

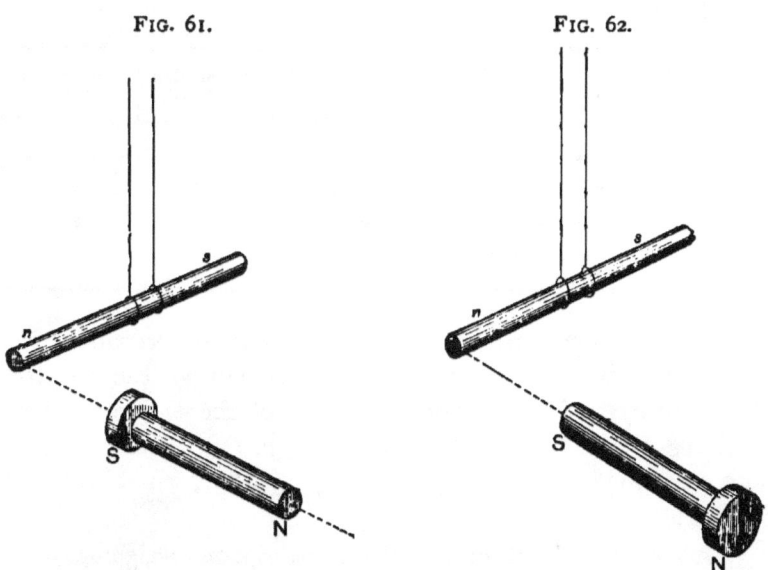

FIG. 61. FIG. 62.

DEFLECTING A STEEL MAGNET HAVING BIFILAR SUSPENSION, POLE-PIECE ON NEAR END.   DEFLECTING STEEL MAGNET, POLE-PIECE ON DISTANT END.

undoubtedly be to promote the leakage of magnetic lines into the air at the front end without much affecting the distribution of those lines in the space in front of the pole, the action was more marked.

Pole-piece used.	Deflexion (degrees).
none	8·5
A	10·0
B	10.3
C	10·3
F	10·1

L

Still confining ourselves to straight electromagnets, I now invite your attention to some experiments made in 1862 by the late Count Du Moncel as to the effect of adding a polar expansion to the iron core. He used as his core a small iron tube, the end of which he could close up with an iron plug, and around which he placed an iron ring which fitted closely on to the pole. He used a special lever arrangement to measure the attraction exercised upon an armature distant in all cases one millimetre from the pole. The results were as follows :—

	Without ring on pole.	With ring on pole.
Tubular core alone	11	10
,, ,, with iron plug	17	14
Core provided with mass of iron at distant end	27	25
,, ,, ,, iron plug	38	33

After hunting up these researches, it was extremely interesting to find that so important a fact had not escaped the observant eye of the original inventor of the electromagnet. In Sturgeon's *Experimental Researches*, p. 113, there is a footnote, written apparently about the year 1832, which runs as follows :—

"An electromagnet of the above description, weighing three ounces and furnished with one coil of wire, supported fourteen pounds. The poles were afterwards made to expose a larger surface by welding to each end of the cylindric bar a square piece of good soft iron; with this alteration only, the lifting power was reduced to about five pounds, although the magnet was annealed as much as possible."

Effect of Jacketing an Electromagnet.

We saw that this straight electromagnet, whether used broadside-on or end-on, could act on the compass needle at some distance from it, and deflect it. In those experiments there was no return path for the magnetic lines that flowed through the iron core, save that afforded by the surrounding air. The lines flowed round in wide-sweeping curves from one end to the other, as in Fig. 17, the magnetic field being quite extensive. Now, what will happen if we provide a return path? Suppose I surround the electromagnet with an iron tube of the same length as itself, the lines will flow along in one direction through the core, and will find an easy path back along the outside of the coil. Will the magnet thus jacketed pull more powerfully or less on that little suspended magnet? We should expect it to pull less powerfully, for if the magnetic lines have a good return path here through the iron tube, why should they force themselves in such a quantity to a distance through air in order to get home? No; they will naturally return short back from the end of the core into the tubular iron jacket. That is to say, the action at a distance ought to be diminished by putting on that iron tube outside. The matter is readily put to the test of experiment by placing a straight electromagnet, either in the end-on position, or in the broadside-on position, near an indicating magnetic needle. Let the deflexion of the latter be observed when the exciting current is turned on, first when there is no external jacket, secondly when an external iron jacket is placed around the electromagnet. In the latter case it will be seen that when the current is turned on the indicating needle is scarcely affected at all. The iron jacket causes that magnet to have much *less* action at a distance. Yet it has actually been proposed to use jacketed magnets of this sort in telegraph instruments, and in electric motors, on the ground that they give a bigger pull.

Iron-clad electromagnets such as this produce less action at a distance across air than do the ordinary forms, but there yet remains the question whether they give a bigger pull in contact? Yes, undoubtedly they do; because everything that is helping the magnetism to get round to the other end increases the goodness of the magnetic circuit, and therefore increases the total magnetic flux.

L 2

This experiment may be tried upon a piece of apparatus similar to one which has been used for some years at the Finsbury Technical College. It consists of a straight electro-magnet M set upright in a base-board, over which is erected a gallows of wood. Across the frame of the gallows goes a winch W, on the axle of which is a small pulley with a cord knotted to it. To the lower end of the cord is hung a common spring balance, from the hook of which depends a small horizontal disk of iron A, to act as an armature. By means of the winch I lower this disk down to the top of the electro-magnet. The current is turned on; the disk is attracted. On winding up the winch I increase the upward pull until the disk is detached. See, it required about 9 lb. to pull it off.

FIG. 63.

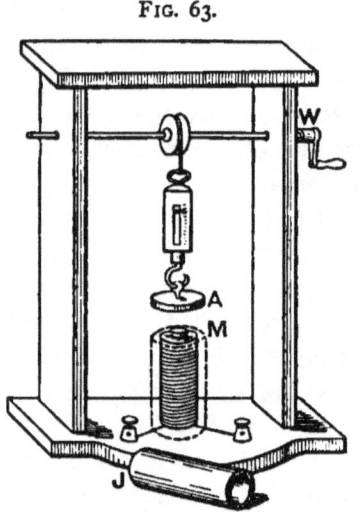

ELECTROMAGNET AND IRON JACKET.

I now slip over the electromagnet, without in any way attaching it, this loose jacket J of iron—a tube, the upper end of which stands flush with the upper polar surfaee. Once more I lower the disk, and this time it attaches itself at its middle to the central pole, and at its edges to the tube. What force will now be required to detach it? The tube weighs about ½ lb., and it is not fixed at the bottom. Will 9½ lb. suffice to lift the disk? By no means. My balance only measures up to 24 lb., and even that pull will not suffice to detach the disk. I know of one case where the pull of the straight core was increased sixteen-fold by the mere addition of a good return-path of iron to complete the magnetic circuit. But the jacketed form is not good for anything except increasing the tractive power. Jacketing an electromagnet which already possesses a return circuit of iron is an absurdity. For this reason the proposal made by one inventor to put iron tubes outside the coils of a horse-shoe electromagnet is one to be avoided.

We will take another paradox, which equally can be explained by the principle of the magnetic circuit. Suppose you take an iron tube as an interior core; suppose you cut a little piece off the end of it—a mere ring of the same size. Take that little piece and lay it down on the end. It will be stuck with a certain amount of pull. It will pull off easily. Take that same round piece of iron, put it on edgewise, where it only touches one point of the circumference, and it will stick on a good deal tighter, because it is there in a position to increase the magnetic flow of the magnetic lines. By concentrating the flow of magnetic lines over a small surface of contact increases B at that point, and B², integrated over the lesser area of the contact, gives a total bigger pull than is the case when the edge is touched all round against the edge of the tube.

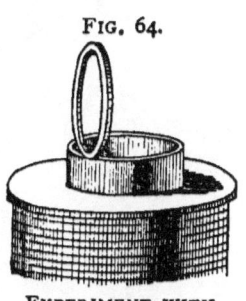

FIG. 64.

EXPERIMENT WITH TUBULAR CORE AND IRON RING.

Here is a still more curious experiment. I use a cylindrical electromagnet set up on end, the core of which has at the top a flat circular polar surface about two inches in dimeter. I now take a round disk of thin iron—ferrotype or tin-plate will answer quite well—which is a little smaller than the polar face. What will happen when this disk is laid down flat and centrally on the polar face? Of course you will say that it will stick tightly on. If it does so, the magnetic lines which come in through its under surface will pass through it and come out on its upper surface in large quantities. It is clear that they cannot all, or even any considerable proportion of them, emerge sideways through the edges of the thin disk, for there is not substance enough in the disk to carry so many magnetic lines. As a matter of fact the magnetic lines do come through the disk, and emerge on its upper surface, making indeed a magnetic field over its upper surface that is nearly as intense as the magnetic field beneath its under surface. If the two magnetic fields were

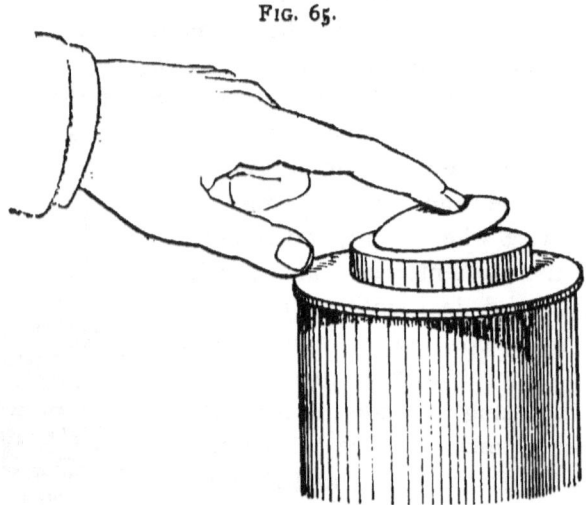

FIG. 65.

EXPERIMENT WITH IRON DISK ON POLE OF THE ELECTROMAGNET.

exactly of equal strength, the disk ought not be attracted either way. Well, what is the fact? The fact, as you see now that the current has been turned on, is that the disk absolutely refuses to lie down on the top of the pole. If I hold it down with my finger, it actually bends itself up, and requires force to keep it down. I lift my finger and over it flies. It will go anywhere in its effort to better the magnetic circuit rather than lie flat on the top of the pole.

Distribution on Polar Surface.

Next I invite your attention to some experiments, originally due to Vom Kolke, published in the *Annalen* forty years ago, respecting the distribution of the magnetic lines where they emerge from the polar surface of an electromagnet. The first one now described relates to a straight electromagnet with a cylindrical, flat-ended core (Fig. 66). In what way will the magnetic lines be distributed over at the end? Fig. 17, p. 41, illustrates roughly the way in which, when there is no return-path of iron, the magnetic lines leak through the air. The main leakage is through the ends, though there is some at the sides also. Now the question of the end distribution we shall try by using a small bullet of iron, which will be placed at different points from the middle to the edge, a spring balance being employed to measure the force required to detach it. The pull at the edge is much stronger than at the middle, at least four or five times as great. There is a regular increase of pull from the middle to the edge.

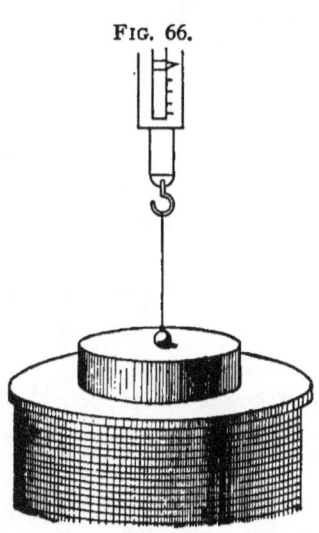

FIG. 66.

EXPLORING POLAR DISTRIBUTION WITH SMALL IRON BALL.

The magnetic lines, in trying to complete their own circuit, flow most numerously in that direction where they can go furthest through iron on their journey. They leak out more strongly at all edges and corners of a polar surface. They do not flow out so strongly at the middle of the end surface, otherwise they would have to go through a larger air-circuit to get back home. The iron is consequently more saturated round the edge than at the middle; therefore, with a very small magnetizing force, there is a great disproportion between the pull at the middle and that at the edges. With a very large magnetizing force you do not get the same disproportion, because if the edge is already far saturated you cannot by applying higher magnetizing power increase its magnetization much, but you can still force more lines through the middle. The consequence is, if you plot out the results of a succession of experiments of the pull at different points, the curves obtained are, with larger magnetizing forces, more nearly straight than are those obtained with small magnetizing force.

The results obtained by Vom Kolke upon a solitary cylindrical pole 12 cm. in diameter, which he examined at every half cm. from centre to edge, are given below. The pull required to detach a small iron ball 3 mm. in diameter was almost six times as great at the edge as in the middle.

Radial distance from centre (centimetres).	0	½	1	1½	2	2½	3
Pull exerted on small sphere in contact	8·75	8·75	8·88	9·16	9·40	10·19	10·83

Radial distance from centre (centimetres).	3½	4	4½	5	5½	6
Pull exerted on small sphere in contact	11·34	12·38	13·52	17·30	25·00	52·20

In this experiment the current was furnished by a single Grove's cell; but with increased battery power, though the pull at the edge and that at the centre were both increased, yet the ratio of them diminished.

Number of Grove's cells used.	Pull at edge.	Pull at centre.	Ratio of edge-pull to centre-pull.
1	52·20	8·75	5·96
2	81·85	19·80	4·20
3	126·50	27·7	5·10
6	227·50	52·25	4·35

Distribution on Polar Surface. 153

In another of Vom Kolke's experiments,* using Plücker's great horse-shoe electromagnet (p. 28), a series of observations were made across the polar face, 10·2 cm. in diameter, both in a direction *c d* at right angles to the line joining the poles, and in a diameter *a b* along that line. A short pointed iron wire 2·6 cm. long was used for detaching. In

FIG. 67.

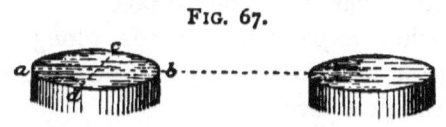

POLES OF ELECTROMAGNET EXPLORED BY VOM KOLKE.

the measurements made along *c d*, (Fig. 67) the radial distance was divided into eight equal parts; and four sets of measurements were made while the electromagnet was excited in four different ways :—

    I. Both coils excited so as to aid one another in magnetizing;
    II. Both coils excited so as to oppose one another;
    III. One coil excited only, on the limb examined;
    IV. One coil excited only on the other limb.

The forces required to detach the exploring rod were found in these four cases to be as follows :—

Distance from the Centre.	I.	II.	III.	IV.
8	54·2	30·8	45·2	22·5
7	45·5	27·0	40·0	18·5
6	40·4	22·9	34·0	16·6
5	38·0	21·5	32·0	15·4
4	37·0	19·0	30·0	13·8
3	35·5	17·9	29·2	13·2
2	35·0	17·4	28·1	12·6
1	35·0	17·0	28·1	12·5
0	35·0	16·6	28·0	12·5

* *Pogg. Ann.*, lxxxi., 321, 1850.

The great distance between the cores, 28·4 cm., may account for the figures in Case II. being as large as they are; the coils being excited so as to oppose one another, there must have been a consequent pole at the yoke joining the cores, the whole of the magnetic lines emerging at either polar surface leaking back externally to this region instead of leaking over from pole to pole.

In the further explorations made along the diameter $a\,b$, the strongest pull was always found at the point b, the inner edge of the pole face.

It is easy to observe such varieties of distribution by merely putting a polished iron ball upon the end of the electromagnet. The magnetic behaviour of iron balls is very curious. A small round piece of iron does not tend to move at all in the most powerful magnetic field if that magnetic field is uniform. All that a small ball of iron tends to do is to move from a place where the magnetic field is weak to a place where the magnetic field is strong. Let the iron ball be placed down anywhere near the middle of the polar surface. The ball at once rolls to the edge as in Fig. 68, and will not stay at the middle. If I take a larger two-pole electromagnet (like Fig. 12) what will the case now be? Clearly the shortest path of the magnetic lines through the air is the path just across from the edge of one polar surface to the edge of the other between the poles. The lines are most dense in the region where they arch over in as short an arch as possible, and they will be less dense along the longer paths, which arch more widely over.

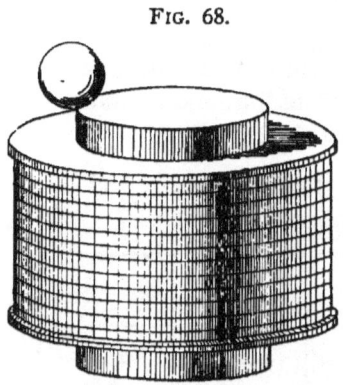

Fig. 68.

Iron Ball attracted to Edge of Polar Face.

Therefore, as there is a greater tendency to flow from the inner edge of one pole to the inner edge of the other, and less tendency to flow from the outer edge of one to the outer edge of the other, the biggest pull ought to be on the inner edges of the pole. We will now try it. On putting the iron ball anywhere on the pole it immediately rolls until it stands perpendicularly over the inner edge.

CHAPTER V.

EXTENSION OF THE LAW OF THE MAGNETIC CIRCUIT TO CASES OF ATTRACTION OF AN ARMATURE AT A DISTANCE. CALCULATION OF MAGNETIC LEAKAGE.

I NOW pass to the consideration of the attraction of a magnet on a piece of iron at a distance. And here I come to a very delicate and complicated question. What is the law of force of a magnet—or electromagnet—acting at a point some distance away from it? I have a very great controversy to wage against the common way of regarding this. The usual thing that is proper to say is that it all depends on the law of inverse squares. In academical examinations they always expect you to give the law of inverse squares. What is the law of inverse squares? We had better understand what it is before we condemn it. It is a statement to the following effect :—That the action of the magnet (or of the pole some people say), at a point at a distance away from it, varies inversely as the square of the distance from the pole. There is a certain action at 1 inch away. Double the distance; the square of that will be four, and, inversely, the action will be one-fourth; at double the distance the action is one-fourth; at three times the distance the action is one-ninth, and so on. You just try it with any electromagnet; nay, take any magnet you like, and unless you hit upon the particular case, I believe you will find it to be universally untrue. Experiment does not prove it. Coulomb, who was supposed to establish the law of inverse squares by means of the torsion balance, was working with long thin needles of specially hard steel, carefully magnetized so that the only leakage of magnetism from the magnet might be as nearly as possible leakage in radiating tufts at the very ends. He practically had point-poles. When the only surface magnetism is at the end faces, the magnet lines leak out like rays from a centre, in radial lines. Now the law of inverse squares is never true except for the action of points; it is a *point* law. There has been a lively discussion going on quite lately whether sound varies as the square of the distance—or rather, whether the intensity of it does—and the people who dispute on both sides of the case do not seem to know what the law of inverse squares means. I have also seen the statement by one who is supposed to be an eminent authority on eyesight, that the intensity of the colour of a scarlet geranium varies inversely with the square of the distance from which you see it. More utter nonsense was never written. The fact is, the law of inverse squares, which is a perfectly true mathematical law, is true not only for electricity, but for light, for sound, and for everything else, provided it is applied to the one case to which a law of inverse squares is applicable. That law is a law expressing the way in which action at a distance falls off when the thing from which the action is proceeding is so small compared with the distance in question that it may be regarded as *a point*. The law of inverse squares is the law universal of action proceeding from a point. The music of an orchestra at 10 feet distance is not four times as loud as at 20 feet distance; for the size of an orchestra cannot be regarded as a mere point in comparison with these distances. If you can conceive of an object giving out a sound, and the object being so small, in relation to the distance at which you are away from it, that it is a point, the law of inverse squares is all right for that, not for the intensity of your hearing, but for the intensity of that to which your sensation is directed. When the magnetic action proceeds from something so small that it may be regarded as a point compared with the distance, then the law of inverse squares is necessarily and mathematically true. If you could get an electromagnet or a magnet, with poles so small in proportion to its length that you can consider the end face of it as the only place through which magnetic lines leak up into the air, and the ends themselves so small as to be relatively mere points; if, also, you can regard those end faces as something so far away from whatever they are going to act upon that the distance between them shall be large compared with their size, and the end itself so small as to be a point, then, and then only, is the law of inverse squares true. It is a law of the action of points. What do we find with electromagnets? We are dealing with pieces of iron which are not infinitely long with respect to their cross-section, and generally possessing round or square end-faces of definite magnitude, which are quite close to the armature; and which are not so infinitely far away that you can consider the polar face a point as compared with its distance away from the object upon which it is to act. Moreover, with real electromagnets there is always lateral leakage; the magnetic lines do not all emerge from the iron through the end face. Therefore, the law of inverse squares is not applicable to that case. What do we mean by a pole, in the first place? We must settle that before we can even begin to apply any law of inverse squares. When leakage occurs all over a great region, as shown in this diagram, every portion of the region is polar; the word polar simply means that you have a place somewhere on the surface of the magnet where filings will stick on; and if filings will stick on to a considerable way down toward the middle all that region must be considered polar, though more strongly at some parts than at others. There are some cases where you can say that the polar distribution is such that the magnetism leaking through the surface acts as if there were a magnetic centre of gravity a little way down, not actually at the end; but cases where you can say there is such a distribution as to have a magnetic centre of gravity are strictly few. When Gauss had to make up his magnetic measurements of the earth, to describe the earth's magnetism, he found it absolutely impossible to assign any definite centre of gravity to the observed distribution of magnetism over the northern regions of the earth; that, indeed, there was not in this sense any definite magnetic pole to the earth at all. Nor is there to our magnets. There is a polar region, but not a pole; and if there is no centre of gravity of the surface magnetism that you can call a pole from which to measure distance, how about the law of inverse squares? Allow me to show you an apparatus (Fig. 69), the only one I ever heard of in which the law of inverse squares is true.

FIG. 69.

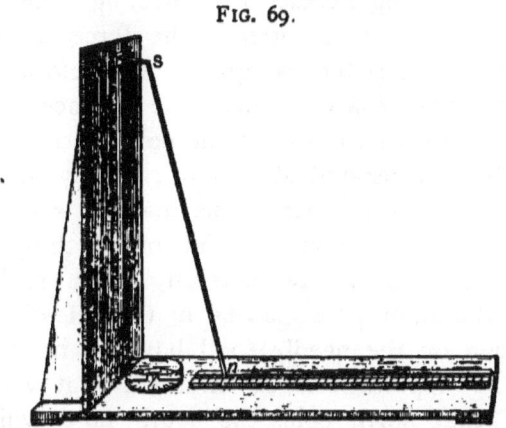

APPARATUS TO ILLUSTRATE THE LAW OF INVERSE SQUARES.

Here is a very long thin magnet of steel, about 3 feet long, very carefully magnetized so as to have no leakage until quite close up to the end. The consequence is that for practical purposes you may treat this as a magnet having point poles, about an inch away from the ends. The south pole is upwards, and the north pole is below, resting in a groove in a base-board which is graduated with a scale, and is set in a direction east and west. I use a long magnet, and keep the south pole well away, so that it shall not perturb the action of the north pole, which, being small, I ask to be allowed to consider as a point. I am going to consider this point as acting on a small compass needle suspended over a card under this glass case constituting a little magnetometer.

If this were properly arranged in a room free from all other magnets, and set so that that needle shall point north, what will be the effect of having the north pole of the long magnet at some distance eastwards? It will repel the north end and attract the south, producing a certain deflexion which we can read off; reckoning the force which causes it by calculating the tangent of the angle of the deflexion. Now, let us move the north pole (regarded as a point) nearer or farther, and study the effect. Suppose we halve the distance from the pole to the indicating needle, the deflecting force at half the distance is four times as great; the force at double the distance is one quarter as great. Wherefore? Because, firstly, we have taken a case where the distance apart is very great compared with the size of the pole; secondly, the pole is practically concentrated at a point; thirdly, there is only one pole acting; and, fourthly, this magnet is of hard steel, and its magnetism in no way depends on the thing it is acting on, but is constant. I have carefully made such arrangements that the other pole shall be in the axis of rotation, so that its action on the needle shall have no horizontal component. The apparatus is so arranged that whatever the position of that north pole, the south pole, which merely slides perpendicularly up and down on a guide, is vertically over the needle, and therefore does not tend to turn it round in any direction whatever. With this apparatus one can approximately verify the law of inverse squares. But this is not like any electromagnet ever used for any useful purpose. You do not make electromagnets long and thin, with point poles a very large distance away from the place where they are to act; no, you use them with large surfaces close up to their armature.

There is yet another case which follows a law that is not a law of inverse squares. Suppose you take a bar magnet, not too long, and approach it broadside-on towards a small compass needle, Fig. 70. Of course, you know as soon as you get anywhere near the compass needle it turns round. Did you ever try whether the effect is inversely proportional to the square of the distance reckoned from the middle of the compass needle to the middle of the magnet. Do you think that the deflexions will vary inversely with the squares of the distances? You will find they do not. When you place the bar magnet like that, broadside-on to the needle, the deflexions vary inversely as the cube of the distance, not the square.

FIG. 70.

DEFLEXION OF NEEDLE CAUSED BY BAR MAGNET BROADSIDE-ON.

Now, in the case of an electromagnet pulling at its armature at a distance, it is utterly impossible to state the law in that misleading way. The pull of the electromagnet on its armature is not proportional to the distance, nor to the square of the distance, nor to the cube, nor to the fourth power, nor to the square root, nor to the three-halfth root, nor to any other power of the distance whatever, direct or inverse, because you find, as a matter of fact, that as the distance alters something else alters too. If your poles were always of the same strength, if they did not act on one another, if they were not affected by the distance in between, then some such law might be stated. If we could always say, as we used to say in the old language, "at that pole," or "at that point," there are to be considered so many "units of magnetism," and at that other place so many units, and those are going to act on one another; then you could, if you wished, calculate the force by the law of inverse squares. But that does not correspond to anything in fact, because the poles are not points, and further, the quantity of magnetism on them is not a fixed quantity. As soon as the iron armature is brought near the pole of the electromagnet there is a mutual interaction; more magnetic lines flow out from the pole than before, because it is easier for magnetic lines to flow through iron than through air. Let us consider a little more narrowly that which happens when a layer of air is introduced into the magnetic circuit of an electromagnet. Here we have (Fig. 71) a closed magnetic circuit, a ring of iron, uncut, such as that used in the experiments on p. 70. The only reluctance in the path of the magnetic lines is that of the iron, and this reluctance we know to be small. Compare

FIG. 71.

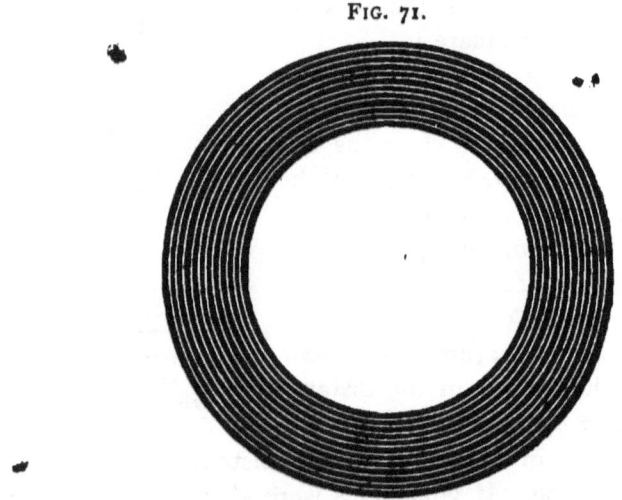

CLOSED MAGNETIC CIRCUIT.

Fig. 71 with Fig. 72, which represents a divided ring with air-gaps in between the severed ends. Now air is a less permeable medium for magnetic lines than iron is, or, in other words, it offers a greater magnetic reluctance. The magnetic permeability of iron varies, as we know, both with its quality and with the degree of magnetic saturation. Reference to Table IV. shows that if the iron has been magnetized up so as to carry 16,000 magnetic lines per sq. cm., the permeability at that stage is about 320. Iron at that stage conducts magnetic lines 320 times better than air does; or air offers 320 times as much reluctance to magnetic lines as iron (at that stage) does. So then the reluctance in the gaps to magnetization is 320 times as great as it would have been if the gaps had been filled up with iron. Therefore, if you have the

Effect of Gaps in Circuit.

same magnetizing coil with the same battery at work, the introduction of air-gaps into the magnetic circuit will, as a first effect, have the result of decreasing the number of magnetic lines that flow round the circuit. But this first effect itself produces a second effect. There are fewer magnetic lines

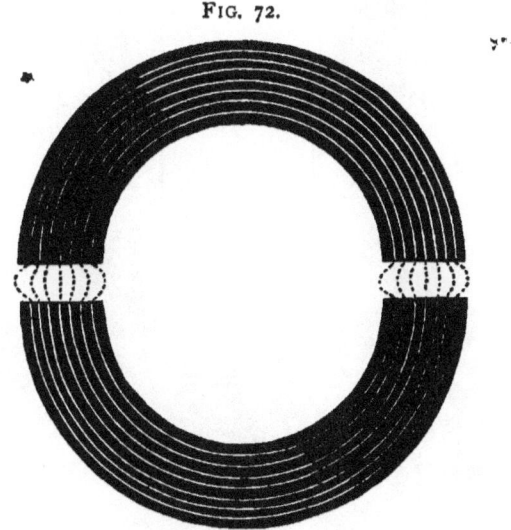

FIG. 72.

DIVIDED MAGNETIC CIRCUIT.

going through the iron. Consequently, if there were 16,000 lines per sq. cm. before there will now be fewer—say only 12,000 or so. Now refer back to Table IV., and you will find that when B is 12,000 the permeability of the iron is not 320 but 1400 or so. That is to say, at this stage, when the magnetization of the iron has been pushed only so far, the magnetic reluctance of air is 1400 times greater than that of iron, so that there is a still greater relative throttling of the magnetic circuit by the reluctance so offered by the air-gaps.

Apply that to the case of an actual electromagnet. Here is a diagram, Fig. 73, representing a horse-shoe electromagnet with an armature of equal section in contact with it. The actual electromagnet used was of the size shown. You can calculate out, from the section, the length of iron, and the table of permeability, how many ampere-turns of excitation will produce any required pull. But now consider that

FIG. 73.

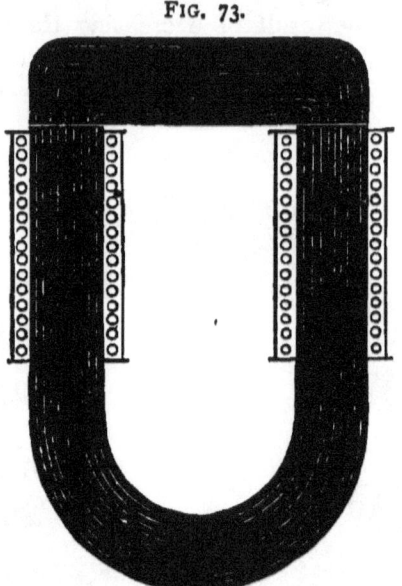

ELECTROMAGNET WITH ARMATURE IN CONTACT.

same electromagnet, as in Fig. 74, with a small air-gap between the armature and the polar faces. The same circulation of current will not now give you as much magnetism as before, because you have interposed air-gaps; and by the very fact of putting in reluctance there, the number of magnetic lines is reduced.

Try, if you like, to interpret this in the old way by the old notion of poles. The electromagnet has two poles, and these excite induced poles in the opposite surface of the armature, resulting in attraction. If you double the distance from the pole to the iron, the magnetic force (always supposing the poles are mere points) will be one-quarter, hence the induced pole on the armature will only be one-quarter as strong. But the pole of the electromagnet is itself weaker. How much weaker? The law of inverse squares does not give you the slightest clue to this all-important fact. If you cannot say how much weaker the primary pole is, neither can you say

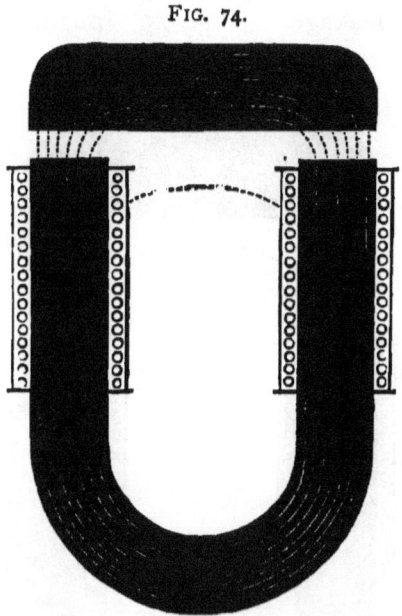

FIG. 74.

ELECTROMAGNET WITH NARROW AIR-GAPS.

how much weaker the induced pole will be, for the latter depends upon the former. The law of inverse squares in a case like this is absolutely misleading.

Moreover, a third effect comes in. Not only do you cut down the magnetism by making an air-gap, but you have a new consideration to take into account. Because the magnetic lines, as they pass up through one of the air-gaps along the armature, down the air-gap at the other end, encounter a considerable reluctance; the whole of the magnetic lines will not go that way, a lot of them will take some shorter cut, although it may be all through air, and you will have some leakage across from limb to limb. I do not say you never have leakage under other circumstances; even with an armature in apparent contact there is always a certain amount of sideway leakage. It depends on the goodness of the contact. And if you widen the air-gaps still farther, you will have still more reluctance in the path, still less magnetism, and still more leakage. Fig. 75 roughly indicates this

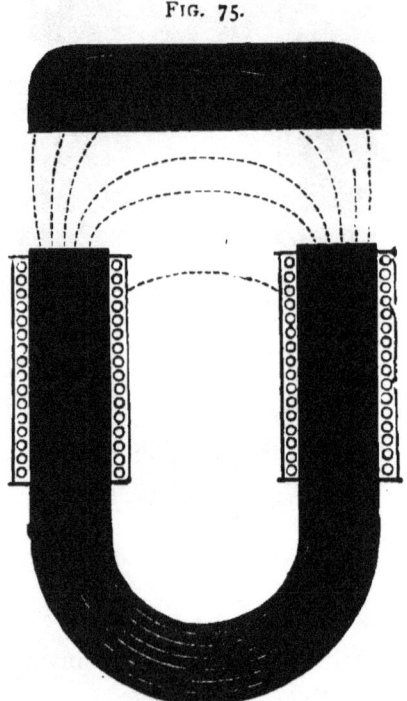

FIG. 75.

ELECTROMAGNET WITH AMMATURE AT DISTANCE.

further stage. The armature will be far less strongly pulled, because, in the first place, the increased reluctance strangles the flow of magnetic lines, so that there are fewer of them in the magnetic circuit; and, in the second place, of this lesser number only a fraction reach the armature, because of the increased leakage. When you take the armature entirely away, the only magnetic lines that go through the iron are those that flow by leakage across the air from one limb to the

Leakage of Lines. 167

other. This is roughly illustrated by.Fig. 76, the last of this set.

Leakage across from limb to limb is always a waste of the magnetic lines, so far as useful purposes are concerned. Therefore it is clear that, in order to study the effect of intro-

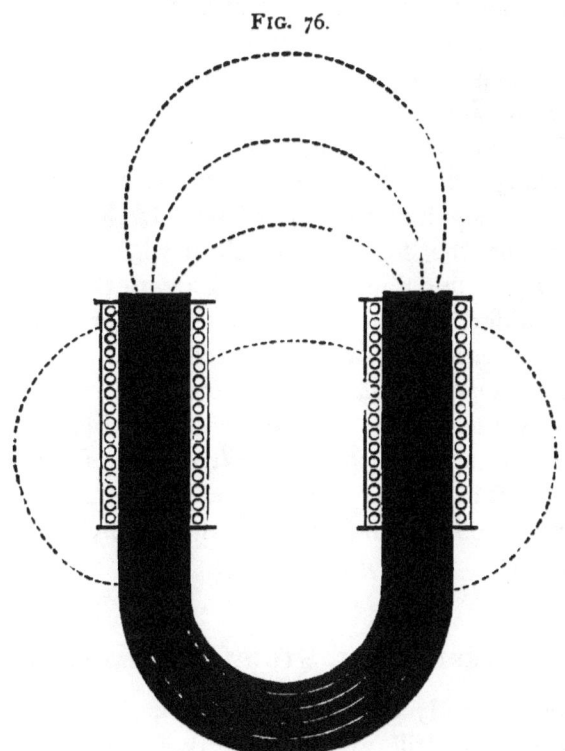

FIG. 76.

ELECTROMAGNET WITHOUT ARMATURE.

ducing the distance between the armature and the magnet, we have to take into account the leakage; and to calculate the leakage is no easy matter. There are so many considerations that occur as to that which one has to take into account, that it is not easy to choose the right ones and leave the wrong ones. Calculations we must make by-and-bye—they will be added in Appendix C of this book—but for the moment experiment seems to be the best guide.

I will therefore refer, by way of illustrating this question of leakage, to some experiments made by Sturgeon. Sturgeon had a long tubular electromagnet made of a piece of old musket-barrel of iron wound with a coil; he put a compass needle about a foot away, and observed the effect. He found the compass needle deflected about 23°; then he got a rod of iron of equal length and put it in at the end, and found that on putting it in so that only the end was introduced the deflection increased from 23° to 37°; but when he pushed the iron right home into the gun-barrel it went back to nearly 23°. How do you account for that? He had unconsciously increased its facility for leakage when he lengthened out the iron core. And when he pushed the rod right home into the barrel, the extra leakage which was due to the added surface could not and did not occur. There was additional cross-section, but what of that? The additional cross-section is practically of no account. You want to force the magnetism across some 20 inches of air, which resists from 300 to 1000 times as much iron. What is the use of doubling the section of the iron? You want to reduce the air reluctance, and you have not reduced the air by putting a core into the tube.

EXPERIMENTAL STUDY OF LEAKAGE.

In order to study this question of leakage, and the relation of leakage to pull, still more incisively, I devised some time ago a small experiment with which a group of my students at the Technical College have been diligently experimenting. Here (Fig. 77) is a horse-shoe electromagnet. The core is of soft wrought iron, wound with a known number of turns of wire. It is provided with an armature. We have also wound on three little exploring coils, each consisting of five turns of wire only, one, C, right down at the bottom, on the bend; another, B, right round the pole, close up to the armature; and a third, A, around the middle of the armature. The object of these is to ascertain how much of the magnetism which was created in the core by magnetizing power of these coils ever got into the armature. If the armature is at a considerable distance away, there is naturally a great deal of leakage. The coil C, around the bend at the bottom, is to catch all the magnetic lines that go through the iron; the coil B, at the poles, is to catch all that have not leaked outside before the magnetism has crossed the joint; while the coil A, right around the middle of the armature, catches all the lines that actually pass into the armature and pull at it. We measure, by means of the ballistic galvanometer and these three exploring coils, how much magnetism gets into the armature at different distances, and are able thus to determine the leakage and compare these amounts with the calculations made, and with the attractions at different distances. The amount of magnetism that gets into the armature does not go by a law of inverse squares, I can assure you, but by quite other laws. It goes by laws which can only be expressed as particular cases of the law of the magnetic circuit. The most important element of the calculations, indeed, in many cases is the amount of percentage of leakage that must be allowed for. Of the magnitude of this matter you will get a very good idea by the result of these experiments following.

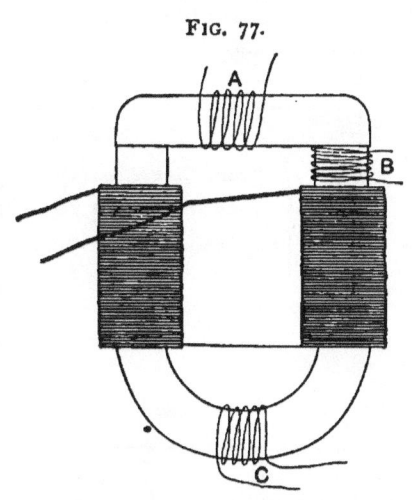

FIG. 77.

EXPERIMENT ON LEAKAGE OF ELECTROMAGNET.

The iron core is 13 mm. in diameter, and the coil consists of 178 turns. The first swing of the galvanometer when the current was suddenly turned on or off measures the number of magnetic lines thereby sent through, or withdrawn from, the exploring coil that is at the time joined to the galvanometer.

The currents used varied from 0·7 of an ampere to 5·7 amperes. Six sets of experiments were made, with the armature at different distances. The numerical results are given below:—

I.—WITH WEAK CURRENT (0·7 AMPERE).

	A	B	C
In contact	12,506	13,870	14,190
Armature distance. { 1 mm. ..	1,552	2,163	3,786
2 mm. ..	1,149	1,487	2,839
5 mm. ..	1,014	1,081	2,028
10 mm. ..	676	1,014	1,690
Removed	—	675	1,352

II.—STRONGER CURRENT (1·7 AMPERE).

	A	B	C
In contact	18,240	19,590	20,283
Armature distance. { 1 mm. ..	2,570	3,381	5,408
2 mm. ..	2,366	2,839	5,073
5 mm. ..	1,352	2,299	5,949
10 mm. ..	811	1,352	3,381
Removed	—	1,308	3,041

III.—STILL STRONGER CURRENT (3·7 AMPERES).

	A	B	C
In contact	20,940	22,280	22,960
Armature distance. { 1 mm. ..	5,610	7,568	11,831
2 mm. ..	4,597	6,722	9,802
5 mm. ..	2,569	3,245	7,436
10 mm. ..	1,149	2,704	7,098
Removed	—	2,366	6,427

Study of Leakage.

IV.—STRONGEST CURRENT (5·7 AMPERES).

		A	B	C
In contact		21,980	23,660	24,040
Armature distance.	1 mm.	8,110	10,810	17,220
	2 mm.	5,611	8,464	15,886
	5 mm.	4,056	5,273	12,627
	10 mm.	2,029	4,057	10,142
Removed		—	3,581	9,795

These numbers may be looked upon as a kind of numerical statement of the facts roughly depicted in Figs. 73 to 76, on pp. 164–7. The numbers themselves, so far as they relate to the measurements made (1) in contact, (2) with gaps of one mm. breadth, are plotted out on Fig. 78; there being three curves, A, B, and C, for the measurements made when the armature was in contact, and three others, A_1, B_1, and C_1, made at the 1 mm. distance. A dotted line gives the plotting of the numbers for the coil C, with different currents, when the armature was removed.

On examining the numbers in detail we observe that the largest number of magnetic lines forced round the bend of the iron core, through the coil C, was 24,040 (the cross section being a little over 1 sq. cm.), which was when the armature was in contact. When the armature was away, the same magnetizing power only evoked 9795 lines. Further, of those 24,040, 23,660 (or 98½ per cent.) came up through the polar surfaces of contact, and of those again 21,980 (or 92½ per cent. of the whole number) passed through the armature. There was leakage, then, even when the armature was in contact, but it amounted to only 7½ per cent. Now, when the armature was moved but 1 mm. (*i.e.* $\frac{1}{25}$ in.) away, the presence of the air-gaps had this great effect, that the total magnetic flux was at once choked down from 24,040 to 17,220. Of that number only 10,810 (or 61 per cent.) reached the polar surfaces, and only 8110 (or 47 per cent. of the total number) succeded in going through the armature. The leakage in this case was 53 per cent! With a 2 mm. gap, the leakage was 65 per cent. when the strongest current was used. It was 68 per cent. with a 5 mm. gap, and 80 per cent. with a 10 mm.

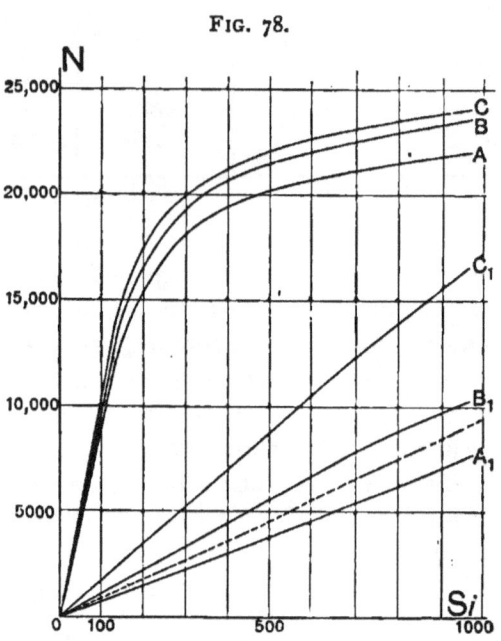

FIG. 78.

CURVES OF MAGNETIZATION PLOTTED FROM PRECEDING.

gap. It will further be noticed that whilst a current of 0·7 ampere sufficed to send 12,506 lines through the armature when it was in contact, a current eight times as strong could only succeed in sending 8,110 lines when the armature was distant by a single millimetre.

Such an enormous diminution in the magnetic flux through the armature consequent upon the increased reluctance and increased leakage occasioned by the presence of the air-gaps, proves how great is the reluctance offered by air, and how essential it is to have some practical rules for calculating reluctances and estimating leakages to guide us in designing electromagnets to do any given duty.

CALCULATION OF LEAKAGE.

The calculation of magnetic reluctances of definite portions of a given material is now comparatively easy, and, thanks to the formulæ of Professor Forbes, given in Appendix C, it is now possible in certain cases to estimate leakages. Of these methods of calculation some examples will also be given in the Appendix. I have, however, found Forbes's rules, which were intended to aid the design of dynamo-machines, not very convenient for the common cases of electro-magnets, and have therefore cast about to discover some more apposite mode of calculation. To predetermine the probable percentage of leakage one must first distinguish between those magnetic lines which go usefully through the armature (and help to pull it) and those which go astray through the surrounding air and are wasted so far as any pull is concerned.

Having set up this distinction, one then needs to know the relative magnetic conductance, or *permeance*, along the path of the useful lines and that along the innumerable paths of the wasted lines of the stray field. For (as every electrician accustomed to the problems of shunt circuits will recognise) the quantity of lines that go respectively along the useful and wasteful paths will be directly proportional to the conductances (or permeances) along those paths, or will be inversely proportional to the respective resistances along those paths. It is customary in electromagnetic calculations to employ a certain coefficient of allowance for leakage, the symbol for which is v, such that when we know the number of magnetic lines that are wanted to go through the armature we must allow for v times as many in the magnetic core. Now, if u represents permeance along the useful path, and w the permeance of all the waste paths along the stray field, the total flux will be to the useful flux as $u + w$ is to u. Hence the coefficient of allowance for leakage v, is equal to $u + w$ divided by u. The only real difficulty is to calculate u and w. In general u is easily calulated, it is the reciprocal of the sum of all the magnetic reluctances along the useful path from pole to pole.

In the case of the electromagnet used in the experiments last described, the magnetic reluctances along the useful path are three in number, that of the iron of the armature and those of the two air-gaps. The following formula is applicable,

$$\text{Reluctance} = \frac{l_1}{A_1 \mu_1} + \frac{2 l_2}{A_2}$$

if the data are specified in centimetre measure; the suffixes 1 and 2 relating respectively to the iron and to the air. If the data are specified in inch measures, the formula becomes

$$\text{Reluctance} = 0 \cdot 3132 \left\{ \frac{l'_1}{A'_1 \mu_1} + \frac{2 l''_2}{A''_2} \right\}$$

But it is not so easy to calculate the reluctance (or its reciprocal, the permeance) for the waste lines of the stray field, because the paths of the magnetic lines spread out so extraordinarily, and bend round in curves from pole to pole.

Fig. 79 gives a very fair representation of the spreading of the lines of the stray field that leaks across between the two limbs of a horse-shoe electromagnet made of round iron. And for square iron the flow is much the same, except that it is concentrated a little by the corners of the metal. Forbes's rules do not help us here. We want a new mode of considering the subject.

The problems of flow, whether of heat, electricity, or of magnetism, in space of three dimensions, are not amongst the most easy of geometrical exercises. However, some of them have been worked out, and may be made applicable to our present need. Consider, for example, the electrical problem of finding the resistance which an indefinitely extended liquid (say a solution of sulphate of copper of given density) offers when acting as a conductor of electric currents flowing across between two indefinitely long parallel cylinders of copper. Fig. 79 may be regarded as representing a transverse section of such an arrangement, the sweeping curves representing lines of flow of current. In a simple case like this it is possible to find an accurate expression for the resistance (or for the conductance) of a layer or stratum of unit thickness. It depends on the diameters of the cylinders, on their distance apart, and on the specific conductivity of the medium. It is not by any means proportional to the distance between them, being, in fact, almost independent of the distance, if that is greater than twenty times the perimeter of either cylinder. Neither is it even approximately proportional to the perimeter of the cylinders except in those cases when the shortest distance between them is less than a tenth part of the perimeter of either. The resistance, for unit length of the cylinders, is, in fact, calculated out by the rather complex formula :—

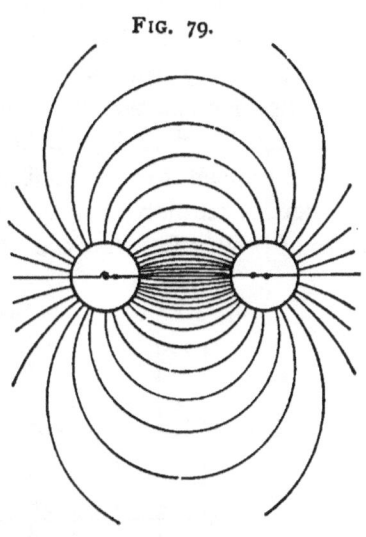

FIG. 79.

CURVES OF FLOW OF MAGNETIC LINES IN AIR FROM ONE CYLINDRICAL POLE TO ANOTHER.

$$R = \frac{1}{\pi \mu} \log \text{nat.} \; h;$$

Where

$$h = \frac{2a}{b + 2a - \sqrt{b^2 + 4ab}};$$

the symbol a standing for the radius of the cylinder; b for the shortest distance separating them; μ for the permeability, or in the electric case the specific conductivity of the medium.

Now, I happened to notice, as a matter that greatly simplifies the calculation, that if we confine our attention to a transverse layer of the medium of given thickness, the resistance between the two bits of the cylinders in that layer depends on the ratio of the shortest distance separating them to their periphery, and is independent of the absolute size of the system. If you have the two cylinders an inch round, and an inch between them, then the resistance of the slab of medium (of given thickness) in which they lie will be the same as if they were a foot round and a foot apart. Now that simplifies matters very much, and, thanks to my friend and former chief assistant, Dr. R. Mullineux Walmsley, who devoted himself to this troublesome calculation, I am able to give you, in tabular form, the magnetic resistances within the limits of proportion that are likely to occur.

FIG. 80.

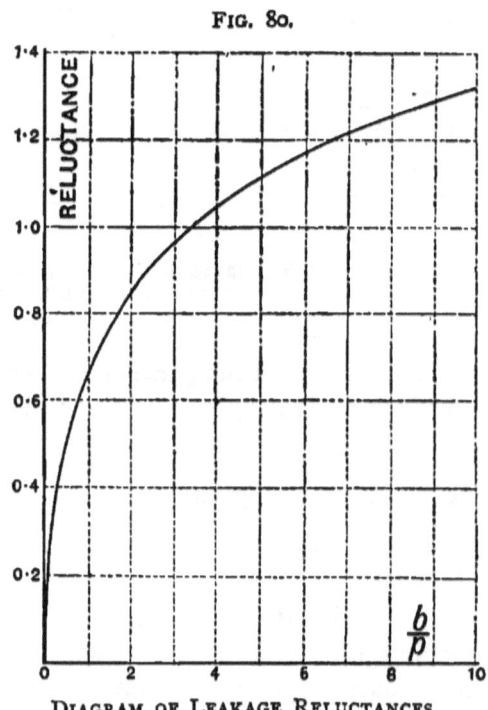

DIAGRAM OF LEAKAGE RELUCTANCES.

The numbers from columns 1 and 2 of the following Table are plotted out graphically in Fig. 80 for more convenient reference. As an example of the use of the Table we will take the following :—

Calculation of Leakage.

Example.—Find the magnetic reluctance and permeance between two parallel iron cores of 1 inch diameter and 9 inches long, the least distance between them being $2\frac{3}{8}$ inches. Here $b = 2·375$, $p = 3·1416$; $b \div p = 0·756$. Reference to the Table shows (by interpolation) that the reluctance and permeance for unit thickness of slab are respectively 0·183 and 5·336. For 9 inches thickness they will, therefore, be 0·021 and 48·02 respectively.

TABLE XIV.—MAGNETIC RELUCTANCE OF AIR BETWEEN TWO PARALLEL CYLINDRICAL LIMBS OF IRON.

$\dfrac{b}{p}$ Ratio of least distance apart to perimeter.	Magnetic reluctance in C.G.S. units = the magneto-motive force ÷ total magnetic flux. Slab = 1 cm. thick.		Magnetic reluctance in inch units = the ampere-turns ÷ the total magnetic flux. Slab 1 inch thick.	
	Reluctance.	Permeance.	Reluctance.	Permeance.
0·1	0·2461	4·063	0·0771	12·968
0·2	0·3404	2·938	0·1066	9·377
0·3	0·4084	2·449	0·1280	7·815
0·4	0·4628	2·161	0·1450	6·897
0·5	0·5084	1·967	0·1593	6·278
0·6	0·5479	1·825	0·1717	5·825
0·8	0·6140	1·629	0·1924	5·198
1·0	0·6681	1·497	0·2003	4·777
1·2	0·7144	1·400	0·2238	4·571
1·4	0·7550	1·324	0·2365	4·228
1·6	0·7903	1·265	0·2476	4·039
1·8	0·8220	1·217	0·2575	3·883
2·0	0·8511	1·202	0·2667	3·750
4·0	1·0500	0·952	0·3290	3·040
6·0	1·1710	0·854	0·3669	2·726
8·0	1·2624	0·792	0·3955	2·528
10·0	1·3250	0·755	0·4151	2·409

NOTE.—In the above Table, unit length of cylinders is assumed (1 centimetre in columns 2 and 3; 1 inch in columns 4 and 5); the flow of magnetic lines being reckoned as in a slab of infinite extent, and of unit thickness. Symbols : p = perimeter of cylinder ; b = shortest distance between cylinders. In columns 2 and 3 the unit reluctance is that of a centimetre cube of air. In columns 4 and 5 the unit reluctance is so chosen (as in the rest of these chapters wherever such measures are used) that the reduction of ampere-turns to magneto-motive force by multiplying by $4\pi \div 10$ is avoided. This will make the reluctance of the inch cube of air equal to $10 \div 4\pi \div 2·54 = 0·3132$; and its permeance as 3·1931.

When the permeance across between the two limbs is thus approximately calculable, the waste flux across the space is estimated by multiplying the permeance so found by the average value of the difference of magnetic potential between the two limbs; and this—if the yoke which unites the limbs at their lower end is of good solid iron, and if the parallel cores offer little magnetic reluctance as compared with the reluctance of the useful paths, or of that of the stray field—may be simply taken as half the ampere-turns (or, if centimetre measures are used, multiply by $1 \cdot 2566$).

The method here employed in estimating the reluctance of the waste field is, of course, only an approximation; for it assumes that the leakage takes place only in the planes of the slabs considered. As a matter of fact, there is always some leakage out of the planes of the slabs. The real reluctance is always therefore somewhat less, and the real permeance somewhat greater, than that calculated from Table VIII.

For the electromagnets used in ordinary telegraph instruments the ratio of b to p is not usually very different from unity, so that for them the permeance across from limb to limb per inch length of core is not very far from $5 \cdot 0$, or nearly twice the permeance of an inch cube of air.

APPLICATION TO SPECIAL CASES.

We are now in a position to see the reason for a curious statement of Count Du Moncel which for long puzzled me. He states that he found, using distance apart of 1 mm., that the attraction of a two-pole electromagnet for its armature was less when the armature was presented laterally than when it was placed in front of the pole ends, in the ratio of 19 to 31. He does not specify in the passage referred to what was the shape of either the armature or the cores. If we assume that he was referring to an electromagnet with cores of the usual sort—round iron with flat ends, presumably like Fig. 23—then it is evident that the air-gaps, when the armature is presented sideways to the magnet, are really greater than when the armature is presented in the usual way, owing to the cylindric curvature of the core. So, if at equal measured distance the reluctance in the circuit is greater, the magnetic flux will be less and the pull less.

It ought also now to be evident why an armature made of iron of a flat rectangular section, though when in contact it sticks on tighter edgeways, is at a distance attracted more powerfully if presented flatways. The gaps when it is presented flatways (at an equal least distance apart) offer a lesser magnetic reluctance.

Another obscure point also becomes explainable, namely, the observation by Lenz, Barlow, and others, that the greatest amount of magnetism which could be imparted to long iron bars by a given circulation of electric current was (nearly) proportional, not to the cross-sectional area of the iron, but to its surface! The explanation is this. Their magnetic circuit was a bad one, consisting of a straight rod of iron and of a return path through air. Their magnetizing force was being in reality expended not so much on driving magnetic lines through iron (which is readily permeable) but on driving the magnetic lines through air (which is, as we know, much less permeable), and the reluctance of the return paths through the air is—when the distance from one to the other of the exposed end parts of the bar is great compared with its periphery—very nearly proportional to that periphery, that is to say, to the exposed surface.

Another opinion on the same topic was that of Professor Müller, who laid down the law that for iron bars of equal length, and excited by same magnetizing power, the amount of magnetism was proportional to the square root of the periphery. A vast amount of industrious scientific effort has been expended by Dub, Hankel, Von Feilitzsch, and others on the attempt to verify this "law." Not one of these experimenters seems to have had the faintest suspicion that the real thing which determined the amount of magnetic flow was not the iron but the reluctance of the return-path through air.

Von Feilitzsch plotted out the accompanying curves (Fig. 81), from which he drew the inference that the law of the square root of the periphery was established. The very straightness of these curves shows that in no case had the iron become so much magnetized as to show the bend that indicates approaching saturation. Air, not iron, was offering the main part of the resistance to magnetization in the whole of these experiments. I draw from the very same curves the conclusion that the magnetization is not proportional to the square root of the periphery, but is more nearly proportional to the periphery itself; indeed, the angles at which the different curves belonging to the different peripheries rise show that the amount of magnetism is very nearly as the surface.

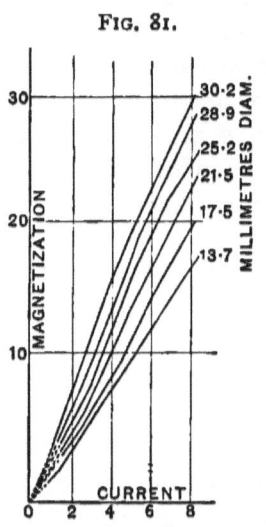

FIG. 81.

VON FEILITZSCH'S CURVES OF MAGNETIZATION OF RODS OF VARIOUS DIAMETERS.

Observe here we are not dealing with a closed magnetic circuit where section comes into account; we are dealing with a bar in which the magnetism can only get from one end to the other by leaking all round into the air. If, therefore, the reluctance of the air-path from one end of the bar to the other is proportional to the surface, we should get some curves very like these; and that is exactly what happens. If you have a solid, of a certain given geometrical form, standing out in the middle of space, the permeance which the space around it (or rather the medium filling that space) offers to the magnetic lines flowing through it, is practically proportional to the surface. It is distinctly so for similar geometrical solids, when they are relatively small as compared with the distance between them.

Electricians know that the resistance of the liquid between two small spheres, or two small disks of copper immersed in a large bath of sulphate of copper, is practically independent of the distance between them, provided they are not within ten diameters, or so, of one another. In the case of a long bar we may treat the distance between the protruding ends as sufficiently great to make an approximation to this law hold good. Von Feilitzsch's bars were, however, not so long that the average value of the length of path from one end surface to the other end surface, along the magnetic lines, was infinitely great as compared with the periphery. Hence the departure from exact proportionality to the surface. His bars were 9·1 cm. long, and the peripheries of the six were respectively 94·9, 90·7, 79·2, 67·6, 54·9, and 42·9 mm.

DIFFERENCES BETWEEN LONG AND SHORT CORES.

It has long been a favourite idea with telegraph engineers that a long-legged electromagnet in some way possessed a greater "projective" power than a short-legged one; that, in brief, a long-legged magnet could attract an armature at a greater distance from its poles than could a short-legged one made with iron cores of the same section. The reason is not far to seek. To project or drive the magnetic lines across a wide intervening air-gap requires a large magnetizing force, on account of the great reluctance and the great leakage in such cases; and the great magnetizing force cannot be got with short cores, because there is not, with short cores, a sufficient length of iron to receive all the turns of wire that are in such a case essential. The long leg is wanted simply to carry the wire necessary to provide the requisite circulation of current.

We now see how, in designing electromagnets, the length of the iron core is really determined; it must be long enough to allow of the winding upon it of the wire which, without overheating, will carry the ampere-turns of exciting current which will suffice to force the requisite number of magnetic lines (allowing for leakage) across the reluctances in the useful path. We may come back to this matter after we have settled, as in the next chapter, the mode of calculating the quantity of wire that is required.

There is one other way in which the difference of behaviour between long and short magnets—I am speaking of horse-shoe shapes—comes into play. So far back as in 1840, Ritchie found that it was more difficult to magnetize steel magnets (using for that purpose electromagnets to stroke them with) if those electromagnets were short than if they were long. He was, of course, comparing magnets which had the same tractive power, that is to say, presumably had the same section of iron magnetized up to the same degree of magnetization. This difference between long and short cores is obviously to be explained on the same principle as the greater projecting power of the long-legged magnets. In order to force magnetism not only through an iron arch but through whatever is beyond, which has a lesser permeability for magnetism, whether it be an air-gap or an arch of hard steel destined to retain some of its magnetism, you require magnetomotive force enough to drive the magnetism through that resisting medium; and, therefore, you must have turns of wire—that implies that you must have length of leg on which to wind those turns. Ritchie also found that the amount of magnetism remaining behind in the soft iron arch, after turning off the current, at the first removal of the armature, was a little greater with long than with short magnets; and, indeed it is what we should expect now, knowing the properties of iron, that long pieces, however soft, retain a little more—have a little more memory, as it were, of having been magnetized—than short pieces. Later on I shall have specially to draw your attention to the behaviour of short pieces of iron which have no magnetic memory.

FURTHER EXPERIMENTAL DATA.

I have several times referred to experimental results obtained in past years, principally by German and French workers, buried in obscurity in the pages of foreign scientific journals. Too often, indeed, the scattered papers of the German physicists are rendered worthless or unintelligible reason of the omission of some of the data of the experiments. They give no measurements perhaps of their currents, or they used an uncalibrated galvanometer, or they do not say how many windings they were using in their coils; or perhaps they give their results in some obsolete phraseology. They are extremely addicted to informing you about the "magnetic moments" of their magnets. Now the magnetic moment of an electromagnet is the one thing that one never wants to know. Indeed the magnetic moment of a magnet of any kind is a useless piece of information, except in the case of bar-magnets of hard steel that are to be used in the determination of the horizontal component of the earth's magnetic force. What one does want to know about an electromagnet is the number of magnetic lines flowing through its circuit, and this the older researches rarely afford the means of ascertaining. Nevertheless, there are some investigations worthy of study, of which some account must be given.

Hollow Cores versus Solid Cores.

I can only now describe some experiments of Von Feilitzsch upon the vexed question of tubular cores, a matter touched by Sturgeon, Pfaff, Joule, Picklès, and later by Du Moncel. To examine the question whether the inner part of the iron really helps to carry the magnetism, Von Feilitzsch prepared a set of thin iron tubes which could slide inside one another. They were all 11 cm. long, and their peripheries varied from 6·12 cm. to 9·7 cm. They could be pushed within a magnetizing spiral, to which either small or large currents could be applied, and their effect in deflecting a magnetic needle was noted, and balanced by means of a compensating steel magnet, from the position of which the forces were reckoned and the magnetic moments calculated out. As the tubes were of equal lengths, the magnetization is approximately proportional to the magnetic moment. The outermost tube was first placed in the spiral, and a set of observations made; then the tube of next smaller size was slipped into it and another set of observations made; then a third tube was slipped in until the whole of the seven were in use. Owing to the presence of the outer tube in all the experiments, the reluctance of the air return paths was alike in every case. The curves given in Fig. 82 indicate the results.

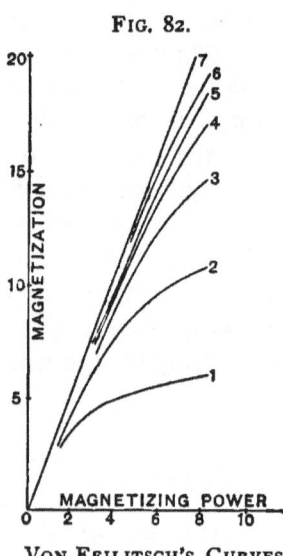

FIG. 82.

VON FEILITSCH'S CURVES OF MAGNETIZATION OF TUBES.

The lowest curve is that corresponding to the use of the first tube alone. Its form, bending over and becoming nearly horizontal, indicates that with large magnetizing power it became nearly saturated. The second curve corresponds to the use of the first tube with the second within it. With greater section of iron saturation sets in at a later stage. Each successive tube adds to the capacity for carrying magnetic lines, the beginning of saturation being scarcely perceptible, even with the highest magnetizing power, when all seven tubes were used. All the curves have the same initial slope. This indicates that with small magnetizing forces, and when even the least quantity of iron was present, when the iron was far from saturation, the main resistance to magnetization was that of the air-paths, and it was the same whether the total section of iron in use was large or small.

EFFECT OF SHAPE OF SECTION.

So far as the carrying capacity for magnetic lines is concerned, one shape of section of cores is as good as another; square or rectangular is as good as round if containing equal sectional area, so long as one is dealing with closed or nearly closed magnetic circuits. But there are two other reasons, both of which tell in favour of round cores. First, the leakage of magnetic lines, from core to core is, for equal mean distances apart, proportional to the surface of the core ; and the round core has less surface than square or rectangular of equal section. All edges and corners, moreover, promote leakage. Secondly, the quantity of copper wire that is required for each turn will be less for round cores than for cores any other shape, for of all geometrical figures of equal area the circle is the one of the least periphery.

The preceding experiments of Von Waltenhofen relate to a case wherein the magnetic circuit is not a closed one, but in which, on the contrary, the magnetic lines have to find their way, by leaking through the surface of the core, through the air from one pole to the other ; and in such cases, at low degrees of saturation it is surface presented to air rather than internal cross section, which is the determining consideration.

Some experiments described on p. 75 in the chapter on the properties of iron are in entire accordance with these results. The further experiments of Bosanquet on rings of different thickness, are conclusive upon the point. The question may therefore be disposed of once for all, with the following remark. In all cases where the magnetic circuit is a closed, or nearly closed one, the quantity of magnetic lines generated by a given magnetizing power will be proportional to the area of cross-section of iron in the core, and does not depend on the shape of that area : whereas in all cases—such as those of bar electromagnets, straight plungers, and the like—where the permeance of the magnetic circuit depends chiefly on the return path of the magnetic lines through air, the magnetic flux generated by a given magnetizing power will depend scarcely at all on the area of cross-section of the iron, but will depend mainly on the facility which it offers for the emergence of magnetic lines into the air, and therefore is nearly proportional to the surface.

EFFECT OF DISTANCE BETWEEN POLES.

Another matter that Du Moncel experimented upon, and Dub and Nicklès likewise, was the distance between the poles. Dub considered that it made no difference how far the poles were apart. Nicklès had a special arrangement made which permitted him to move the two upright cores or limbs, 9 cm. high, to and fro on a solid bench or yoke of iron. His armature was 30 cm. long. Using very weak currents he found the effect best when the shortest distance between the poles was 3 cm.; with a stronger current, 12 cm.; and with his strongest current, nearly 30 cm. I think leakage must have a deal to do with these results. Du Moncel tried various experiments to elucidate this matter, and so did Professor Hughes, in an important, but too little known, research which came out in the *Annales Télégraphiques* in the year 1862.

RESEARCHES OF PROFESSOR HUGHES.

His object was to find out the best form of electromagnet, the best distance between the poles, and the best form of armature for the rapid work required in Hughes's printing telegraphs. One word about Hughes's magnets. This diagram (Fig. 83) shows the form of the well-known Hughes's electromagnet. I feel almost ashamed to say those words "well-known," because, although on the Continent everybody knows what you mean by a Hughes's electromagnet, in England scarcely any one knows what you mean. Englishmen do not even know that Professor Hughes has invented a special form of electromagnet. Hughes's special form is this:—A permanent steel magnet, generally a compound one, having soft iron pole-pieces, and a couple of coils on the pole-pieces only. As I have to speak of Hughes's special contrivance amongst the mechanisms that will occupy our attention next week, I only now refer to this magnet in one particular. If you wish a magnet to work rapidly, you will secure the most rapid action, not when the coils are distributed all along, but when they are heaped up near, not necessarily entirely on, the poles.

Hughes made a number of researches to find out what the right length and thickness of these pole-pieces should be. It was found an advantage not to use too thin pole-pieces, otherwise the magnetism from the permanent magnet did not pass through the iron without considerable reluctance, being choked by insufficiency of section ; also not to use too thick pieces, otherwise they presented too much surface for leakage across from one to the other. Eventually a particular length was settled upon, in proportion about six times the diameter, or rather longer.

Fig. 83.

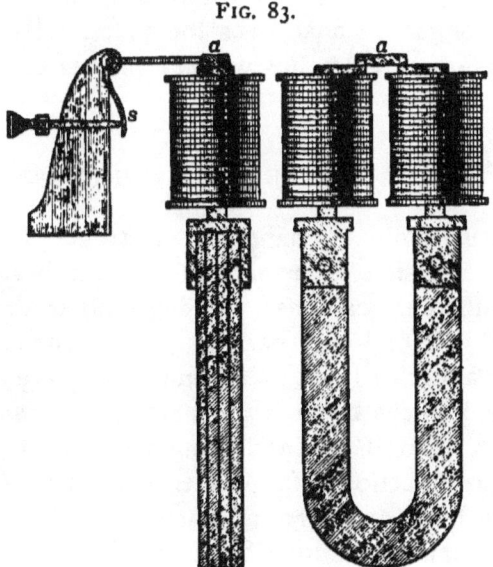

Hughes's Electromagnet.

In the further researches that Hughes made he used a magnet of shorter form, not shown here, more like those employed in relays, and with an armature from 2 to 3 mm. thick, 1 cm. wide, and 5 cm. long. The poles were turned over at the top towards one another. Hughes tried whether there was any advantage in making those poles approach one another, and whether there was any advantage in having as long an armature as 5 cm. He tried all different kinds, and plotted out the results of observations in curves, which could be compared and studied. His object was to ascertain the conditions which would give the strongest pull, not with a steady current but with such currents as were required for operating his printing telegraph instruments; currents which lasted but one to twenty hundredths of a second. He found it was decidedly an advantage to shorten the length of the armature, so that it did not protrude far over the poles. In fact he got a sufficient magnetic circuit to secure all the attractive power that he needed, without allowing as much chance of leakage as there would have been had the armature extended a longer distance over the poles. He also tried various forms of armature having very various cross-sections.

Position and Form of Armature.

In one of Du Moncel's papers on electromagnets[*] you will also find a discussion on armatures, and the best forms for working in different positions. Amongst other things in Du Moncel you will find this paradox; that whereas using a horse-shoe magnet with flat poles, and a flat piece of soft iron for armature, it sticks on far tighter when put on edgeways; on the other hand, if you are going to work at a distance, across air, the attraction is far greater when it is set flatways. I explained the advantage of narrowing the surfaces of contact by the law of traction; B^2 coming in. Why should we have for an action at a distance the greater advantage from placing the armature flatway to the poles? It is simply that we thereby reduce the reluctance offered by the air-gap to the flow of the magnetic lines. Du Moncel also tried the difference between round armatures and flat ones, and found that a cylindrical armature was only attracted about half as

[*] *La Lumière Électrique*, vol. ii.

strongly as a prismatic armature, having the same surface when at the same distance. Let us examine this fact in the light of the magnetic circuit. The poles are flat. You have at a certain distance away a round armature; there is a certain distance between its nearest side and the polar surfaces. If you have at the same distance away a flat armature having the same surface, and, therefore, about the same tendency to leak, why do you get a greater pull in this case than in that? I think it is clear that if they are at the same distance away, giving the same range of motion, there is a greater magnetic reluctance in the case of the round armature, although there is the same periphery, because though the nearest part of the surface is at the prescribed distance, the rest of the under surface is farther away; so that the gain found in substituting an armature with a flat surface is a gain resulting from the diminution in the resistance offered by the air-gap.

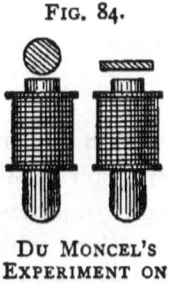

FIG. 84.

DU MONCEL'S EXPERIMENT ON ARMATURES.

POLE-PIECES ON HORSE-SHOE MAGNETS.

Another of Du Moncel's researches[*] relates to the effect of polar projections or shoes—moveable pole-pieces, if you like—upon a horse-shoe electromagnet. The core of this magnet was of round iron 4 cm. in diameter, and the parallel limbs were 10 cm. long, and 6 cm. apart. The shoes consisted of two flat pieces of iron slotted out at one end, so that they could be slid along over the poles and brought nearer together. The attraction exerted on a flat armature across air-gaps 2 mm. thick was measured by counterpoising. Exciting this electromagnet with a certain battery, it was found that the attraction was greatest when the shoes were

[*] *La Lumière Electrique*, vol. iv., p. 129.

pushed to about 15 mm., or about one-fourth of the interpolar distance, apart. The numbers were as follows :—

Distance between shoes. Millimetres.	Attraction, in grammes.
2	900
10	1,012
15	1,025
25	965
40	890
60	550

With a stronger battery the magnet without shoes had an attraction of 885 grammes, but with the shoes 15 mm. apart, 1,195 grammes. When one pole only was employed, the attraction, which was 88 grammes without a shoe, was *diminished,* by adding a shoe, to 39 grammes!

CHAPTER VI.

RULES FOR WINDING COPPER WIRE COILS.

Winding of the Copper.

I NOW take up the question of winding the copper wire upon the electromagnet. How are we to determine beforehand the amount of wire required, and the proper gauge of wire to employ?

The first stage of such a determination is already accomplished; we are already in possession of the formulæ for reckoning out the number of ampere-turns of excitation required in any given case. It remains to show how from this to calculate the amount of bobbin-space, and the quantity of wire to fill it. Bear in mind that a current of 10 amperes (*i.e.*, as strong as that used for a big arc light) flowing once around the iron, produces exactly the same effect magnetically as a current of 1 ampere flowing around 10 times, or as a current of only 100th part of an ampere flowing around a 1,000 times. In telegraphic work the currents ordinarily used in the lines are quite small, usually from 5 to 20 thousandths of an ampere; hence, in such cases, the wire that is wound on need only be a thin one, but it must have a great many turns. Because it is thin and has a great many turns, and is consequently a long wire, it will offer a considerable resistance. That is no advantage, but does not necessarily imply any greater waste of energy than if a thicker coil of fewer turns were used with a correspondingly larger current. . Consider a very simple case. Suppose a bobbin is already filled with a certain number of turns of wire, say 100, of a size large enough to carry 1 ampere, without over-heating; it will offer a certain resistance, it will waste a certain amount of the energy of the current, and it will have a certain magnetizing power. Now suppose this bobbin to be re-wound with a wire of half the diameter; what will the result be? If the wire is half the diameter, it will have one-quarter the sectional area, and the bobbin will hold four times as many turns (assuming insulating materials to occupy the same percentage of the available volume). The current which such a wire will carry will be one-fourth as great. The coil will offer sixteen times as much resistance, being four times as long and of one-fourth the cross-section as the other wire. But the waste of energy will be the same, being proportional to the resistance and to the square of the current; for $16 \times \frac{1}{16} = 1$. Consequently the heating effect will be the same. Also the magnetizing power will be the same, for though the current is only one-quarter of an ampere, it flows around 400 turns; the ampere-turns are 100, the same as before. The same argument would hold good with any other numerical instance that might be given. It therefore does not matter in the least to the magnetic behaviour of the electromagnet whether it is wound with thick wire or thin wire, provided the thickness of the wire corresponds to the current it has to carry, so that the same number of watts of power are spent in heating it. For a coil wound on a bobbin of given volume the magnetizing power is the same for the same heat waste. But the heat waste increases in a greater ratio than the magnetizing power, if the current in a given coil is increased; for the heat is proportional to the square of the current, and the magnetizing power is simply proportional to the current. Hence it is the heating effect which in reality determines the winding of the wire. We must—assuming that the current will have a certain strength—allow enough volume to admit of our getting the requisite number of ampere-turns without over-heating. A good way is to assume a current of one ampere while one calculates out the coil. Having done this, the same volume holds good for any other gauge of wire appropriate to any other current. The terms "long-coil" magnet and "short-coil" magnet are appropriate for those electromagnets which have, respectively, many turns of thin wire and few turns of thick wire. These terms are preferable to "high-resistance" and "low-resistance," sometimes used to designate the two classes of windings, because, as I have just shown, the resistance of a coil has in itself nothing to do with its magnetizing power. Given the volume occupied by the copper, then for any current-density (say, for example, a current-density of 2000 amperes per square inch of cross-section of the copper), the magnetizing power of the coil will be the same for all different gauges of wire. The specific conductivity of the copper itself is of importance; for the better the conductivity, the less the heat-waste per cubic inch of winding. High-conductivity copper is therefore to be preferred in every case.

Now the heat which is thus generated by the current of electricity raises the temperature of the coil (and of the core), and it begins to emit heat from its surface. It may be taken as a sufficient approximation that a single square inch of surface, warmed 1° Fahr. above the surrounding air, will steadily emit heat at the rate of $\frac{1}{225}$ of a watt. Or, if there is provided only enough surface to allow of a steady emission of heat at the rate of 1 watt* per square inch of surface, the temperature of that surface will rise to about 225° Fahr. above the temperature of the surrounding air. This number is determined by the average emissivity of such substances as cotton, silk, varnish, and other materials of which the surfaces of coils are usually composed.

In the specifications for dynamo machines, it is usual to

* The *watt* is the unit of rate of expenditure of energy, and is equal to ten million ergs per second, or to $\frac{1}{746}$ of a horse-power. A current of one ampere, flowing through a resistance of one ohm, spends energy in heating at the rate of one watt. One watt is equivalent to 0·24 calories per second, of heat. That is to say, the heat developed in one second, by expenditure of energy at the rate of one watt, would suffice to warm one gramme of water through 0·24 (Centigrade) degrees. As 252 calories are equal to one British lb. (Fahrenheit) unit of heat, it follows that heat emitted at the rate of one watt would suffice to warm 3·4 pounds of water one degree Fahrenheit in one hour; or one British unit of heat equals 1,058 watt-seconds.

O

TABLE XV.—WIRE GAUGE AND AMPERAGE TABLE.

PERMISSIBLE AMPERAGE, PROBABLE HEATING, AND PERMISSIBLE DEPTH.

S.W.G.	Dimensions				At 1000 Amperes to square inch			At 2000 Amperes to square inch			At 3000 Amperes to square inch			At 4000 Amperes to square inch		
	Diam. (inch).	Section (square inch bare).	Turns to 1 linear inch (covered).	Turns per square inch (covered).	A	F	D	A	F	D	A	F	D	A	F	D
22	·028	·00062	23·81	624	·616	2·28	4·5	1·23	9·12	1·13	1·85	20·52	·50	2·46	36·5	·28
20	·036	·0010	20·00	440	1·018	3·18	3·9	2·036	12·72	·87	3·05	28·62	·43	4·07	50·9	·24
19	·040	·0012	18·52	377	1·26	3·56	3·6	2·52	14·24	·92	3·78	32·04	·41	5·04	57·0	·23
18	·048	·0018	16·13	286	1·81	4·64	3·3	3·62	18·56	·83	5·43	41·76	·37	7·24	74·2	·21
17	·056	·0024	14·28	224	2·4	5·47	3·2	4·8	21·9	·79	7·2	49·2	·35	9·6	87·5	·19
16	·064	·0032	12·83	181	3·2	6·57	3·0	6·4	26·3	·74	9·6	59·1	·33	12·8	105·1	·18
15	·072	·0040	11·63	149	4·0	7·40	2·9	8·0	29·6	·72	12·0	66·6	·32	16·0	118·4	·17
14	·080	·0050	10·64	124	5·0	8·46	2·8	10·0	33·8	·70	15·0	76·3	·31	20·0	135·4	·17
13	·092	·0060	9·44	98·2	6·6	9·97	2·7	13·2	39·9	·67	19·8	89·7	·30	26·4	159·5	·16
12	·104	·0085	8·48	79·1	8·5	11·53	2·6	17·0	46·1	·65	25·5	103·8	·29	34·0	184·4	·16
11	·116	·0105	7·69	65·0	10·5	12·8	2·5	21·0	51·2	·63	31·5	115·2	·28	42·0	204·8	·16
10	·128	·0128	7·04	54·5	12·8	14·3	2·4	25·6	57·2	·61	38·4	128·7	·27	51·2	228·8	·15
9	·144	·0163	6·33	44·1	16·3	16·4	2·4	32·6	65·6	·60	48·9	147·6	·27	65·2	262·4	·15
8	·160	·0201	5·74	36·3	20·1	18·4	2·3	40·2	73·6	·59	60·3	165·6	·26	80·4	294·4	·15
7	·176	·0243	5·26	30·4	24·3	20·4	2·3	48·6	81·6	·58	72·9	183·6	·26	97·2	326·4	·15

Wire Gauge and Amperage.

Stranded.																
7/22	·840	·0043	9·62	101·8	4·3	6·73	4·0	8·6	26·9	·99	12·9	24·6	·44	17·2	107·7	·25
7/20	·108	·0072	7·81	67·1	7·13	8·94	3·7	14·3	35·7	·92	21·4	80·5	·48	28·5	143·0	·23
7/18	·144	·0128	6·09	40·8	12·7	12·4	3·4	25·4	49·6	·83	38·1	111·6	·39	50·8	198·4	·21
7/16	·192	·0229	5·10	28·6	22·9	17·2	3·2	45·8	68·7	·79	68·7	154·5	·35	91·6	274·7	·20
7/15	·216	·0289	4·27	20·1	28·9	19·5	3·1	57·8	78·0	·78	86·7	175·4	·34	115·6	311·8	·20
7/14	·240	·0356	3·87	15·5	35·6	21·8	3·1	71·2	87·1	·76	106·8	195·9	·34	142·4	348·3	·19
7/13	·276	·0462	3·38	12·6	46·2	24·7	3·0	92·4	98·8	·74	138·6	222·3	·33	184·8	395·2	·19
7/12	·312	·0595	3·01	9·97	59·5	28·5	2·9	179·0	114·0	·72	178·5	256·5	·32	238·0	456·0	·18

Figures in columns marked A signify number of amperes that the wire carries.

Figures in columns marked F signify number of degrees (Fahrenheit) that the coil will warm up if there is only one layer of wire, and on the assumption that the heat is radiated only from the outer surface of the coil: they are calculated by the following modifications of Forbes's rule:—

Rise in temperature (Fahrenheit degrees) = 225 × number of watts lost per square inch.

= 159 × sectional area × number of turns to 1 inch (at 1000 amperes per square inch).

Figures in columns marked D are the depths in inches to which wire may be wound if 1 watt be lost by each square inch of radiating surface, the outside radiating surface of the bobbin being only considered.

Rule of calculating a 7-strand cable :—Diameter of cable = 1·134 × diameter of equivalent round wire.

Figures under heading "Turns to 1 linear inch," are calculated for cotton-covered wires of average thicknesses of coverings used for the different gauges, viz. 14 mils additional diameter on round wires (from No. 22), and 20 mils on stranded or square wire.

Figures under heading "Turns per square inch" are calculated from preceding, allowing 10 per cent. for bedding of layers.

Resistance (ohms) of coil of copper wire, occupying v cubic inches of coil-space, and of which the gauge is d mils uncovered, and D mils covered, may be approximately calculated by the rule :—

$$\text{ohms} = 960{,}700 \frac{v}{D^2 d^2}.$$

The data respecting sizes of wires of various gauges are kindly furnished by the London Electric Wire Company.

lay down a condition that the coils shall not heat more than a certain number of degrees warmer than the air. With electromagnets it is a safe rule to say that no electromagnet ought ever to heat up to a temperature more than 100° Fahr. above the surrounding air. In many cases it is quite safe to exceed this limit.

The resistance of the insulated copper wire on a bobbin may be approximately calculated by the following rule:—If d is the diameter of the naked wire, in mils, and D is the diameter, in mils, of the wire when covered, then the resistance per cubic inch of the coil will be :—

$$\text{Ohms per cubic inch} = \frac{960,700}{D^2 \times d^2}.$$

We are therefore able to construct a wire gauge and amperage table which will enable us to calculate readily the degree to which a given coil will warm when traversed by a given current, or conversely what volume of coil will be needed to provide the requisite circulation of current without warming beyond any prescribed excess.

Accordingly, I here give a *Wire Gauge and Amperage Table* (pp. 194, 195), which we have been using for some time at the Finsbury Technical College. It was calculated out under my instructions by one of the demonstrators of the College, Mr. Eustace Thomas, to whom I am indebted for the great care bestowed upon the calculations.

For many purposes, such as for use in telegraphs and electric bells, smaller wires than any of those mentioned in the table are required. The table is, in fact, intended for use in calculating magnets in larger engineering work.

A rough and ready rule sometimes given for the size of wire is to allow $\frac{1}{1000}$ of a square inch per ampere. This is an absurd rule, however, as the figures in the table show. Under the heading 1000 amperes to square inch, it appears that if a No. 18 S.W.G. wire is used, it will at that rate carry 1·81 amperes; that if there is only one layer of wire, it will only warm up 4·64° Fahr., consequently one might wind layer after layer to a depth of 3·3 inches, without getting up to the limit of allowing one square inch per watt for the emission of heat. In very few cases does one want to wind a coil so thick as 3·3 inches. For very few electromagnets is it needful that the layer of coil should exceed ½ an inch in thickness; and if the layer is going to be only ½ an inch thick, or about one-seventh of the 3·3, one may use a current density $\sqrt{7}$ times as great as 1000 amperes per square inch, without exceeding the limit of safe working. Indeed, with coils only ½ inch thick, one may safely employ a current density of 3000 amperes per square inch, owing to the assistance which the core gives for the dissipation and emission of heat.

Suppose, then, we have designed a horse-shoe magnet with a core 1 inch in diameter, and that after considering the work it has to do, it is found that a magnetizing power of 2400 ampere-turns is required; suppose also that it is laid down that the coil must not warm up more than 50° Fahr. above the surrounding air—what volume of coil will be required? Assume first that the current will be 1 ampere; then there will have to be 2400 turns of a wire which will carry 1 ampere. If we took a No. 20 S.W.G. wire, and wound it to a depth of ½ an inch, that would give 220 turns per inch length of coil; so that a coil 11 inches long, and a little over ½ inch deep (or 10 layers deep) would give 2400 turns. Now Table XV. shows that if this wire were to carry 1·018 ampere, it would heat up 225° Fahr., if wound to a depth of 3·9 inches. If wound to ½ inch, it would therefore heat up about 30° Fahr.; and with only 1 ampere would of course heat less. This is too good; try the next thinner wire. No. 22, S.W.G. wire, at 2000 amperes to square inch, will carry 1·23 ampere; and heats 225° if wound up 1·13 inch. If it is only to heat 50° it must not be wound more than ¼ inch deep; but if it only carries current of 1 ampere it may be wound a little deeper—say to 14 layers. There will then be wanted a coil about 7 inches long to hold the 2400 turns. The wire will occupy about 3·85 square inches of total cross section; and the volume of the space occupied by the winding will be

26·95 cubic inches. Two bobbins, each 3½ inches long and 0·65 deep, to allow for 14 layers, will be suitable to receive the coils.

By the light of the knowledge one possesses as to the relation between emissivity of surface, rate of heating by current, and limiting temperatures, it is seen how little justification there is for such empirical rules as that which is often given, namely, to make the depth of coil equal to the diameter of the iron core. Consider this in relation to the following fact; that in all those cases where leakage is negligible, the number of ampere-turns that will magnetize up a thin core to any prescribed degree of magnetization will magnetize up a core of any section whatever, and of the same length, to the same degree of magnetization. A rule that would increase the depth of copper proportionately to the diameter of the iron core is absurd.

Where less accurate approximations are all that is needed, more simple rules can be given. Here are two cases :—

Case 1. *Leakage assumed to be negligible.* — Assume $B = 16,000$, then $H = 50$ (see Table IV). Hence the ampere-turns per centimetre of iron will have to be 40, or per inch of iron, 102; for H is equal to 1·2566 times the ampere-turns per centimetre. Now if the winding is not going to exceed ½ inch in depth, we may allow 4000 amperes per square inch without serious over-heating. And the 4000 ampere-turns will require a 2-inch length of coil, or each inch of coil carries 2000 ampere-turns without over-heating. Hence each inch of coil ½ inch deep will suffice to magnetize up 20 inches length of iron to the prescribed degree.

Case 2. *Leakage assumed to be* 50 *per cent.*—Assume B in air gap $= H = 8000$, then to force this across requires ampere-turns 6400 per centimetre of air, or 16,250 per inch of air. Now if winding is not going to exceed ½ inch depth, each inch length of coil will carry 2000 ampere-turns. Hence, 8 inches length of coil ½ inch deep will be required for 1 inch length of air, magnetized up to the prescribed degree.

WINDINGS FOR CONSTANT PRESSURE AND FOR CONSTANT CURRENT.

In winding coils for magnets that are to be used on any electric light system, it should be carefully borne in mind that there are separate rules to be considered according to the nature of the supply. If the electric supply is at *constant pressure*, as usual for glow lamps, the winding of coils of electromagnets follows the same rule as the coils of voltmeters. If the supply is with *constant current*, as usual for arc lighting in series, then the coils must be wound with due regard to the current which the wire will carry when lying in layers of suitable thickness, the number of turns being in this case the same whether thin or thick wire is used.

If we assume that a safe limit of temperature is 90° Fahr. higher than the surrounding air, then the largest current which may be used with a given electromagnet is expressed by the formula:—

$$\text{Highest permissible amperes} = 0.63 \sqrt{\frac{s}{r}}.$$

where s is the number of square inches of surface of the coils and r their resistance in ohms.

Similarly, for coils to be used as shunts, we have:—

$$\text{Highest permissible volts} = 0.63 \sqrt{sr}.$$

The magnetizing power of a coil, supplied at a given number of volts of pressure, is independent of its length, and depends only on its gauge, but the longer the wire the *less* will be the heat waste. On the contrary, when the condition of supply is with a constant number of amperes of current, the magnetizing power of a coil is independent of the gauge of the wire, and depends only on its length; but the larger the gauge the less will be the heat waste.

Miscellaneous Rules about Winding.

To reach the same limiting temperature with bobbins of equal size wound with wires of different gauge, the cross-section of the wire must vary with the current it is to carry; or, in other words, the current-density (amperes per square inch) must be the same in each. Table XV. shows the amperages of the various sizes of wires, at four different values of current-density.

To raise to the same temperature two similarly-shaped coils, differing in size only, and having the gauges of the wires in the same ratio (so that there are the same number of turns on the large coil as on the small one), the currents must be proportional to the square roots of the cubes of the linear dimensions.

Sir William Thomson has given a useful rule for calculating windings of electromagnets of the same type, but of different sizes. Similar iron cores, similarly wound with lengths of wire proportional to the squares of their linear dimensions, will, when excited with equal currents, produce equal intensities of magnetic field at points similarly situated with respect to them.

Similar electromagnets of different sizes must have ampere-turns proportional to their linear dimensions if they are to be magnetized up to an equal degree of saturation.

Various Modes of Winding.

A variety of winding has been suggested, namely, to employ in the coils a wire of graduated thickness. It has been shown by Sir William Thomson to be advantageous in the construction of the coils of galvanometers to use for the inner coils of small diameter a thin wire; then, as the diameter of the windings increase, a thicker wire; the thickest wire being used on the outermost layers; the gauge being thus proportioned to the diameter of the windings. But it by no means follows that the plan of using *graded wire*, which is satisfactory for galvanometer coils, is necessarily good for electromagnets. In designing electromagnets it is necessary to consider the means of getting rid of heat; and it is obvious that the outer layers are those which are in the most favourable position for getting rid of this heat. Experience shows that the under layers of coils of electromagnets always attain a higher temperature than those at the surface. If, therefore, the inner layers were to be wound with finer wire, offering higher resistance, and generating more heat than the outer layers, this tendency to over-heating would be still more accentuated. Indeed, it would seem wise rather to reverse the galvanometer plan, and wind electromagnets with wires that are stouter on the inner layers, and finer on the outer layers.

Yet another mode of winding is to employ several wires united in parallel, a separate wire being used for each layer, their anterior extremities being all soldered together at one end of the coil, and their posterior extremities being all soldered together at the other. Magnetically, this mode of winding presents not the slightest advantage over winding with a single stout wire of equivalent section. But it has lately been discovered that this mode of winding with *multiple wire* possesses one incidental advantage, namely that its use diminishes the tendency to sparking which occurs at break of circuit.

Another mode, proposed by Victor Serrin in 1876, consists in winding an iron core, insulated by enamelling, with flat spirals of sheet copper also protected by coatings of enamel.

Winding in Cloisons.

For one particular case there is an advantage in winding a coil in sections; that is to say, in placing partitions or *cloisons* at intervals along the bobbin, and winding the wire so as to fill up each of the successive spaces between the partitions before passing on from one space to the next. The case in which this construction is advantageous is the unusual case of coils that are to be used with currents supplied at very high potentials. For when currents are supplied at very high potentials there is a very great tension * exerted on the insulating material, tending to pierce it with a spark. By winding a coil in *cloisons*, however, as originally suggested by Ritchie, there is never so great a difference of potentials between the windings on two adjacent layers as there would be if the layers were wound from end to end of the whole length of coil. Consequently, there is never so great a tension on the insulating material between the layers, and a coil so wound is less likely to be injured by the occurrence of a spark.

FALLACIES ABOUT WINDING.

It is curious what erroneous notions crop up from time to time about winding electromagnets. In 1869, a certain Mr. Lyttle took out a patent for winding the coils in the following way. Wind the first layer as usual, then bring the wire back to the end where the winding began and wind a second layer, and so on. In this way all the windings will be right-handed, or else all left-handed, not alternately right and left as in the ordinary winding. Lyttle declared that this method of winding a coil gave more powerful effects; so did M. Brisson, who re-invented the same mode of winding in 1873, and solemnly described it. Its alleged superiority was at once disproved by Mr. W. H. Preece, who found the only difference to be that there was more difficulty in carrying out this mode of winding.

Another popular error is that electromagnets in which the wires are badly insulated are more powerful than those in which they are well insulated. This arose from the ignorant use of electromagnets having long thin coils (of high resistance) with batteries consisting of a few cells (of low

---

* The tension on the insulating material, tending to pierce it with a spark, is proportional to the square of the difference of potentials (per unit thickness) to which the insulating material is subject. It is incorrect to talk about the tension of the conductor or about the tension of the current; for the tension or electric stress is always an action affecting the di-electric or insulating material.

electromotive force). In such cases, if some of the coils are short-circuited, more current flows, and the magnetizing power may be greater. But the scientific cure is either to re-wind the magnet with an appropriate coil of thick wire, or else to apply another battery having an electromotive force that is greater.

INSULATION OF WIRE.

Instructions concerning the proper insulation of the wires, and between contiguous layers, are given under the heading of Materials, on p. 59 to 64.

SPECIFICATIONS OF ELECTROMAGNETS.

One frequently comes across specifications for construction which prescribe that an electromagnet shall be wound so that its coil shall have a certain resistance. This is an absurdity. Resistance does not help to magnetize the core. A better way of prescribing the winding is to name the ampere-turns and the temperature limit of heating. Another way is to prescribe the number of watts of energy which the magnet is to take. Indeed it would be well if electricians could agree upon some sort of figure of merit by which to compare electromagnets, which should take into account the magnetic output —i.e. the product of magnetic flux into magnetomotive force—the consumption of energy in watts, the temperature rise, and the like.

AMATEUR RULE ABOUT RESISTANCE OF ELECTROMAGNET AND BATTERY.

In dealing with this question of winding copper on a magnet core, I cannot desist from referring to that rule which is so often given, which I often wish might disappear from our text-books—the rule which tells you in effect that you are to waste 50 per cent. of the energy you employ. I refer to the rule which states that you will get the maximum effect out of an electromagnet if you so wind it that the resistance is equal to the resistance of the battery you employ; or that if you have a magnet of a given resistance you ought to employ a battery of the same resistance. What is the meaning of this rule? It is a rule which is absolutely meaningless unless in the first case the volume of the coil is prescribed once for all, and you cannot alter it, or unless once for all the number of battery elements that you can have is prescribed. If you have to deal with a fixed number of battery elements, and you have to get out of them the biggest effect in your external circuit, and cannot beg, buy, or borrow any more cells, it is perfectly true that, for steady currents, you ought to group them so that their internal resistance is equal to the external resistance that they have to work; and then, as a matter of fact, half the energy of the battery will be wasted, but the output will be a maximum. Now that is a very nice rule indeed for amateurs, because an amateur generally starts with the notion that he does not want to economise in his rate of working; it does not matter whether the battery is working away furiously, heating itself, and wasting a lot of power; all he wants is to have the biggest possible effect for a little time out of the fewest cells. It is purely an amateur's rule, therefore, about equating the resistance inside to the resistance outside. But it is absolutely fallacious to set up any such rule for serious working; and not only fallacious, but absolutely untrue if you are going to deal with currents that are going to be turned off and on quickly. For any apparatus like an electric bell, or rapid telegraph, or induction coil, or any of those things where the current is going to vary up and down rapidly, it is a false rule, as we shall see presently. What is the real point of view from which one ought to start? I am often asked questions by, shall I say, amateurs, as well as by those who are not amateurs, about prescribing the battery for a given electromagnet, or prescribing an electromagnet for a given battery. Again, I am often told of cases of failure in which a very little common sense rightly directed might have made a success. What one ought to think about in every case is not the battery, not the electromagnet, but *the line*. If you have a line, then you must have a battery and electromagnet to correspond. If the line is short and thick (a few feet of good copper wire), you should have a short thick battery (a few big cells, or one big cell,) and a short thick coil on your electromagnet. If you have a long thin line, miles of it, say, you want a long thin battery (small cells, and a long row of them) and a long thin coil. That is then our rule; for a short thick line, a short thick battery, and a short thick coil; for a long thin line, a long thin battery, and electromagnet coils to match. You smile: but it is really a good rule that I am giving you; vastly better than the worn-out amateur rule.

But, after all, my rule does not settle the whole question, because there is something more than the whole resistance of the circuit to be taken into account. Whenever you come to rapidly-acting apparatus, you have to think of the fact that the current, while varying, is governed not so much by the resistance as by the inertia of the circuit—its electromagnetic inertia. As this is a matter which will claim our especial attention hereafter, I will leave battery rules for the present, and proceed with the question of design.

EFFECT OF SIZE OF COILS.

Seeing that the magnetizing power which a coil exerts on the magnetic circuit which it surrounds is simply proportional to the ampere-turns, it follows that those turns which lie on the outside layers of the coil, though they are further away from the iron core, possess precisely equal magnetizing power. This is strictly true for all closed magnetic circuits; but in those open magnetic circuits where leakage occurs it is only true for those coils which encircle the leakage lines also. For example, in a short bar-electromagnet, of the wide turns on the outer layer, those which encircle the middle part of the bar do enclose all the magnetic lines, and are just as operative as the smaller turns that underlie them; whilst those wide turns which encircle the end portions of the bar are not so efficient, as some of the magnetic lines leak back past these coils.

Effect of Position of Coils.

Among the other researches which Du Moncel made with respect to electromagnets, was one on the best position for placing the coil upon the iron core. This is a matter that other experimenters have examined. In Dub's book, 'Electromagnetismus,' to which I have several times referred, you will also find many experiments on the best position of a coil; but it is perhaps sufficient to narrate a simple example. Du Moncel had four pairs of bobbins made of exactly the same length, and with 50 m. of wire on each; one pair was 16 cm. long, another pair 8 cm., or half the length, with not quite so many turns, because of course the diameter of the outer turns was larger, one 4 cm. in length, and another 2 cm. These were tried both with bar magnets and horse-shoes. It will suffice perhaps to give the result of the horse-shoe. The horse-shoe was made long enough— 16 cm. only, a little over 6 in., long—to carry the longest coil. Now when the compact coils 2 cm. long were used, the pull on the armature at a distance away of 2 mm. (it was always the same of course in the experiments) was 40 gm. Using the same weight of wire, but distributed on the coils twice as long, the pull was 55 gm. Using the coils 8 cm. long, it was 75 gm.; and using the coils 16 cm. long, covering the length of each limb, the pull was 85, clearly showing that, where you have a given length of iron, the best way of winding a magnet to make it pull with its greatest pull is not to heap the coil up against the poles, but to wind it uniformly, for this mode of winding will give you more turns, therefore more ampere-turns, therefore more magnetization. An exception might, however, occur in some case where there is a large percentage of leakage. With club-footed magnets results of the same kind are obtained. It was found in every case that it was well to distribute the coil as much as possible along the length of the limb. All these experiments were made with a steady current. It does not follow, however, because winding the wire over the whole length of core is best for steady currents that it is the best winding in the case of a rapidly varying current; indeed, it will be shown that it is not.

MISTAKEN METHODS OF CONSTRUCTION.

It is sometimes of advantage to consider faulty or mistaken methods of construction, and to grasp the reasons why they are defective. Fig. 85 furnishes an example. It is one of many forms suggested by Roloff, and consists of three similar cylindrical cores, the ends of which form the corners of an equilateral triangle; the armature being an equilateral plate of iron with rounded corners. It would obviously be absurd to wind the coils so as to make all three polar ends north poles. If one is a north pole the other two would be south poles.

In such a construction—a tripolar magnet, in fact—there is no *a priori* objection; a tripolar magnet of good design being described on p. 317, Fig. 169a. But to carry out the idea properly, one of the cores ought to have double the cross-section of either of the other two; so that the magnetic lines which flow down two of the cores may return up the other core, and find adequate section of iron.

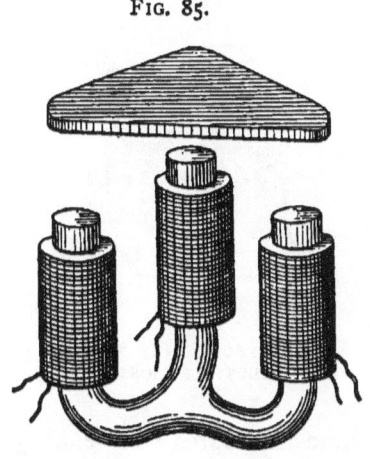

FIG. 85.

THREE-BRANCH ELECTROMAGNET.

Another mistaken mode of construction, used at one time in the field-magnets of Edison's dynamo machines, and also in some of Gramme's earlier patterns, is to employ several parallel cores, each separately wound with wire, united to a common pole-piece at one end and to a common yoke at the other, as in Fig. 86. To divide up the iron limb thus is worse than useless, for if the wires are wound so as to produce the same polarity in the parallel cores, much of the wire is wasted; for in the intermediate space between two such cores there are two sets of currents flowing in opposite directions, mutually destroying one another's magnetizing action. It would be far better to wrap the wire round the outside of all the cores, and fill up the gaps with iron, or else use one solid core. In Fig. 87 are shown three ways of winding; *a* corresponding with the magnet shown in Fig. 86. If these

FIG. 86.   FIG. 87.

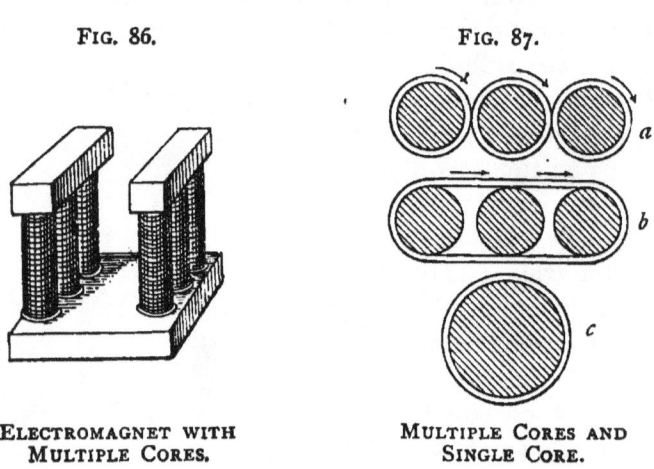

ELECTROMAGNET WITH MULTIPLE CORES.   MULTIPLE CORES AND SINGLE CORE.

three iron cores were each 3 inches in diameter, and the layer of winding ½ inch thick, the (average) length of wire required to put one turn round one core will be almost exactly 11 in.; or, to put one turn round the whole section of iron, the length required will be 33 in. If, however, the wire is simply carried once round all three cores, as in Fig. 87*b*, the length needed will be only 27 in. If the same amount of iron were fused into one large cylinder of the same section, having therefore a diameter of 5·196 in., the (average) length of wire required for one turn will be only 19·05 in.

It becomes, therefore, in order to save copper, an important question what is the best form of section to give to cores. This point is readily answered by considering the geometrical fact that of all possible forms enclosing equal area the one with least periphery is the circle. For facilitating comparison, the following table exhibits the relative lengths of wire required to wind round various forms of section enclosing equal area; the area of the simple circular form being taken as unity:—

Circle	3·54
Square	4·00
Rectangle, 2 : 1	4·24
Rectangle, 3 : 1	4·62
Rectangle, 10 : 1	6·91
Oblong made of one square between two semicircles	3·76
Oblong made of two squares between two semicircles	4·28
Two circles side by side	4·997
Two circles, but wire wound round both together	4·10
Three circle „ „ all „	6·13
Four circles „ „ „ „	7·09

RELATION OF RESISTANCE TO VOLUME OF COIL AND GAUGE OF WIRE.

If it be assumed that the thickness of the insulation is proportional to the thickness of the wire on which it is wound, it follows that the weight of copper in a coil filling a bobbin of given dimensions will be the same, whether a thick wire or a thin one be employed. Further, for a given volume to be filled with coils, the number of *ohms* of resistance of the coil will vary *directly as the square of the number of turns* in the coil. For if a coil wound with 100 turns of a given gauge be rewound with 200 turns of a wire having half the sectional area, the resistance of this new winding will obviously be four times as great as that of the original winding. Also, by a similar argument, it follows that the resistance of a coil of given volume will vary *inversely as the square of the sectional area* of the wire used. And as this area is proportional to the square of the diameter of the wire, it follows that the

P

resistance will vary *inversely as the fourth power of the diameter of the wire.* These rules are only approximate, because in the case of thin wires the layer of insulation occupies a greater relative thickness than in the case of thick wires. The formula given on p. 193 will here be found of service.

Brough's Formulæ.

A more complete formula is that given by Brough,* in which the only assumption made is that the turns of the coil will lie in square order, instead of bedding in between one another.

To find diameter d of a wire to fill bobbin of given dimensions (outer diameter A, inner diameter a, length b,) and produce given resistance R ohms. Radial depth of insulating layer is called u. Let the resistance of a wire (of the quality to be used) of 1 unit length and 1 unit diameter be called $= c$ (ohm), then

$$d = \sqrt{u^2 + \sqrt{\frac{\pi b c (A^2 - a^2)}{4 R}}} - u;$$

and the total length of the wire l will be

$$l = \frac{\pi b (A^2 - a^2)}{4 (2u + d)^2}$$

These formulæ can be used either for inch measures or for centimetres by inserting the proper value of c.

* *Journal Society Telegraph Engineers*, vol. v., p. 256.

CHAPTER VII.

SPECIAL DESIGNS. RAPID-ACTING ELECTROMAGNETS. RELAYS AND CHRONOGRAPHS.

WE are now prepared to consider many details of design that are of importance in the production of electromagnets adapted for special purposes.

CONTRAST BETWEEN ELECTROMAGNETS AND PERMANENT MAGNETS.

It will not be inappropriate here to enter a caution against the idea that all the results obtained in the two preceding chapters from electromagnets are equally applicable to permanent magnets of steel; they are not, for this simple reason. With an electromagnet, when you put the armature near, and make the magnetic circuit better, you not only get more magnetic lines going through that armature, but you get more magnetic lines going through the whole of the iron. You get more magnetic lines round the bend when you put an armature on to the poles, because you have a magnetic circuit of less reluctance, with the same external magnetizing power in the coils acting around it. Therefore, in that case, you will have a greater magnetic flux all the way round. The data obtained with the electromagnet (Fig. 77, p. 169), with the exploring coil, C, on the bend of the core, when the armature was in contact, and when it was removed, are most significant. When the armature was present it multiplied the total magnetic flow tenfold for weak currents, and nearly threefold for strong currents. But with a steel horse-shoe, magnetized once for all, the magnetic lines that flow around the bend of the steel are a fixed quantity, and however much you diminish the reluctance of the magnetic circuit you do not create or evoke any more. When the armature is away the magnetic lines arch across, not at the ends of the horseshoe only, but from its flanks; the whole of the magnetic lines leaking somehow across the space. When you have put the armature on, these lines, instead of arching out into space as freely as they did, pass for the most part along the steel limbs and through the iron armature. You may still have a considerable amount of leakage, but you have not made one line more go through the bent part. You have absolutely the same number going through the bend with the armature off as with the armature on. You do not add to the total number by reducing the magnetic reluctance, because you are not working under the influence of a constantly impressed magnetizing force. By putting the armature on to a steel horse-shoe magnet you only *collect* the magnetic lines, you do not *multiply* them. This is not a matter of conjecture.

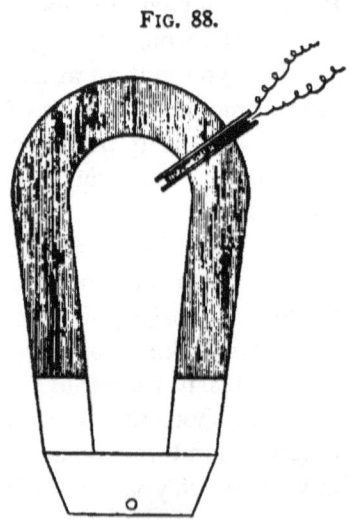

FIG. 88.

EXPERIMENT WITH PERMANENT MAGNET.

A group of my students have been making experiments in the following way. They took this large steel horseshoe magnet (Fig. 88), the length of which from end to end through the steel is 42½ in.; a light narrow frame was constructed so that it could be slipped on over the magnet, and on it were wound 30 turns of fine wire, to serve as an exploring coil. The ends of this coil were carried to a distant part of the laboratory, and connected to a sensitive ballistic galvanometer. The mode of experimenting is as follows:—The coil is slipped on over the magnet (or over its armature) to any desired position. The armature of the magnet is placed gently upon the poles and time enough is allowed to elapse for the galvanometer needle to settle to zero; the armature is then suddenly detached. The first swing measures the change due to removing the armature, in the number of magnetic lines that pass through the coil in the particular position.

I will roughly repeat the experiment before you: the spot of light on the screen is reflected from my galvanometer at the far end of the table. I place the exploring coil just over the pole, and slide on the armature; then close the galvanometer circuit. Now I detach the armature, and you observe the large swing. I shift the exploring coil right up to the bend; replace the armature; wait until the spot of light is brought to rest at the zero of the scale. Now, on detaching the armature, the movement of the spot of light is quite imperceptible. In our careful laboratory experiments, the effect was noticed inch by inch all along the magnet. The effect when the exploring coil was over the bend was not as great as 1-3000th part of the effect when the coil was hard up to the pole. We are therefore justified in saying that the number of magnetic lines in a permanently magnetized steel horseshoe magnet is not altered by the presence or absence of the armature.

You will have noticed that I always put on the armature gently. It does not do to slam on the armature: every time you do so, you knock some of the so-called permanent magnetism out of it. But you may pull off the armature as suddenly as you like. It does the magnet good rather than harm. There is a popular superstition that you ought never to pull off the keeper of a magnet suddenly. On investigation, it is found that the facts are just the other way. You may pull off the keeper as suddenly as you like; but you should never slam it on.

Another point of difference between a permanent magnet of steel and an electromagnet is to be found in the following fact:—Suppose two such magnets, both of horse-shoe form, are taken of equal size; and that the electromagnet is excited just so much that its traction on an iron keeper in contact is equal to that of the permanent magnet. Then if the attraction of the same magnets on an armature at a little distance away is tried, it will be found that the permanent magnet will exert a considerably greater pull than that of the electromagnet. For equal tractive power the steel magnet has the greater range.

ELECTROMAGNETS FOR MAXIMUM TRACTION.

These have already been dealt with in Chapter IV.; the characteristic feature of all the forms suitable for traction being the compact magnetic circuit.

Several times it has been proposed to increase the power of electromagnets by constructing them with intermediate masses of iron between the central core and the outside, between the layers of windings. All these constructions are founded on fallacies. Such iron is far better placed either right inside the coils or right outside them, so that it may properly constitute a part of the magnetic circuit. The constructions known as Camacho's and Cance's, and one patented by Mr. S. A. Varley in 1877, belonging to this delusive order of ideas, are now entirely obsolete.

Another construction which is periodically brought forward as a novelty, is the use of iron windings of wire or strip in place of copper winding. The lower electric conductivity of iron, as compared with copper, makes such a construction wasteful of exciting power. To apply equal magnetizing power by means of an iron coil implies the expenditure of about six times as many watts as need be expended if the coil is of copper.

The latest specimen of this order of construction is Ricco's[*] electromagnet, constructed of a single long sheet of flexible iron rolled around a central iron rod, with intermediate sheets of oiled paper to insulate the successive convolutions; the whole being held together by insulated rings.

[*] *Mem. R. Acad. Sci. Modena*, ser. 2, vol. iv., p. 27, 1886.

ELECTROMAGNETS FOR MAXIMUM RANGE OF ATTRACTION.

We have already laid down, in Chapter V., the principles which will enable us to design electromagnets to act at a distance. We want our magnet to project, as it were, its force across the greatest length of air gap. Clearly, then, such a magnet must have a very large magnetizing power, with many ampere-turns upon it, to be able to make the required number of magnetic lines pass across the air resistance. Also it is clear that the poles must not be too close for its work, otherwise the magnetic lines at one pole will be likely to curl round and take short cuts to the other pole. There must be a wider width between the poles than is desirable in electromagnets for traction.

ELECTROMAGNETS WITHOUT IRON.

If the iron core of an electromagnet be omitted, the coil will by itself attract pieces of iron when excited by the electric current; its action is, however, much less powerful than when an iron core is present. The special case of a long tubular coil attracting an iron plunger into its interior is discussed separately in Chapter VIII., as this arrangement presents some important characteristic properties; possessing, in particular, an extended range of motion.

ELECTROMAGNETS OF MINIMUM WEIGHT.

In designing an apparatus to put on board a boat or a balloon, where weight is a consideration of primary importance, there is again a difference. There are three things that come into play—iron, copper, and electric current. The current weighs nothing, therefore if you are going to sacrifice everything else to weight, you may have comparatively little iron, but you must have enough copper to be able to carry the electric current; and under such circumstances you must not mind heating your wires nearly red hot to pass the biggest possible current. Provide as little copper as you conveniently can, sacrificing economy in that case to the attainment of your object; but of course you must use fire-proof material, such as asbestos, for insulating, instead of cotton or silk.

ELECTROMAGNETS OF MINIMUM COST.

In order to reduce the first cost of construction, some manufacturers have used cast iron instead of wrought iron This is a saving in the case of large electromagnets such as those used for the field-magnets of dynamo-machines, at least in those cases where the extra bulk and weight are admissible. But for small forms, such as those used in electric bells, cast iron is inadmissible. *Malleable iron* has indeed been used for these purposes; and the author was the first to recommend that *mitis iron* should be tried for cores. There is little gain even in these cases, because the saving in using an iron of a cheaper and inferior magnetic quality is practically counterbalanced by the resulting necessity of using a greater weight of copper. But labour counts also in the cost of production. Hence forms have been sought which shall need the minimum of labour in their construction. One such form has established itself in commerce. It is that called by Count du Moncel the *aimant boiteux*, or club-footed magnet (Fig. 24, p. 51). It is a horse-shoe in fact, with a coil upon one pole and no coil upon the other. The advantage of that construction is simply, I suppose, that you will save labour—you will only have to wind the wire on one pole instead of two. Whether that is an improvement in any other sense is a question for experiment to determine; but on which theory perhaps might now be able to say something. Count du Moncel, who made many experiments on this form of magnet, ascertained that there was for an equal weight of copper a slight falling off in power with the club-footed magnet. Indeed one might almost predict, for a given weight of copper, if you wound all in one coil only, you will not make as many turns as if you wound it in two; the outer turns on the coil being so much larger than the average turn when wound in two coils. Consequently the number of ampere-turns with a given weight of copper would be rather smaller, and you would require more current to bring the magnetizing power up to the same value as with the two coils. At the same time the one coil may be produced a little more cheaply than the two; and indeed such electromagnets are really quite common, being largely used, for the sake of cheapness and compactness, in indicators of electric bells. In this form the yoke may be made a little shorter, and the two cores a little nearer together than is practicable in electromagnets with two bobbins: so shortening slightly the length of the magnetic circuit.

Du Moncel tried various experiments about this form to find whether it acted better when the armature was pivoted over one pole or over the other, and found it worked best

FIG. 89.

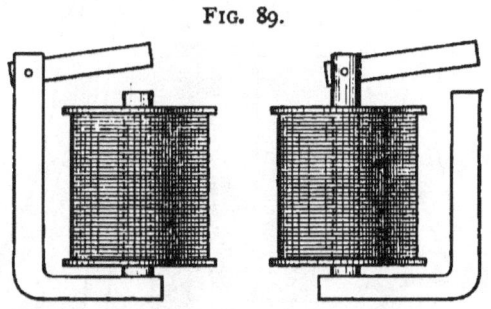

CLUB-FOOT ELECTROMAGNETS WITH HINGED ARMATURES.

when the armature was actually hinged on to that pole which comes up up through the coil. He made two experiments, trying coils on one or other limb, the armature being in each case set at an equal distance. In one experiment he found the pull was 35 grammes, with an armature hinged on to the idle pole, and 40 grammes when it was hinged on to the pole which carried the coil.

Another form of electromagnet, having but one coil, is used in the electric bells of church-bell pattern, of which

218   *The Electromagnet.*

Mr. H. Jensen is the designer. In Jensen's electromagnet a straight cylindrical core receives the bobbin for the coil, and, after this has been pushed into its place, an oval pole-piece, or in some forms two pole-pieces, are screwed upon its ends, serving thus to bring the magnetic circuit across the ends of

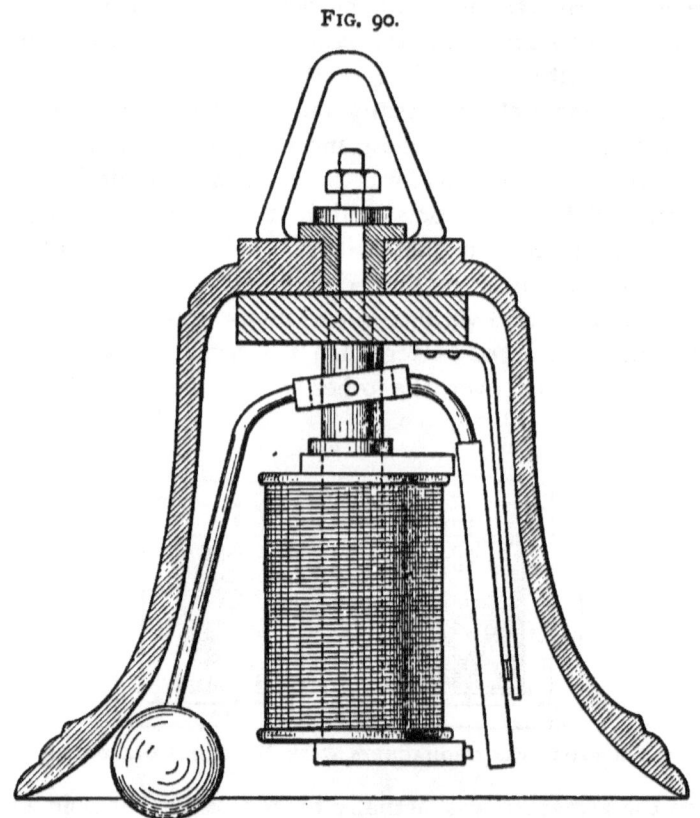

Fig. 90.

Jensen's Electric Bell.

the bobbin, and forming a magnetic gap along the side of the bobbin. The armature is a rectangular strip of soft iron about the same length as the core, and is attracted at one end by one pole-piece, and at the other end by the other.

Electromagnets for use with Rapidly Changing Currents.

When you are designing electromagnets for use with rapidly changing currents, whether intermittent or alternating, it is necessary to make a change in one respect, namely, you must so laminate the iron that internal eddy-currents shall not occur; indeed, for all rapid acting electromagnetic apparatus it is a good rule that the iron must not be solid. It is not usual with telegraphic instruments to laminate them by making up the core of bundles of iron plates or wires, but they are often made with tubular cores, that is to say, the cylindrical iron core is drilled with a hole down the middle, and the tube so formed is slit with a saw-cut to prevent the circulation of currents in the substance of the tube. When electromagnets are to be employed with rapidly alternating currents, such as are used for electric lighting, the frequency of the alternations being usually about 100 periods per second, slitting the cores is insufficient to guard against eddy-currents; nothing short of completely laminating the cores is a satisfactory remedy.

All the necessary instructions on this head are given in Chapter XI. Meantime it is essential to understand that there exists a sort of electromagnetic inertia, or self-inductive property of the electric circuit, which causes the currents to rise and fall later in time than the electromotive forces by which they are occasioned. In all such cases the impedance which the circuit offers is made up of two things—resistance and inductance. Both these causes tend to diminish the amount of current that flows, and the reluctance also tends to delay the flow. As Chapter XI. is devoted to alternating current apparatus, it will be sufficient here to deal with the case of intermittent currents, such as are used in telegraphic signalling.

ELECTROMAGNETS FOR QUICKEST ACTION: LAW OF HELMHOLTZ.

Professor Hughes's researches on the form of electro-magnet best adapted for rapid signalling, have already been mentioned. Mention has also incidentally been made of the fact that where rapidly varying currents are employed, the strength of the electric current that a given battery can yield is determined not so much by the resistance of the electric circuit, as by its electric inertia. It is not a very easy task to explain precisely what happens to an electric circuit when the current is turned on suddenly. The current does not suddenly rise to its full value, being retarded by inertia. The ordinary law of Ohm in its simple form no longer applies; one needs to apply that other law which bears the name of the law of Helmholtz, the use of which is to give us an expression, not for the final value of the current, but for its value at any short time, t, after the current has been turned on. The strength of the current after a lapse of a short time, t, cannot be calculated by the simple process of taking the electromotive force and dividing it by the resistance, as you would calculate steady currents.

In symbols, Helmholtz's law is :—

$$i_t = \frac{E}{R}\left(1 - e^{-\frac{R}{L}t}\right).$$

In this formula i_t means the strength of the current after the lapse of a short time t (seconds); E is the electromotive force (volts); R the resistance of the whole circuit (in ohms); L its co-efficient of self-induction (in quads*); and e the number 2·7183, which is the base of the Napierian logarithms. Let us look at this formula; in its general form it resembles Ohm's law, but with a new factor, namely, the expression contained within the brackets. This factor is necessarily a

---

* The *quad* or *quadrant*, formerly called the *secohm*, is the unit in which to express co-efficients of self-induction. Some American electricians call it the *henry*. See Appendix B, on Units.

fractional quantity, for it consists of unity less a certain negative exponential, which we will presently further consider. If the factor within brackets is a quantity less than unity, that signifies that i_t will be less than $E \div R$. Now the exponential of negative sign, and with negative fractional index, is rather a troublesome thing to deal with in a popular lecture.

Our best way is to calculate some values, and then plot it out as a curve. When once you have got it into the form of a curve, you can begin to think about it, for the curve gives you a mental picture of the facts that the long formula expresses in the abstract. Accordingly we will take the following case. Let $E = 10$ volts; $R = 1$ ohm; and let us take a relatively large self-induction, so as to exaggerate the effect; say let $L = 10$ quads. This gives us the following :—

t (sec.)	$e^{+\frac{R}{L}t}$	i_t
0	1	0
1	1·105	0·950
2	1·221	1·810
5	1·649	3·936
10	2·718	6·343
20	7·389	8·646
30	20·08	9·501
60	403·4	9·975
120	162800·0	9·999

In this case the value of the steady current, as calculated by Ohm's law, is 10 amperes; but Helmholtz's law shows us that with the great self-induction, which we have assumed to be present, the current, even at the end of 30 seconds, has only risen up to within 95 per cent. of its final value; and only at the end of two minutes has practically attained full strength. These values are set out in the highest curve in Fig. 91, in which, however, the further supposition is made that the number of spirals S in the coils of the electromagnet is 100, so that when the current attains its full value of 10 amperes, the full magnetizing power will be $Si = 1000$. It will be noticed that the curve rises from zero at first steeply and nearly in a straight line, then bends over, and then becomes nearly straight again, as it gradually rises to its limiting value. The first part of the curve—that relating to the strength of the current after a *very small* interval of time—is the period within which the strength of the current is governed by inertia (*i. e.*, the self-induction) rather than by resistance. So that for very small values of t the formula might be approximately written as :—

$$i_t = \frac{E\,t}{L}$$

Or, in words, the effect of self-induction during the first small intervals of time is such as though the only resistance in the circuit was equal to the quads of self-induction divided by the small fraction of a second that has elapsed. Now dividing by a small fraction is equivalent to multiplying by a large number. The effect then during the first instants is the same as if there was an enormous resistance which diminished as time went on.

Time-Constants of Electromagnets.

It was remarked above that, in the first period after the current is turned on, its strength is governed rather by the self-induction of the circuit than by its resistance—by the quads, rather than by the ohms. In reality the current is not governed either by the self-induction or by the resistance alone, but by the ratio of the two. This ratio is sometimes called the "time-constant" of the circuit, for it represents *the time* which the current takes in that circuit to rise to a definite fraction of its final value. This definite fraction is the fraction $\frac{e-1}{e}$; or in decimals, 0·634. All curves of rise-of-current are alike in general shape—they differ only in scale;

that is to say, they differ only in the height to which they will ultimately rise, and in the time they will take to attain this fraction of their final value.

Example 1.—Suppose $E = 10$; $R = 400$ ohms; $L = 8$. The final value of the current will be 0·025 amperes or 25 milliamperes. Then the time-constant will be $8 \div 400 = $ 1-50th sec.

Example 2.—The P.O. Standard "A" relay has $R = 400$ ohms; $L = 3·25$. It works with 0·5 milliampere current, and therefore will work with 5 Daniell cells through a line of 9600 ohms. Under these circumstances the time-constant of the instrument on short circuit is 0·0081 sec.

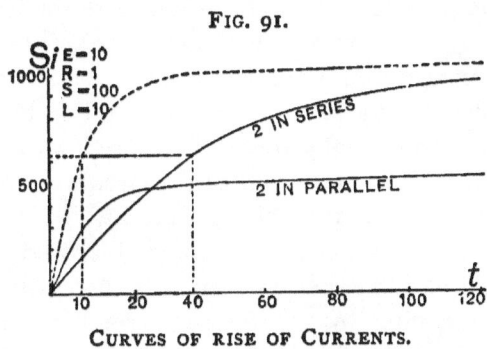

FIG. 91.

CURVES OF RISE OF CURRENTS.

It will be noted that the time-constant of a circuit can be reduced either by diminishing the self-induction or by increasing the resistance. In Fig. 91 the position of the time-constant for the top curve is shown by the vertical dotted line at 10 seconds. The current will take 10 seconds to rise to 0·634 of its final value. This retardation of the rise of current is simply due to the presence of coils and electromagnets in the circuit, the current as it grows being retarded because it has to create magnetic fields in these coils, and so sets up opposing electromotive forces that prevent it from growing all at once to its full strength. Many electricians, unacquainted with Helmholtz's law, have been in the habit of accounting for this by saying that there is a lag in the iron of the electromagnet cores. They tell you that an iron core cannot be magnetized suddenly; that it takes time to acquire its magnetism. They think it is one of the properties of iron. But we know that the only true time-lag in the magnetization of iron—that which is properly termed "viscous hysteresis"—does not amount to any great percentage of the whole amount of magnetization, takes comparatively a long time to show itself, and cannot therefore be the cause of the retardation which we are considering.

There are also electricians who will tell you that when magnetization is suddenly evoked in an iron bar, there are induction currents set up in the iron which oppose and delay its magnetization. That they oppose the magnetization is perfectly true; but if you carefully laminate the iron so as to eliminate eddy-currents, you will find, strangely enough, that though the magnetism rises faster than before, the current rises still more slowly to its final value. For by laminating the iron you have virtually increased the self-inductive action, and increased the time-constant of the circuit. The lag is not in the iron, but in the magnetizing current.

Continental electricians are in the habit of distinguishing between the "variable period" and the "steady period" of the current, meaning by the former term the period during which the current is still rising, and by the latter the period after the current has so risen. It is impossible, however, to draw a hard and fast line between the two periods: for the current attains its final value by imperceptible degrees. Ohm's law applies only to the state of things when this steady value has been attained. So long as the strength of the current falls perceptibly short of its steady value, so long must it be considered as still in the variable period, in which the formula of Helmholtz expresses the law.

It is sometimes asked whether an electromagnet which is sluggish can be quickened in its action by rewinding it with a coil of some other resistance. As will be seen, the answer must depend on the other resistances and inductances in the circuit. Assuming perfect insulation, and a given volume to be filled by the coil, then it is impossible to alter the time-constant of the electromagnet *itself*, by changing the winding.

For in a coil of given volume the resistance and the self-induction are each proportional to the square of the number of turns; and the ratio of the two is therefore a constant, independent of the gauge of the wire. But if an electromagnet is to be used in a long line without other coils or electromagnets in the circuit which, though it has little self-induction, has considerable resistance, it may be worth while to rewind the electromagnet with thicker wire in fewer turns so as to diminish the time-constant of the circuit as a whole. It was shown by Von Beetz* that, though the strength of the current and the final value of the magnetism be the same in each case, the magnetism of the core of an electromagnet is more rapidly established by a large electromotive force working through a large resistance, than by a smaller electromotive force working through a small resistance. In other words, if due precautions were taken to vary the battery so as to keep the current up to the same final value, the time-constant of the circuit was diminished by *increasing* the resistance.

CONNECTING COILS FOR QUICKEST ACTION.

Now let us apply these most important, though rather intricate considerations to the practical problems of the quick working of the electromagnet. Take the case of an electromagnet forming some part of the receiving apparatus of a telegraph system, in which it is desired to secure very rapid working. Suppose the two coils that are wound upon the horseshoe core are connected together in series. The co-efficient of self-induction for these two is four times as great as that of either separately; co-efficients of self-induction being proportional to the square of the number of turns of wire that surround a given core. Now if the two coils, instead of being put in series, are put in parallel, the co-efficient of self-induction will be reduced to the same value as if there were only one coil, because half the line current (which is

---

* For some earlier experiments of Hipp, see *Mittheilungen der Berner naturforschenden Gesellschaft*, 1855, p. 190.

practically unaltered) will go through each coil. Hence the time constant of the circuit when the coils are in parallel will be a quarter of that which it is when the coils are in series; on the other hand, for a given line-current, the final magnetizing power of the two coils in parallel is only half what it would be with the coil in series. The two lower curves in Fig. 91 illustrate this, from which it is at once plain that the magnetizing power for very brief currents is greater when the two coils are put in parallel with one another than when they are joined in series.

Now this circumstance has been known for some time to telegraph engineers. It has been patented several times over. It has formed the theme of scientific papers, which have been read both in France and in England. The explanation generally given of the advantage of uniting the coils in parallel is, I think, fallacious; namely, that the "extra-currents" (*i. e.*, currents due to self-induction), set up in the two coils are induced in such directions as tend to help one another when the coils are in series, and to neutralise one another when they are in parallel. It is a fallacy, because in neither case do they neutralize one another. Whichever way the current flows to make the magnetism, it is opposed in the coils while the current is rising, and helped in the coils while the current is falling, by the so-called extra-currents. If the current is rising in both coils at the same moment, then, whether the coils are in series or parallel, the effect of self-induction is to retard the rise of the current. The advantage of parallel grouping is simply that it reduces the time-constant.

BATTERY GROUPING FOR QUICKEST ACTION.

One may consider the question of grouping the battery cells from the same point of view. How does the need for rapid working, and the question of time-constant, affect the best mode of grouping the battery cells? The amateur's rule, which tells you to so arrange your battery that its internal resistance should be equal to the external resistance, gives you a result wholly wrong for rapid working. The supposed best arrangement will not give you (at the expense even of economy) the best result that might be got out of the given number of cells. Let us take an example and calculate it out, and place the results graphically before our eyes in the form of curves. Suppose the line and electromagnet have together a resistance of 6 ohms, and that we have 24 small Daniell's cells each of electromotive force, say, 1 volt, and of internal resistance, 4 ohms. Also let the coefficient of self-induction of the electromagnet and circuit be 6 quadrants. When all the cells are in series the resistance of the battery will be 96 ohms, the total resistance of the circuit 102 ohms, and the full value of the current 0·235 ampere. When all the cells are in parallel the resistance of the battery will be 0·133 ohm, the total resistance 6·133 ohms, and the full value of the current 0·162 ampere. According to the amateur rule of grouping cells so that internal resistance equal external, we must arrange the cells in 4 parallels, each having 6 cells in series, so that the internal resistance of the battery will be 6 ohms, total resistance of circuit 12 ohms, full value of current 0·5 ampere. Now the corresponding time-constants of the circuit in the three cases (calculated by dividing the co-efficient of self-induction by the total resistance) will be respectively—in series, 0·06 sec.; in parallel, 0·96 sec.; grouped for maximum steady current, 0·5 sec. From these data we may now draw the three curves, as in Fig. 92, wherein the abscissæ are the values of time in seconds, and the ordinates the current. The faint vertical dotted lines mark the time-constants in the three cases. It will be seen that when rapid working is required the magnetizing current will rise, during short intervals of time, more rapidly if all the cells are put in series than it will do if the cells are grouped according to the amateur rule.

When they are all put in series so that the battery has a much greater resistance than the rest of the circuit, the current rises much more rapidly, because of the smallness of the time-constant, although it never attains the same ultimate maximum as when grouped in the other way. That is to say, if there is self-induction as well as resistance in the circuit, the amateur rule does not tell you the best way of arranging the battery. There is another mode of regarding the matter which is helpful. Self-induction, while the current is growing, acts as if there were a sort of spurious addition to the resistance of the circuit; and while the current is dying away it acts of course in the other way, as if there were a subtraction from the resistance. Therefore you ought to arrange the battery so that the internal resistance is equal to the real

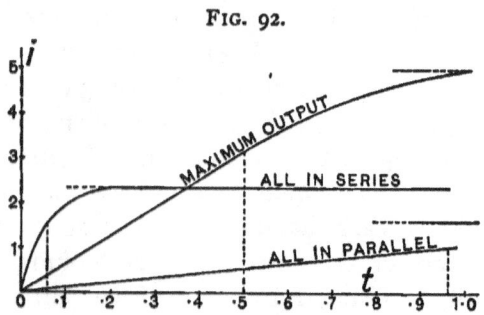

FIG. 92.

CURVES OF RISE OF CURRENT WITH DIFFERENT GROUPINGS OF BATTERY.

resistance of the circuit, plus the spurious resistance during that time. But how much is the spurious resistance during that time? It is a resistance proportional to the time that has elapsed since the current was turned on. So then it comes to a question of the length of time for which you want to work it. What fraction of a second do you require your signal to be given in? What is the rate of the vibrator of your electric bell? Suppose you have settled that point, and that the short time during which the current is required to rise is called t; then the apparent resistance at time t after the current is turned on is given by the formula:—

$$R_t = R \times e^{\frac{R}{L} t} \div \left(e^{\frac{R}{L} t} - 1 \right).$$

TIME-CONSTANTS OF ELECTROMAGNETS.

Reference may here be made to some determinations made by M. Vaschy,* respecting the co-efficients of self-induction of the electromagnets of a number of pieces of telegraphic apparatus. Of these I must only quote one result, which is very significant; it relates to the electromagnet of a Morse receiver of the pattern habitually used on the French telegraph lines.

	L, in quadrants.
Bobbins separately, without iron cores	0·233 and 0·265
,, separately, with iron cores	1·65 and 1·71
,, with cores joined by yoke, coils in series	6·37
,, with armature resting on poles	10·68

It is interesting to note how the perfecting of the magnetic circuit increases the self-induction.

Thanks to the kindness of Mr. Preece, I have been furnished with some most valuable information about the co-efficients of self-induction, and the resistance of the standard pattern of relays, and other instruments which are used in the British Postal Telegraph service, from which data one is able to say exactly what the time-constants of those instruments will be on a given circuit, and how long in their case the current will take to rise to any given fraction of its final value. Here let me refer to a very capital paper by Mr. Preece in an old number of the 'Journal of the Society of Telegraph Engineers,' a paper "On Shunts," in which he treats this question, not as perfectly as it could now be treated with the fuller knowledge we have in 1890 about the co-efficients of self-induction, but in a very useful and practical way. He showed most completely that the more perfect the magnetic circuit is—though, of course, you are getting more magnetism from your current—the more is that current retarded. Mr. Preece's mode of experiment was extremely simple; he observed the throw of the galvanometer, when the circuit which contained the battery and the electromagnet

* *Bulletin de la Société Internationale des Électriciens*, 1886.

was opened by a key which at the same moment connected the electromagnet wires to the galvanometer. The throw of the galvanometer was assumed to represent the extra-current which flowed out. Fig. 93 represents a few of the results of Mr. Preece's paper. Take from an ordinary relay a coil, with its iron core, half the electromagnet, so to speak, without any yoke or armature. Connect it up as described, and observe the throw given to the galvanometer. The amount of throw obtained from the single coil was taken as unity, and all others were compared with it. If you join up two such coils as they are usually joined in series, but without any iron yoke across the cores, the throw was 17. Putting the iron yoke

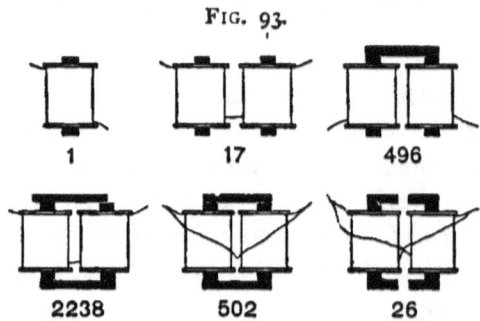

ELECTROMAGNETS OF RELAY, AND THEIR EFFECTS.

across the cores, to constitute a horseshoe form, 496 was the throw; that is to say, the tendency of this electromagnet to retard the current was 496 times as great as that of the simple coil. But when an armature was put over the top, the effect ran up to 2238. But the mere device of putting the coils in parallel, instead of in series, the 2238 came down to 502, a little less than the quarter value which would have been expected. Lastly, when the armature and yoke were both of them split in the middle, as is done in fact in all the standard patterns of the British Postal Telegraph relays, the throw of the galvanometer was brought down from 502 to 26. Relays so constructed will work excessively rapid. Mr. Preece states that with the old pattern of relay having so much self-induction as to give a galvanometer throw of 1688, the speed of signalling was only from 50 to 60 words per minute; whereas, with the standard relays constructed on the new plan, the speed of signalling is from 400 to 450 words per minute. It is a very interesting and beautiful result to arrive at from the experimental study of these magnetic circuits.

BRITISH POSTAL TELEGRAPH RELAYS.*

As examples of the construction of rapid-working telegraphic instruments are given four standard patterns of relays as used in the Postal Telegraph service of the British Government.

The "A" pattern of standard relay has an electromagnet which is depicted half-size in Fig. 94, one of the bobbins being represented as unwound, the other wound. The official specification runs as follows :—" The iron forming the cores of the electromagnets, to be turned from a solid bar of the best Swedish iron, and when properly annealed must not retain a trace of residual magnetism, after a current due to an electromotive force of 50 volts has been sent through the coils on short circuit." There are four wires 5 mils ($= 0·005$ inch) in diameter upon the pair of electromagnets, each of which must have a resistance of 100 ohms, at a temperature of 60° F. (a deviation of *not more than* one per cent. either way will be accepted). These wires must be of equal lengths, so that they are electromagnetically differential when a current due to an E.M.F. of 50 volts is passed through them. Each pair of wires to have ends of different colours (green and white). And in the official memorandum respecting the examination of apparatus it is added :—" The insulation of the parts electrically separated must also be practically perfect. This

FIG. 94.

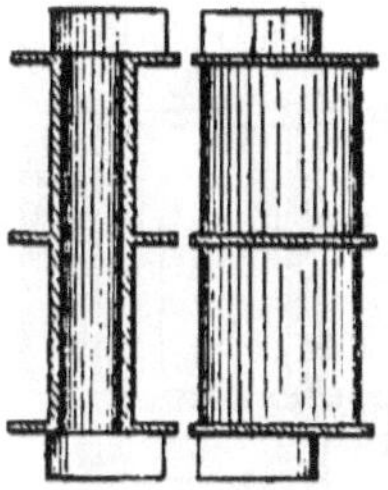

P.O. RELAY, "A" PATTERN.

---

* The author is indebted to the Chief Electrician of the British Postal Telegraphs, W. H. Preece, Esq., F.R.S., for particulars about these instruments.

condition must especially be observed between the coils and the cores of all electromagnets, and between the two wires of differentially-wound instruments, which should not be approved if the insulation resistance is less than one megohm." The total length of wire employed in the "A" relay is 320 yards; its resistance when all coils are in series is 400 ohms; and its co-efficient of self-induction 3·25 quads.

The "B" pattern of standard relay, given half-size in Fig. 95, is likewise wound with wire of 5 mils diameter. There are two wires, each of 320 yards length, and resistance

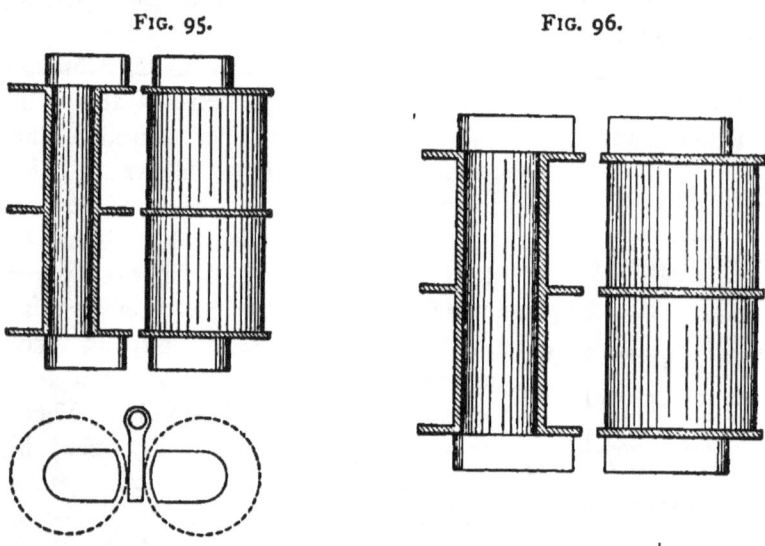

FIG. 95.   FIG. 96.

P.O. RELAY, "B" PATTERN.   P.O. RELAY, "C" PATTERN.

400 ohms; but the two, after being wound, are permanently joined in parallel, making the resistance of the instrument 200 ohms, and its self-induction 2·14 quads.

The "C" pattern (Fig. 96, half-size), the largest of the P. O. standard relays, is wound with 5 mil wire 960 yards long, and has a resistance of 1200 ohms, and a co-efficient of self-induction of 26·4 quads. It has three times the length of wire, and about nine times the self-induction of the "A" pattern. The same instructions quoted above respecting the cores used for the "A" pattern relate also to the "B" and "C" patterns. In each of these instruments there is also used a curved permanent magnet (of tungsten steel) to polarize the tongue armatures, which are two in number in each instrument; one being placed between the upper pole-pieces and the other between the lower.

Fig. 97 depicts (half size) the form of Siemens' relay adopted by the British Postal Telegraphs. It is wound with 646 yards of 7 mil wire; and has resistance 400 ohms, and self-induction 7·09 quads.

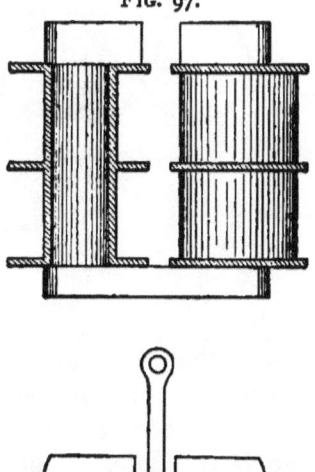

FIG. 97.

SIEMENS' RELAY.

HIGGINS'S ELECTROMAGNETS.

The electromagnets used in the automatic type-printing telegraph instruments of the Exchange Telegraph Co. furnish another example of special design for rapid work. Mr. Higgins states[*] that cores which are short in proportion to their diameter are more rapid than long slender cores; also that projecting ends and attached pole-pieces are abandoned because they retard action, and that metal checks or bobbins also retard the magnetism. The cores of these electromagnets (Fig. 98), are made from a very pure kind of charcoal-iron manufactured in Switzerland. One-third of their diameter is bored out from the bottom to within a short distance of the poles. They are then slit lengthways and annealed out of contact with the air. Flat armatures are used.

Some information about the special electromagnet used in d'Arlincourt's relay, is to be found in Chapter IX. p. 305.

[*] *Journal of Society of Telegraph Engineers*, vol. vi., p. 129, 1877.

SHORT CORES *versus* LONG CORES.

In considering the forms that are best for rapid action, it ought to be mentioned that the effects of hysteresis in retarding changes in the magnetization of iron cores are much more noticeable in the case of nearly-closed magnetic

FIG. 98.

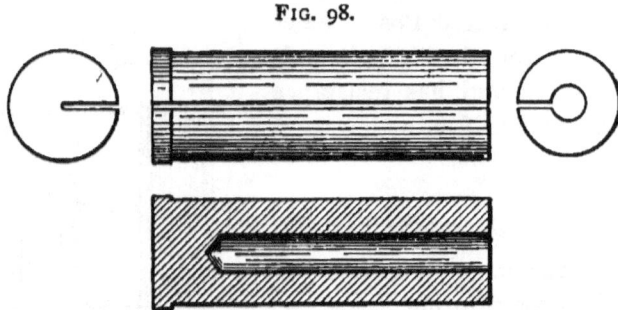

CORES OF EXCHANGE TELEGRAPH CO'S. ELECTROMAGNETS.

circuits than in short pieces. Electromagnets with iron armatures in contact across their poles will retain, after the current has been cut off, a very large part of their magnetism, even if the cores be of the softest of iron. But so soon as the armature is wrenched off the magnetism disappears. An air-gap in a magnetic circuit always tends to hasten demagnetizing. A magnetic circuit composed of a long air-path and a short iron-path demagnetizes itself much more rapidly than one composed of a short air-path and a long iron-path. In long pieces of iron the mutual action of the various parts tends to keep in them any magnetization that they may possess; hence they are less readily demagnetized. In short pieces, where these mutual actions are feeble or almost absent, the magnetization is less stable, and disappears almost instantly on the cessation of the magnetizing force. Short bits and small spheres of iron have no "magnetic memory." Hence the cause of the commonly received opinion amongst telegraph engineers that for rapid work electromagnets must have short cores. As we have seen, the only reason for employing long cores is to afford the requisite length for winding the wire which is necessary for carrying the needful circulation of current to force the magnetism across the air-gaps. If, for the sake of rapidity of action, length has to be sacrificed, then the coils must be heaped up more thickly on the short cores. The electromagnets in American patterns of telegraphic apparatus usually have shorter cores, and a relatively greater thickness of winding upon them, than those of European patterns.

Rapid Electromagnets for use in Chronographs.

Though the electromagnets designed for telegraphic instruments, and particularly for relays, must be rapid in their action, yet those required for actuating the recording styluses of chronographs must be still more rapid. When chronographs are used for such experimental work as measuring the speed of projectiles or the velocity of a sound wave in the air, the usual mode of effecting the record is that an electric current should be automatically interrupted by the motion which is to be recorded, thereby releasing the armature of an electromagnet in the circuit. A spring or other device draws back the armature and causes a stylus attached to it to make a record upon a moving surface usually a piece of paper coiled over a uniformly rotating drum. There are several causes of delay between the act of breaking the current and the completion of the record by the stylus. Firstly, the electromagnet, and the circuit in which it is interposed will have a definite time-constant, so that the current takes time to fall. Secondly, the magnetism may be retarded behind the current if the core or bobbins are so arranged that eddy-currents can occur, or if the iron parts constitute a very nearly closed magnetic circuit. Thirdly, the inertia of the moving parts may be considerable, causing the actual motion to be retarded even after the magnetism has disappeared. The total lapse of time between the rupture of the circuit and the formation of the record, is called the *latency* of the apparatus. The Rev. F. J. Smith, of Oxford, who has given this name, has succeeded in constructing styluses having a latency of less than 0·0003 of a second.

To diminish the time-constant of the circuit, the electromagnet must be made small, and with as little copper wire as possible; preferably placed only on the polar ends of the cores, as in Hughes's form (Fig. 83, p. 187). There ought also to be a considerable additional inductionless resistance in the circuit. The iron ought to be laminated, and divided across at the yoke. The moving parts should be as light and compact as possible, and the controlling springs of such form as to act with rapidity.

The production of such electromagnets has been studied by Hipp,[*] Schneebeli,[†] Marcel Deprez,[‡] Mercadier,[§] F. J. Smith,[‖] and others. It will suffice to describe two of the forms found best by M. Deprez, and that devised by Mr. Smith for use in his laboratory at Oxford.

Fig. 99 gives in its natural size one of the forms used by Deprez. There two electromagnets E, E, each with a straight core built of thin laminæ. The armature is lozenge-shaped, pivoted; limited in its movement by the stop I J, and governed by a spring B K, the stiffness of which is controlled by a lever F. The stylus is at C D; and there is a clamping-screw M by which it can be fixed to a support. By limiting the motion of the armature to 2 mm. the latency can be reduced to 0·00016 sec. on opening the circuit, or 0·00048 sec. on closing it.

The second form used by Deprez is a polarised apparatus containing a powerful permanent magnet, (Fig. 100) set with its limbs upwards, and furnished at the top with two adjustable pole-pieces B D, B D. Between the limbs is a single flat

---

[*] *Mittheilungen der naturforschenden Gesellschaft in Bern,* 1853, p. 113; and 885, p. 190.

[†] *Bulletin de la Société des Sciences naturelles de Neuchâtel,* June 1874, and Feb. 1876.

[‡] *La Lumière Électrique,* iv., p. 282, 1881.

[§] *Ibid.,* iv., p. 404, 1881.

[‖] *Phil. Mag.,* May 1890, p. 377; and Aug. 1890.

coil E with a laminated core. Above the top of this, and magnetized by it, is a little armature of soft iron of triangular

FIG. 99.

FIG. 100.

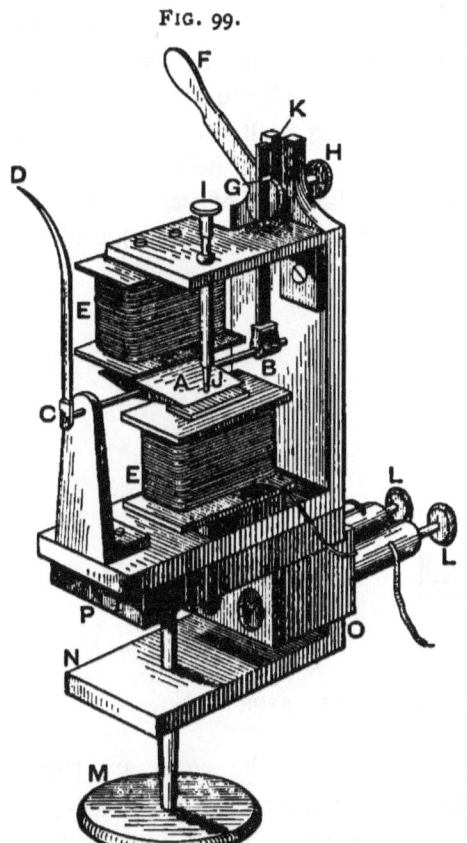

ELECTROMAGNET OF DEPREZ'S CHRONO-GRAPH. NO. 1, FULL SIZE.

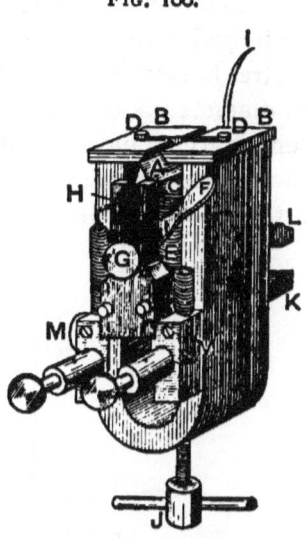

POLARIZED ELECTROMAGNET FOR DEPREZ'S CHRONOGRAPH.

section, pivoted on a knife-edge on the end of the core, and governed, as the armature of the preceding form, by a spring controlled by a lever F. The apparatus is used in a similar way to Hughes's electromagnet (Fig. 83, p. 187); that is to say, the armature is placed in contact with one of the two pole-pieces, and the spring is adjusted so that its force is almost sufficient to detach the armature. When the current is sent around the central core in the proper direction the armature is released, and flies over. An electric current produced by closing the circuit for only $\frac{1}{40000}$ sec. suffices to work the apparatus; but according to Deprez about 0·001 sec. is occupied in the detaching of the armature.

The form of electromagnet used by the Rev. F. J. Smith is depicted in Fig. 101, full size. The yoke consists of a small rectangular block, 18 millimetres long, and 22 square millimetres in section. The cores are thin cylinders, 1·5 millimetre in diameter, and 9·5 millimetres long, well annealed at a low temperature, and not subsequently touched with hammer or file. The armature A is a triangular tube of very thin iron, attached to the aluminium lever L. The contact edge of the

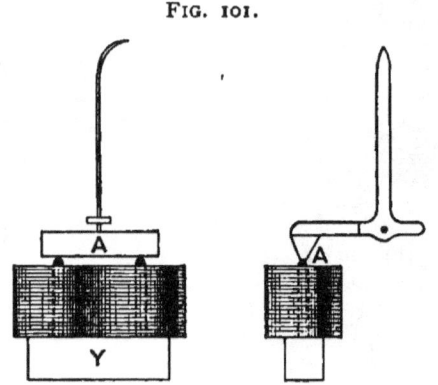

FIG. 101.

ELECTROMAGNET OF SMITH'S CHRONOGRAPH.

armature is rounded, and the ends of the cores are turned truly hemispherical. After the cores are fitted into the yoke the whole is annealed in a gas furnace. The two coils are joined in parallel with one another. Mr. Smith finds that for the same ampere-turns of excitation the length of the cores greatly affects the latency. In these electromagnets the latency is 0·0003 of a second.

RATE OF GAIN OR LOSS OF MAGNETISM.

When the exciting battery is turned on, time is required, as we have seen, for the current to rise to its full strength, and the magnetism consequently takes time to rise: this it does gradually at first, then more rapidly, then again more slowly as the maximun value is approached. So also, an electromagnet, particularly if massive and of such a form as nearly to constitute a closed circuit, takes time to lose its magnetism when the current is turned off. If the circuit of a large electromagnet is suddenly broken, a long thin spark follows the parting ends of the wires, and may last several seconds. In some cases, after the circuit has been parted and the spark has been broken, the magnet still goes on losing its magnetism, and its inductive action may charge the surrounding coil so that it can emit sparks and give shocks to a person touching it.

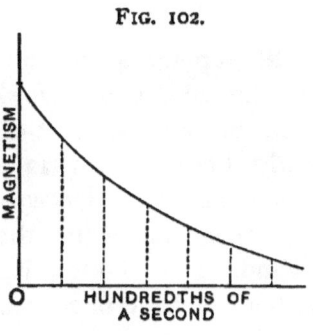

FIG. 102.

CURVE OF LOSS OF MAGNETISM.

The rate at which an electromagnet loses its magnetism is by no means uniform, the curves of loss of magnetism presenting features like those shown in Fig. 102; the rate of loss being greatest at first. Massive electromagnets, such as are used in dynamo-machines, may take some minutes in the operation of losing their magnetism.

CHAPTER VIII.

COIL-AND-PLUNGER.

THE apparatus wherein an iron core is attracted into a tubular coil, or *solenoid*, I take the liberty of naming, for the sake of brevity, as the *coil-and-plunger*. Now, from quite early times, from 1822 at any rate, it was known that a coil would attract a piece of iron into it, and that this action resembles somewhat the action of a piston going into a cylinder; resembled it, I mean to say, in possessing an extended range of action.

A simple experiment will render the matter obvious. A tubular coil, or solenoid, A, mounted on a stand, is placed upon a table, its terminals being connected in circuit with a suitable voltaic battery B. A simple spring-key, or switch, S, is provided, wherewith to complete the circuit at will. The iron core C can be introduced as a plunger into the aperture of the hollow coil. If one end of the plunger is introduced into the mouth of the coil, and the switch is then pressed so as to turn on the current, the plunger is at once drawn into the coil, and settles down with its ends protruding equally through the coil, as shown in Fig. 103. In this position it is in stable equilibrium. If it is forcibly pulled out, at either end, and then let go, it is instantly sucked back to this position by an invisible pull. In fact, to a person handling it, it feels as though in attempting to draw it out he were pulling against an internal spring. From its peculiar action it is sometimes termed a *sucking electromagnet*.

The use of such a device as the coil-and-plunger was patented in this country in 1846 under the name of "a new electromagnet." Electromagnetic engines, or motors, were made on this plan by Page, and afterwards by others, and it became generally known as a distinct device. But even now, if you inquire into the literature of the text-books to know what are the peculiar properties of the coil-and-plunger arrangement, you will find that the books give you next to no information. They are content to deal with the thing in very general terms by saying: Here is a sort of sucking magnet; the core is attracted in. Some books go so far as to tell you that the pull is greatest when the core is about half way in, a

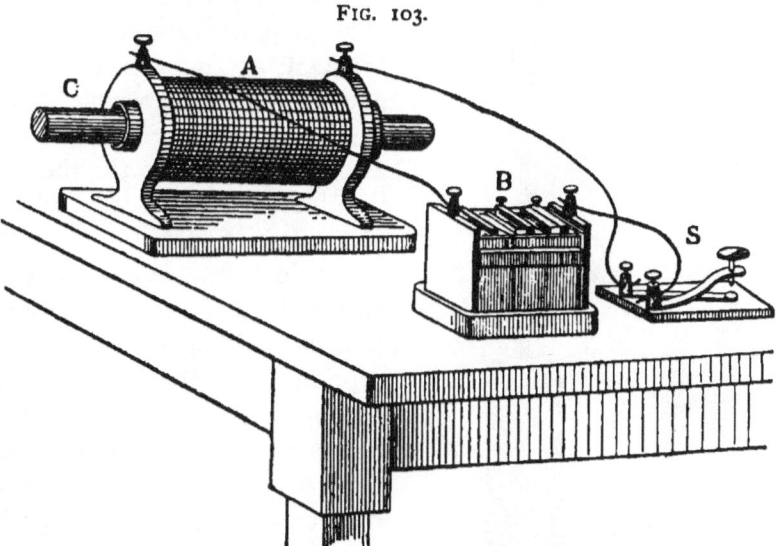

FIG. 103.

EXPERIMENT WITH COIL-AND-PLUNGER.

statement which is true in one particular case, but false in a great many others. Another book tells you that the pull is greatest at a point one centimetre below the centre of the coil, for plungers of all different lengths—which is quite untrue. Another book tells you that a wide coil pulls less powerfully than a narrow one, a statement which is true for some cases and not for others. The books also give you some approximate rules, which, however, are very little to the point. The reason why this ought to receive much more careful consideration is because in this mechanism of coil-and-plunger we have a real means, not only of equalizing, but also of vastly extending the range of the pull of the electromagnet. Let us take a very simple mode of contrasting the range of action of the ordinary electromagnet with the range of action of the coil-and-plunger.

FIG. 104.

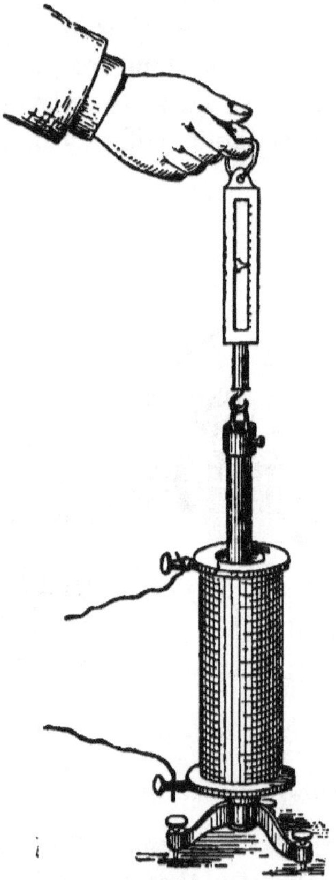

VERTICAL COIL-AND-PLUNGER.

Fig. 104 represents a tubular coil, about 9 inches long, set up vertically, the iron plunger being a rod of the same length, clamped to a loop, whereby it hangs on the hook of an ordinary spring balance. With this simple means it is easy to measure the pull exerted by the coil on the plunger in different positions. If the plunger is first held at a considerable height above the coil, and gradually lowered, it is found that the pull begins to be perceptible when the lower end of the plunger is yet a little distance above the mouth of the coil; and the pull increases as it enters. The pull (in this particular instance, where the plunger is of equal length with the coil) goes on increasing as the plunger descends, and becomes a maximum when the plunger is just a little more than half way immersed; and, from that point onward, though it is still pulled, the pull becomes less and less, until, when the ends of the plunger just coincide with the ends of the coil, the pull ceases. In this instance, then, the pull, though unequally distributed, extends over a range somewhat exceeding the total length of the cylinder.

Here are some numbers which are given in a paper

written by the late Mr. Robert Hunt in 1856, and read before the Institution of Civil Engineers, with that eminent engineer, Robert Stephenson, in the chair. Mr. Hunt discussed the various types of motors, and spoke of this question of the range of action. He named some experiments of his own in which the following was the range of action. There was a horse-shoe electromagnet which at distance zero—that is, when its armature was in contact—pulled with a pull of 220 lb.; when the distance was made only $\frac{4}{1000}$ of an inch (4 mils), the pull fell to 90; and when the distance was increased to 20 mils ($\frac{1}{50}$ of an inch), the pull fell to only 36 lb. The difference from 220 to 36 was within a range of $\frac{1}{50}$ of an inch. He contrasts this with the results given by another mechanism, not quite the simple coil-and-plunger, but a variety of electromagnet brought out about the year 1845 by a Dane living in Liverpool, named Hjörth, wherein a sort of hollow truncated cone of iron (Fig. 105), with coils wound upon it — a hollow electromagnet, in fact — was caused to act on another electromagnet, one being caused to plunge into the other. Now we have no information what the pull was at distance zero with this curious arrangement of Hjörth's, but, at a distance of 1 inch the pull (with a very much larger apparatus than Hunt's) was 160 lb., the pull at 3 inches was 88 lb., at 5 inches 72 lb.

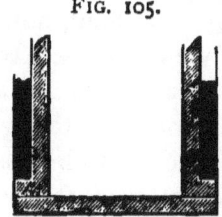

FIG. 105.

HJÖRTH'S ELECTROMAGNETIC MECHANISM.

Here, then, we have a range action going not over $\frac{1}{50}$ of an inch, but over 5 inches, and falling not from 220 to 36, but from 160 to 72, obviously a much more equable kind of range. At the Institution of Civil Engineers on that occasion, a number of the most celebrated men, Joule, Cowper, Sir William Thomson, Mr. Justice Grove, and Professor Tyndall, discussed these matters —discussed them up and down—from the point of view of range of action, and from the point of view of the fact that there was no means of working them at that time except by the consumption of zinc in a primary battery; and they all came to the conclusion that electric motors would never pay. Robert Stephenson summed up the debate at the end in the following words :—" In closing the discussion," he remarked, "there could be no doubt from what had been said that the application of voltaic electricity, in whatever shape it might be developed, was entirely out 'of the question commercially speaking. Without, however, considering the subject in that point of view, the mechanical applications seemed to involve almost insuperable difficulties. The power exhibited by electromagnetism, though very great, extended through so small a space as to be practically useless. *A powerful magnet might be compared, for the sake of illustration, to a steam-engine with an enormous piston, but with an exceedingly short stroke; such an arrangement was well known to be very undesirable.*"

Well, from the discussion in 1856—when this question of the length of range was so distinctly set forth—down to the present time, there have been a large number of attempts to ascertain exactly how to design a long-range electromagnet, and those who have succeeded have, as a general rule, not been the theorists; rather they have been men compelled by force of circumstances to arrive at their result by some kind of—shall we call it—" designing eye," by having a sort of intuitive perception of what was wanted, and going about it in some rough and ready way of their own. Indeed, I am afraid had they tried to get much light from calculations based on orthodox notions respecting the surface distribution of magnetism and all that kind of thing, they would not have been much helped. There is our old friend, the law of inverse squares, which would of course turn up the first thing, and they would be told that it would be impossible to have a magnet that pulled equally through any range, because the pull was certain to vary inversely according to the square of the distance. But then neither this coil nor the plunger can be regarded as a point; and you know that the law of inverse squares is therefore inapplicable.

Now we want to arrive at a true law. We want to know exactly what the law of action of the coil-and-plunger is. It is not a very difficult thing to work out, provided you get hold of the right ideas. We must begin with a simple case, that of a short coil consisting of but one turn, acting on a single point-pole. From this we may proceed to consider the effect on a point-pole of a long tube of coil. Then we may go on to a more complex case of the tube coil acting on a very long iron core; and, last of all, from the very long iron core we may pass to the case of a short core.

You all know how a long tube of coil such as this will act on an iron core. Let us make an experiment with it. I turn on the current so that it circulates around the coil along the tube, and when I hold in front of the aperture of the tube this rod of soft iron, it is sucked into the coil. When I pull it out a little way it runs back, as with a spring. The current happens to be a strong one—about 25 amperes; there are about 700 turns of wire on the coil. The rod is about 1 inch in diameter and 20 inches long. So great is the pull that I cannot pull it entirely out. The pull was very small when the rod was outside, but as soon as it gets in it is pulled actively, runs in, and settles down with the ends equally protruding. The tubular coil I have been using is about 14 inches long; but now let us consider a shorter coil. Here is one only half-an-inch from one end to the other, but I have one somewhat still shorter, so short that the length, parallel to the axis, is very small compared with the diameter of the aperture within. The wire on it consists of but one single turn. Taking such a coil, treating it as only one single ring, with the current going once round, in what way does it act on a magnet that is placed on the axis?

CALCULATION OF EFFECT OF COIL ON PLUNGER.

First of all, take the case of a very long, permanently magnetized steel magnet, so long, indeed, that any action on the more distant pole is so feeble that it may be disregarded altogether, and only one pole, say the north pole, is near the coil. In what way will that single turn of coil act on that single pole? This is the rule, that the pull does not vary inversely as the square of the distance, nor as any power at all of the distance measured straight along the axis, but inversely as the cube of the slant distance. Let the point O in Fig. 106 represent the centre of the ring, its radius being y. The line O P is the axis of the ring, and the distance from O to P we will call x. The slant distance from P to the ring we call a. Then the pull on the axis towards the centre of this coil varies inversely as the cube of a. That law can be plotted out in a curve for the sake of observing the variations of pull at various points along the axis. Allow me to draw your attention to

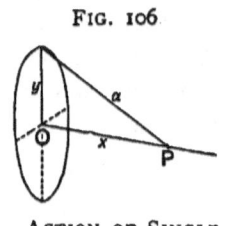

FIG. 106.

ACTION OF SINGLE COIL ON POINT-POLE ON AXIS.

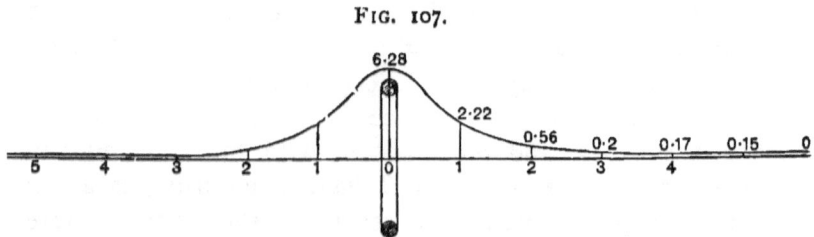

FIG. 107.

ACTION ALONG AXIS OF SINGLE COIL.

Fig. 107, which represents a section or edge view of the coil. At various distances, right and left of the coil, are plotted out vertically the corresponding force, the calculations being made for a current of 10 amperes, circulating once around a ring of 1 cm. radius. The force with which such a current acts on a

Calculation of Effect. 247.

magnetic pole of unit strength placed at the central point is 6·28 dynes. If the pole is moved away down the axis, the pull is diminished. At a distance away equal in length to the radius it has fallen to 2·22 dynes. At a distance equal to twice the radius, or one diameter, it is only 0·56 dynes, less than one-tenth of what it was at the centre. At two diameters it has fallen to 0·17 dynes, or less than 3 per cent.; and the force at three diameters is only about 2 per cent. of that at the centre.

If, then, we could take a *very* long magnet, we may utterly neglect the action on the distant pole. If I had a long steel magnet with the south pole 5 or 6 feet away, and the north pole at a point three diameters (i. e. 6 cm. in this case) distant from the mouth of the coil, then the pull of the current in one spiral on the north pole three diameters away would be practically negligible; it would be less than 2 per cent. of what the pull would be of that single coil when the pole was pushed right up into it. But now, in the case of the tubular coil, consisting of at least a whole layer of turns of wire, the action of all of the turns has to be considered. If the nearest of the turns of wire is at a distance equal to three diameters, all the other turns of wire will be at greater distances, and, therefore, if we may neglect such small quantities as 2 per cent. of the whole amount, we may neglect their action also, for it will be still smaller in amount. Now, for the purpose of arriving at the action of a whole tube of coil, I will adopt a method of plotting devised by Mr. Sayers. Suppose we had a whole tube coiled with copper wire from end to end, its action would be practically the same as though the copper wire were gathered together in small numbers at distant intervals. If, for example, I count the number of turns in a centimetre length of the actual tubular coil which I used in my first experiment, I find there are four. Now if, instead of having four wires distributed over the centimetre, I had one stout wire in the middle of that space to carry four times the current, the general effect would be the same. This diagram (Fig. 108) is calculated out on the supposition that the effect will be not greatly different if the wires were aggregated in that way, and it is easier to calculate. If, beginning at the end of the tube marked A, we take the wires over the first centimetre of length and aggregate them, we can draw a curve, marked 1, for the effect of that lot of wires. For the next lot, we could draw a similar curve, but instead of drawing it on the horizontal line we will add the several heights of the second curve on to those of the first, and that gives the curve marked 2 ; for the third part, add the ordinates of another similar curve, and so gradually build up a final curve for the total action of this tubular coil on a unit pole at different points along the axis. This resultant curve begins about 2½ diameters away from the end, rises

FIG. 108.

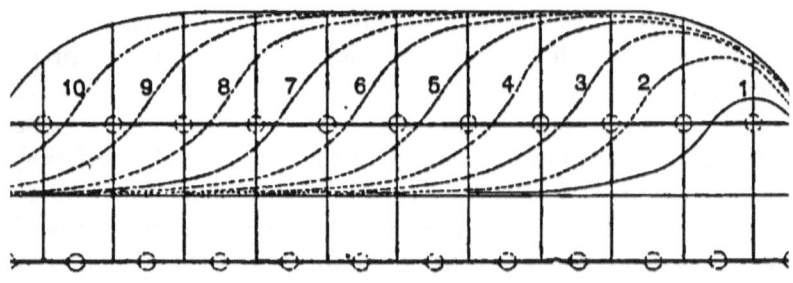

ACTION OF TUBULAR COIL.

gently, and then suddenly, and then turns over and becomes nearly flat with a long level back. It does not rise any more after a point about 2½ diameters along from A ; the curve at that point becomes practically flat, or does not vary more than about 1 per cent., however long the tube may be. For example, in a tubular coil 1 inch in diameter and 20 inches long, there will be a uniform magnetic field for about 15 inches along the middle of the coil. In a tubular coil 3 cm. in diameter and 40 cm. long, there will be a uniform magnetic field for about 32 cm. along the middle of the coil. The meaning of this is that the value of the magnetic forces down the axis of that coil begins outside the mouth of the tube, increases, rises to a certain maximum amount a little within the mouth of the tube, and after that is perfectly constant nearly all the way along the tube, and then falls off symmetrically as you get to the other end. The ordinates drawn to the curve represent the forces at corresponding points along the axis of the tube, and may be taken to represent not simply the magnetizing force, but the pull on a magnetic pole at the end of an indefinitely long, thin steel magnet of fixed strength.

The rule for calculating the intensity of the magnetic force at any point on the axis of the long tubular coil within this region where the force is uniform, is :—

$$H = \frac{4}{10} \pi \times \text{the ampere-turns per cm. of length.}$$

And, as the total magnetizing power of a tubular coil is proportional not only to the intensity of the magnetic force at any point, but also to the length, the integral magnetizing effect on a piece of iron that is inserted into the coil may be taken as practically equal to $\frac{4}{10} \pi \times$ the total number of ampere-turns in that portion of the tubular coil which surrounds the iron. If the iron protrudes as much as three diameters at both ends, the total magnetizing force is simply $\frac{4}{10} \pi \times$ the whole number of ampere-turns.

Now that case is of course not the one we are usually dealing with. We cannot procure steel magnets with unalterable poles of fixed strength. Even the hardest steel magnet, magnetized so as to give us a permanent pole near or at the end of it—quite close up to the end of it when you put it into a magnetizing coil—becomes by that fact further magnetized. Its pole becomes strengthened as it is drawn in, so that the case of an unalterable pole is not one which can actually be realized. One does not usually work with steel; one works with soft iron plungers which are not magnetized at all when at a distance away, but become magnetized in the act of being placed at the mouth of the coil, and which become more highly magnetized the further they go in. They tend, indeed, to settle down with the ends protruding equally, for that is the position where they most nearly complete the magnetic circuit; where, therefore, they are most completely and highly magnetized. Accordingly, we have this fact to deal with, that whatever may be the magnetizing forces all along a tube, the magnetism of the entering core will increase as it goes on. We must therefore have recourse to the following procedure. We will construct a curve in which we will plot, not simply the magnetizing forces of the spiral at different points, but the product of the magnetizing forces into the magnetism of the core which itself increases as the core moves in. The curve with a flat top to it corresponds to an ideal case of a single pole of constant strength. We wish to pass from this to a curve which shall represent a real case, with an iron core. Let us still suppose that we are using a very long core, one so long that when the front pole has entered the coil the other end is still a long way off. With an iron core of course it depends on the size and quality of the iron as to how much magnetism you get for a given amount of magnetizing power. When the core has entered up to a certain point, you have the magnetizing forces up to that point acting on it; it acquires a certain amount of magnetism, so that the pull will necessarily go on increasing and increasing, although the intensity of the magnetic force from point to point along the axis of the coil remains the same, until within about two diameters from the far end. Although the magnetic force inside the long spiral remains the same, because the magnetism of the core is increasing, the pull goes on increasing and increasing (if the iron does not get saturated) at an almost uniform rate all the way up, until the piece of iron has been poked pretty nearly through to the distant end. In Fig. 109, a tubular coil, B A, is represented. Suppose a long iron core is placed on the axis to the right, and that its end is gradually brought up towards B. When it arrives at X, the pull becomes sensible, and increases, at the first rapidly as the core enters the mouth of the tube, then gently as the core travels along, attaining a maximum, C, at about the further end, A, of the tube. When it approaches to the other end, A, it comes to the region where the magnetizing force falls off, but the magnetism is still going on increasing, because something is still being added to the total magnetizing power, and these two effects nearly balance one another, so that the pull arrives at the maximum. This is the highest point, C, on the curve; the greatest pull occurring just as the end of the iron core arrives at the bottom or far end of the tubular coil, from which point there is a very rapid falling off. The question of rapidity of descent from that point depends only on how long the core is. If the core is a very long one, so that its other pole is still very far away,

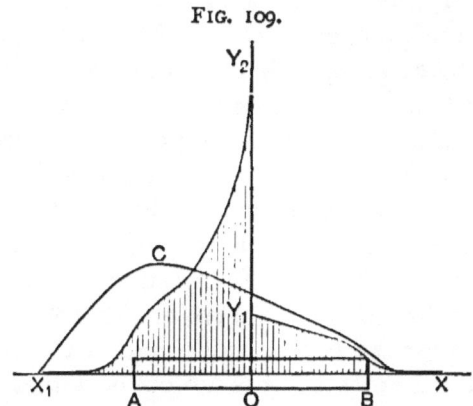

Fig. 109.

DIAGRAM OF FORCE AND WORK OF COIL-AND-PLUNGER.

you have a long, slow descent going on over some three diameters, and gradually vanishing. If, however, the other pole is coming up within measurable distance of B, then the curve will come down more rapidly to a definite point, X. To take a simple case where the iron core is twice as long as the coil, its curve will descend in pretty nearly a straight line down to a point such that the ends of the iron rod stand out equally from the ends of the tube.

Precisely similar effects will occur in all other cases where the plunger is considerably longer than (at least twice as long as) the coil surrounding it. If you take a different case, however, you will get another effect. Take the case of a plunger of the same length as the coil, then this is what necessarily happens. At first the effects are much the same; but as soon as the core has entered about half, or a little more than half its length, you begin to have the action of the other pole that is left protruding outside tending to pull the plunger back; and although the magnetizing force goes on increasing the farther the plunger enters, the repulsion exerted by the coil on the other pole of the plunger keeps increasing still faster as this end nears the mouth of the coil. In that case the maximum will occur at a point a little further than half way along the coil, and from that point the curve will descend and go to zero at A; that is to say, there will be no pull when both ends of the plunger coincide with the two ends of the coil. If you take a plunger that is a little shorter than the coil, then you find that the attraction comes down to zero at an earlier period still. The maximum pull occurs earlier, and so does the reduction of the pull to zero; there being no action at all upon the short core when it lies wholly within that region of the tube within which the intensity of the magnetic force is uniform. That is to say, for any portion of this tube corresponding to the flat top of the curve of Fig. 108, if the plunger of iron is so short as to lie wholly within that region, then there is no action upon it; it is not pulled either way. Now these things can be not only predicted by the help of such a law as that, but verified by experiment. There is a set of tubular coils which is used at the Finsbury Technical College for the purpose of verifying these laws. One of the coils is about 9 inches long, one about half that length, another just a quarter. They are all made alike in this way, that they have exactly the same weight of copper wire, cut from the same hank, upon them. There are, of course, more turns on the long one than on the shorter, because with the shorter ones each turn requires on the average a larger amount of wire, and therefore the same weight of wire will not make the same number of windings. We use that very simple apparatus, a Salter's balance, to measure the pull exerted down to different distances on cores of various lengths. We find that in every case the pull increases and becomes a maximum, then diminishes. We will now make the experiment, taking first a long plunger, roughly about twice as long as the coil. The pull increases as the plunger goes down, and the maximum pull occurs just when the lower end gets to the bottom; beyond that the pull is less. Using the same plunger with these shorter coils, one finds the same thing, in fact more marked, for we have now a core which is more than twice the length of the coil. So we find, taking in all these cases, that the maximum pull occurs, not when the plunger is half way in, as the books say, but when the bottom end of it is just beginning to come out through the bottom of the coil that we are using. If, however, we take a shorter plunger, the result is different. Here is one just the same length as the coil. With this one the maximum pull does occur when the core is about half way in; the maximum pull is just about at the middle. Again, with a very short core—here is one about one-sixth of the length of the coil—the maximum pull occurs as it is going into the mouth of the coil; and when both ends have gone in so far that it gets into the region of equable magnetic field there is no more pull on one end than on the other; one end is trying to move with a certain force down the tube, and the other end is trying to move with exactly equal force up the tube, and the two balance one another. If we carry that to a still more extreme case, and employ a little round ball of iron to explore down the tube, you will find this curious result, that the only place where any pull occurs on the ball is just as it is going in at the mouth. For about half-an-inch in the neck of the coil there is a pull; but there is no pull down the interior of the tube at all, and there is no measurable pull outside.

Now these actions of the coil on the core are capable of being viewed from another standpoint. Every engineer knows that the work done by a force has to be measured by multiplying together the force and the distance through which its point of application moves forward. Here we have a varying force acting over a certain range. We ought, therefore, to take the amount of the force at each point, and multiply that by the adjacent little bit of range, averaging the force over that range, and then take the next value of force with the next little bit of range, and so consider in small portions the work done along the whole length of travel. If we call the length of travel x, the element of length must be called dx. Multiply that by f, the force. The force multiplied by the element of length gives us the work, dw, done in that short range. Now the whole work over the whole travel is made up of the sum of such elements all added together; that is to say, we have to take all the various values of f, multiply each by its own short range dx, and the sum of all those, writing $\int$ for the sum, would be equal to the sum of all the work; that is to say, the whole work done in putting the thing together will be written :—

$$w = \int f\, dx.$$

Now what I want you to think about is this : here, say, is a coil, and there is a distant core. Though there is a current in the coil, it is so far away from the core that practically there is no action; bring them nearer and nearer together; presently they begin to act on one another, there is a pull, which increases as the core enters, then comes to a maximum, then dies away as the end of the core begins to protrude at the other side. There is no further pull at all when the two ends stand out equally.

Now there has been a certain total amount of work done by this apparatus. Every engineer knows that if we can ascertain the force at every point along the line of travel, the work done in that travel is readily expressed by the area of the force-curve. Think of the curve $X C X_1$, in Fig. 109, p. 251, the ordinates of which represent the forces. The whole area underneath this curve represents the work done by the system, and therefore represents equally the work you would have to do upon it in pulling the system apart. The area under the curve represents the total work done in attracting in the iron plunger, with a pull distributed over the range $X X_1$.

Now I want you to compare that with the case of an electromagnet, where, instead of having this distributed pull, you have a much stronger pull over a much shorter range. I have endeavoured to contrast the two in the other curves drawn in Fig. 109. Suppose we have our coil, and suppose the core, instead of being made of one rod such as this, were made in two parts, so that they could be put together with a screw in the middle, or fastened together in any other mechanical way.

Now first treat this rod as a single plunger, screw the two parts together, and begin with the operation of allowing it to enter into the coil, the work done will be the area under the curve which we have already considered. Let us divide the iron core into two. First of all put in one end of it; it will be attracted up in a precisely similar fashion, only, being a shorter bar, the maximum would be a little displaced. Let it be drawn in up to half-way only; we have now a tube half filled with iron, and in doing so we shall have had a certain amount of work done by the apparatus. As the piece of iron is shorter, the force-curve, which ascends from X to Y_1, will lie little lower than the curve $X C X_1$; but the area under that lower curve, which stops half way, will be the work done by the attraction of this half core. Now go to the other end and put in the other half of the iron. You now have not only the attraction of the tube, but that of the piece which is already in place, acting like an electromagnet. Beginning with a gentle attraction, it soon runs up, and draws the force-curve to a tremendously steep peak, becoming a very great force when the distance asunder is very small.

We have therefore in this case a totally different curve made up of two parts, a part for the putting in of the first half of the core, and a steeper part for the second; but the net result is, we have the same quantity of iron magnetized in exactly the same manner by the same quantity of electric current running round the same amount of copper wire—that is to say, the total amount of work done in these two cases is necessarily equal. Whether you allow the entire plunger to come in by a gentle pull over a long range, or whether you put the core in in two pieces—one part with a gentle pull, and the other with a sudden spring up at the end—the total work must be the same; that is to say, the total area under our two new curves must the same as the area under the old curve. The advantage, then, of this coil-and-plunger method of employing iron and copper is, not that it gets any more work out of the same expenditure of energy, but that it distributes the pull over a considerable range. It does not, however, equalise it altogether over the range of travel.

Horse-shoe Plungers.

In 1846 Guillemin suggested the use of a double plunger consisting of a horse-shoe shaped core, the two limbs of which were sucked up into two tubular coils. Devices of this kind are found in several arc lamps.

Experimental Data concerning Coil-and-Plunger.

A number of experimental researches have been made from time to time to elucidate the working of the coil-and-plunger. Hankel, in 1850, examined the relation between the pull in a given portion of the plunger and the exciting power. He found that, so long as the iron core was so thick and the exciting power so small that magnetization of the iron never approached saturation, the pull was proportional to the square of the current, and was also proportional to the square of the number of turns of wire. Putting these two facts together we get the rule—which is true only for an unsaturated core in a given position—that the pull is proportional to the square of the ampere-turns.

This might have been expected, for the magnetism of the iron core will, under the assumptions made above, be proportional to the ampere-turns, and the intensity of the magnetic field in which it is placed being also proportional to the ampere-turns, the pull, which is the product of the magnetism and of the intensity of the field, ought to be proportional to the square of the ampere-turns.

Dub, who examined cores of different thicknesses, found the attraction to vary as the square root of the diameter of the core. His own experiments show that this is inexact, and that the force is quite as nearly proportional to the diameter as to its square root. There is again reason for this. The magnetic circuit consists largely of air-paths by which the magnetic lines flow from one end to the other. As the main part of the magnetic reluctance of the circuit is that of the air, anything which reduces the air reluctance increases the magnetization, and, consequently, the pull. Now, in this case, the reluctance of the air-paths is mainly governed by the surface exposed by the end portions of the iron core. Increasing these diminishes the reluctance, and increases the magnetization by a corresponding amount.

Von Waltenhofen, in 1870, compared the attraction exerted by two equal (short) tubular coils on two iron cores, one of which was a solid cylindrical rod, and the other a tube of equal length and weight, and found the tube to be more powerfully attracted. Doubtless, the effect of the increased surface in diminishing the reluctance of the magnetic circuit explains the cause of the observation.

Von Feilitzsch compared the action of a tubular coil upon a plunger of soft iron with that exerted by the same coil upon a core of hard and magnetized steel of equal dimensions. The plungers (Fig. 110) were each 10·1 cm. long; the coil being 29·5 cm. in length, and 4·2 in diameter. The current, however, which was applied in the experiment with the steel magnet as plunger, was much greater than that used in the case of the soft iron core, which nevertheless gave the more powerful pull. The steel magnet showed a maximum attraction when it had plunged to a depth of 5 cm., whilst the iron core had its maximum at a depth of 7 cm., doubtless because its own magnetization went on increasing more than did that of the steel core. As the uniform field region began at a depth of about 8 cm., and the cores were 10·1 cm. in length, one would expect the attracting force to come to zero when

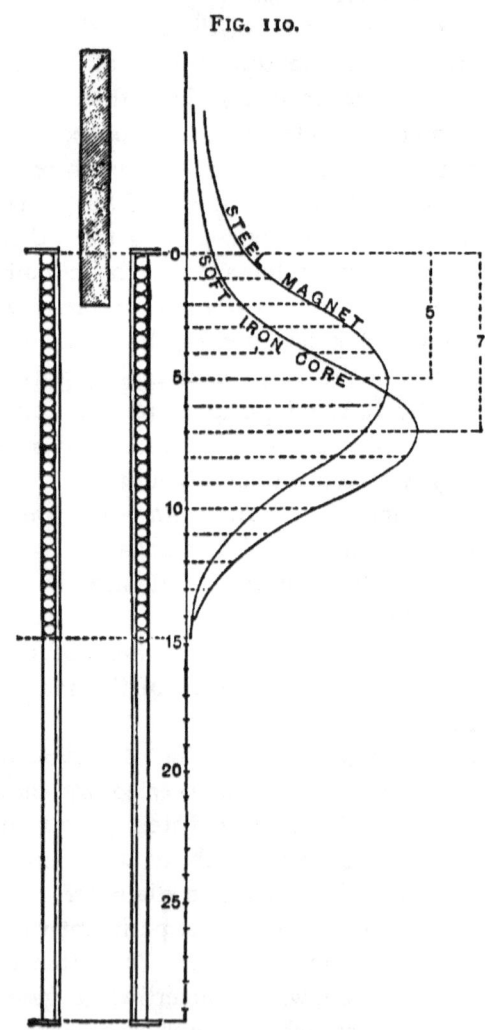

Fig. 110.

Von Feilitzsch's Experiment on Plungers of Iron and Steel.

the cores had plunged in to a depth of about 18 cm. As a matter of fact, the zero point was reached a little earlier. It will be noticed that the pull at the maximum was a little greater in the case of the iron plunger.

The most careful researches of late years are those made by Dr. Theodore Bruger, in 1886. One of his researches, in which a cylindrical iron plunger was used, is represented by two of the curves in Fig. 111. He used two coils, one $3\frac{1}{2}$ cm. long, the other 7 cm. long. These are indicated in the

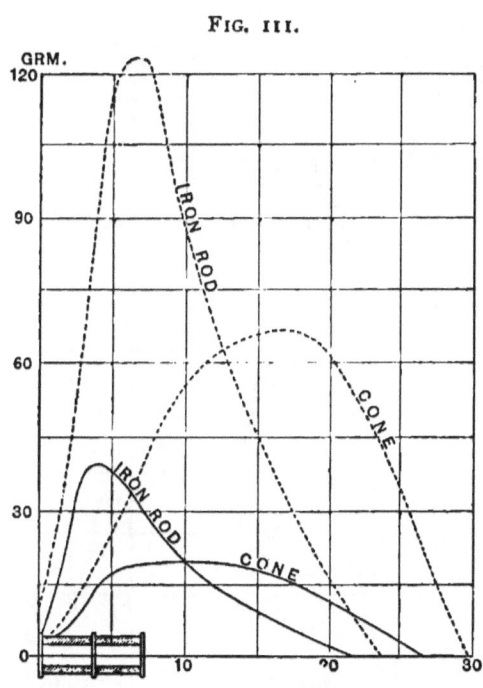

BRUGER'S EXPERIMENTS ON COILS AND PLUNGERS.

bottom left-hand corner. The exciting current was a little over 8 amperes. The cylindrical plunger was 39 cm. long. The plunger is supposed, in the diagram, to enter on the left, and the number of grammes of pull is plotted out opposite the position of the entering end of the plunger. As the two curves show by their steep peaks, the maximum pull occurs just when the end of the plunger begins to emerge through the coil; and the pull comes down to zero when the ends of the core protrude equally. In this figure the dotted curves relate to the use of the longer of the two coils. The height of the peak, with the coil of double length, is nearly four times as great, there being double ampere-turns of excitation. In some other experiments, which are plotted in Fig. 112, the same core was used with a tubular coil 13 cm. long. Using

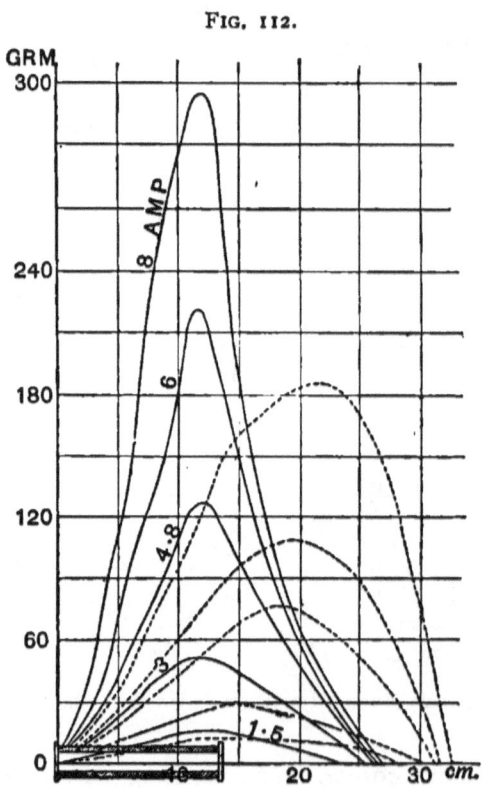

FIG. 112.

BRUGER'S EXPERIMENTS, USING CURRENTS OF VARIOUS STRENGTHS.

currents of various strengths, 1·5 ampere, 3, 4·8, 6, or 8 amperes, the pull is of course different, but broadly you get the same effect, that the maximum pull occurs just where the pole begins to come out at the far end of the tubular coil.

There are slight differences; with the smallest amount of current the maximum is exactly over the end of the tube, but with currents rather larger, the maximum point comes a little farther back. When the core gets well saturated, the force-curve does not go on rising so far; it begins to turn over at an earlier stage, and the maximum place is necessarily displaced a little way back from the end of the tube. That was also observed by Von Waltenhofen when using the steel magnet.

EFFECT OF USING CONED PLUNGERS.

But now, if instead of employing a cylindrical core you employ one that is pointed, you find this completely alters the position of the maximum pull, for now the point is insufficient to carry the whole of the magnetic lines which are formed in the iron rod. They do not come out at the point, but filter through, so to speak, along the sides of the core. The region where the magnetic lines come up through the iron into the air is no longer a definite "pole" at or near the end of the rod, but is distributed over a considerable surface; consequently when the point begins to poke its nose out, you still have a larger portion of iron up the tube, and the pull, instead of coming to a maximum at that position, is distributed over a wider range.

I am now making the experiment roughly with my spring balance and a conical plunger, and I think you will be able to notice a marked difference between this case and that of the cylindrical plunger. The pull increases as the plunger enters, but the maximum is not so well defined with a pointed core as it is with one that is flat-ended. This essential difference between coned plungers and cylindrical ones was discovered by an engineer of the name of Krizik, who applied his discovery in the mechanism of the Pilsen arc lamps. Coned plungers were also examined by Bruger. In Fig. 112 are given the curves that correspond to the use of a coned iron core, as well as those corresponding to the use of the cylindrical iron rod. You will notice that, as compared with the cylindrical plunger, the coned core never gave so big a pull, and the maximum occurred not as the tip emerged, but when it got a very considerable way out on the other side. So it is with both the shorter and the longer coil. The dotted curves in Fig. 112 represent the behaviour of a coned plunger. With the longer coil represented, and with various currents, the maximum pull occurred when the tip had come a considerable way out; and the position of the maximum pull, instead of being brought nearer to the entering end with a high magnetizing current, was actually caused to occur farther down; the range of action became extended with large currents as compared with small ones. Bruger also investigated the case of cores of very irregular shapes, resembling, for example, the shank of a screw-driver, and found a very curious and irregular force-curve. There is a good deal more yet to be done, I fancy, in examining this question of distributing the pull on an attracted core by altering the shape of it; but Bruger has shown us the way, and we ought not to find very much difficulty in following him.

Solid and Hollow Plungers.

It is often supposed that hollow plungers are as good as solid cores. The reason is that when one is dealing with weak magnetizing forces, which do not nearly saturate the iron core, the greater part of the magnetizing force is spent in driving the magnetic lines through the return paths through the air (compare p. 179), and consequently the pull on the core depends much more on the reluctance offered in the air part of the magnetic circuit than on that of the iron part. Hence, for weak magnetizing forces, hollow plungers act just as well as solid ones of the same external diameter and length. But for strong magnetizing forces this is no longer so, because the tubular core becomes practically saturated at an earlier stage than the solid one. In Fig. 113 the behaviour of a tubular core is contrasted with that of a solid core (No. 2) of the same external diameter. In the same figure is given a curve for solid core (No. 1) of smaller size, but slightly greater weight than the tubular one. With small magnetizing forces this core is less strongly magnetized than the tubular one, but with large magnetizing forces it is more strongly magnetized.

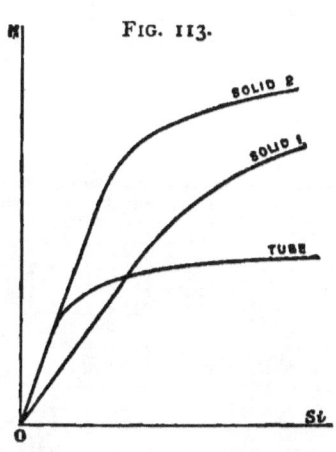

FIG. 113.

HOLLOW *versus* SOLID CORES.

OTHER MODES OF EXTENDING RANGE OF ACTION.

Another way of altering the distribution of the pull is to alter the distribution of the wire on the coil. Instead of having a coned core use a coned coil, the winding being heaped up thicker at one end than at the other. Such a coil, wound in steps of increasing thickness, has been used for some years by Gaiffe, in his arc lamp; it has also been patented in Germany by Leupold. M. Trève has made the suggestion to employ an iron wire coil, so as to utilise the magnetism of the iron that is carrying the current. Trève declares that such coils for an equal current possess four times the pulling power. I doubt whether that is so; but even if it were, we must remember that to drive any given current through an iron wire instead of a copper wire of the same bulk, implies that we must force the current through six times the resistance; and, therefore, we shall have to employ six times the horse-power to drive the same current through the iron wire coil, so that there is really no gain. Again, a suggestion has been made to inclose in an iron jacket the coil employed in this way. Iron-clad solenoids have been employed from time to time, but they do not increase the range of action; what they do is to tend to prevent the falling off of the internal pull at the region within the mouth of the coil. It equalises the internal pull at the expense of all external action. An iron-clad solenoid has practically no attraction at all on anything outside of it, not even on an iron core placed at a distance of half a diameter of the aperture; it is only when the core is inside the tube that the attraction begins, and the magnetizing power is practically uniform from end to end. In 1889 I wished to make use of this property for some experiments on the action of magnetism on light, and for that purpose I had built, by Messrs. Paterson and Cooper, this powerful coil, which is provided with a tubular iron jacket outside, and with a thick iron disk perforated by a central hole covering each end. The magnetic circuit around the exterior of the coil is practically completed with soft iron. With this coil, one may take it, there is an absolutely uniform magnetic field from one end of the tube to the other; not falling off at the ends as it would do if the magnetic circuit had simply an air return. The whole of the ampere-turns of exciting power are employed in magnetizing the central space, in which therefore the actions are very powerful and uniform. The coil and its uses were described in the author's lecture, at the Royal Institution, on "Optical Torque," in 1889.

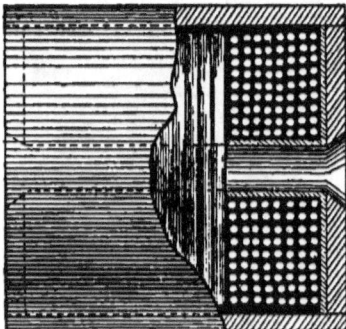

FIG. 114.

IRON-CLAD COIL.

MODIFICATIONS OF THE COIL-AND-PLUNGER.

In one variety of the coil-and-plunger mechanism a second coil is wound on the plunger. In such a case the magnetism of the core is partly due to the current in the coil around it, partly due to the current in the outer coil. It will tend to move into such a position as will make its magnetization a maximum. Hjörth used this modification, and the same thing has been employed in several arc lamps. There is a series of drawings upon this wall depicting the mechanism of about a dozen different forms of arc lamp, all made by Messrs. Paterson and Cooper. In one of these there is a plunger with a coil on it drawn into a tubular coil, the current flowing successively through both coils. In another there are two separate coils in separate circuits, one of thin wire and one of thick, one being connected in series with the arc, and one in shunt.

FIG. 115.

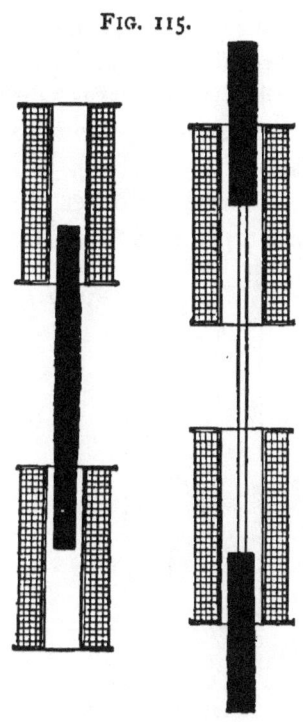

TWO FORMS OF DIFFERENTIAL COIL-AND-PLUNGER.

DIFFERENTIAL COIL-AND-PLUNGER.

There is a drawing here, Fig. 115, showing the arrangement, which was originally introduced by Siemens, wherein a plunger is drawn at one end into the coil that is in the main circuit, and at the other end into a coil that is in shunt. Here, obviously the magnetism of the plunger will depend upon the currents flowing around both of the coils, and will be different according to whether the currents both circulate in the same sense around the iron core, or not. It is also obvious that where one core plunges its opposite ends into two coils, the magnetization will depend on both coils, and the resultant pull will not be simply the difference between the pull of the two coils acting each separately. Fig. 115 shows also another modification of the same arrangement, designed to give greater independence of action to the two coils; for in this case the magnetization of either plunger is practically dependent only upon the current in its own coil. The two plungers are connected mechanically by a rod of brass or other non-magnetic metal. This latter disposition is usually adopted in preference to the former in all cases where alternating currents are used; for with such currents a magnetic-plunger extending from one coil to the other would act like the core of an induction coil, causing the current in one coil to set up induction currents in the other coil. In all cases where alternating currents are used, the cores must be laminated, a split tube being the usual form adopted.

There are other modes of securing differential action. For example, in some forms of the Pilsen arc lamp the (coned)

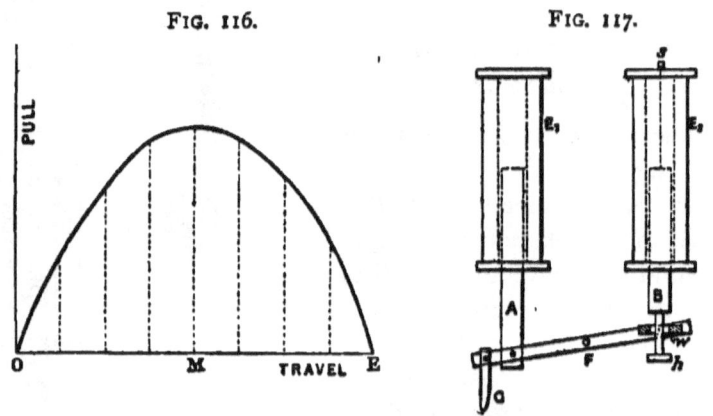

CURVE OF FORCES IN MECHANISM OF PILSEN ARC LAMP.

DIFFERENTIAL PLUNGERS OF THE BROCKIE-PELL ARC LAMP.

iron plungers are connected together by an overhead cord passing over a pulley. In these lamps the coned cores are so shaped, and hung upon a cord of such a length, as to secure that the curve connecting pull and travel (Fig. 116) shall be symmetrical on the two sides of the position of maximum pull.

Fig. 117 depicts a fourth kind of differential arrangement, used in the Brockie-Pell and other arc lamps, in which there are two separate plungers attached to the two ends of a *see-saw* lever. In this case the two magnetizing actions are separate. A is one core, entering into the bobbin E_1, wound with thick wire to carry the main current; while B enters the bobbin E_2, which is wound with fine wire and connected as a shunt.

In a fifth kind of differential arrangement there is but one plunger and one tubular bobbin, upon which are wound the two coils, differentially, so that the action on the plunger is simply due to the difference between the ampere-turns circulating in the two separate wires. This is illustrated by the mechanism of Menges' arc lamp depicted in Fig. 118.

FIG. 118.

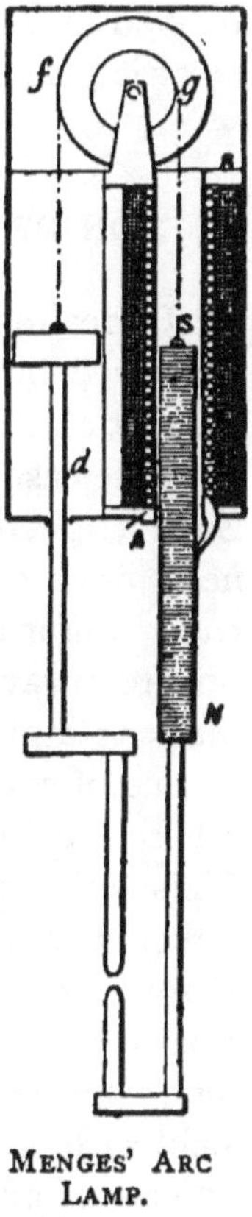

COIL AND COIL-PLUNGER.

When one abandons iron altogether, and merely uses two tubular coils, one of wide diameter, and another of narrower diameter capable of entering into the former, and passes electric currents through both of them, if the currents are circulating in the same fashion through both of them, they will be drawn toward one another. This arrangement has also been used in arc lamps. If the currents circulate in opposite senses in the two coils, then they will tend to push one another apart. In either case the mutual force, in any given position of the coils, will be proportional to the product of their respective ampere-turns.

SECTIONED COIL.

Sectioned Coils, with Plunger.—An important suggestion was made by Page, about 1850, when he designed a form of coil-and-plunger having a travel of indefinitely long range. The coiled tube, instead of consisting merely of one coil excited simultaneously throughout its whole length by the current, was constructed in a number of separate sections or short

MENGES' ARC LAMP.

tubes, associated together end to end, and furnished with means for turning on the electric current into any of the sections separately. Suppose an iron core to be just entering into any section, the current is turned on in that section, and as the end of the core passes through it, the current is then turned on in the section next ahead. In this way an attraction may be kept up along a tube of indefinite length. Page constructed an electric motor on this plan, which was later revived by Du Moncel, and again by Marcel Deprez in his electric "hammer." (Fig. 200, p. 360.)

ACTION OF MAGNETIC FIELD ON SMALL IRON SPHERE.

In dealing with the action of tubular coils upon iron cores, I showed how, when a very short core is placed in a uniform magnetic field, it is not drawn in either direction. The most extreme case is where a small sphere of soft iron is employed. Such a sphere, if placed in even the most powerful magnetic field, does not tend to move in any direction if the field is truly uniform. If the field is not uniform, then the iron sphere always tends to move from the place where the field is weak to a place where the field is stronger. A ball of bismuth or one of copper, tends, on the contrary, to move from a place where the field is strong to a place where the field is weaker. This is the explanation of the actions called "diamagnetic," which were at one time erroneously attributed to a supposed diamagnetic polarity opposite in kind to the ordinary magnetic polarity. A simple way of stating the facts is to say that a small sphere of iron tends to move up the *slope* of a magnetic field with a force proportional to that slope; whilst (in air) a sphere of bismuth or of copper tends, with a feeble force, to move down that slope. Any small piece of soft iron—a short cylinder, for example—shows the same kind of behaviour as a small sphere. In some of Ayrton and Perry's coiled-ribbon amperemeters and voltmeters, and in some of Sir William Thomson's current-meters, this principle is applied.

LONG RANGE ELECTROMAGNETS. INTERMEDIATE FORMS.

The coil-and-plunger apparatus considered in the preceding chapter does not, for a given weight of copper and iron, exert at any part of the extended range of motion a force so great as can be obtained over a very short range when the fixed-core forms of electromagnet are used. Many inventors have therefore sought to devise electromagnets having a range of motion of their armature longer than that of the ordinary bar or horse-shoe form, but without sacrificing altogether the powerful pull afforded by the action of the fixed iron core. For certain purposes it is desirable to construct an electromagnet which, while having the powerful pull of the electromagnet, should have over its limited range of action a more equable pull, resembling in this respect the equalizing of range of the coil-and-plunger. Amongst the various suggestions for long-range electromagnets, the foremost place must be given to those forms which occupy an intermediate place between the coil-and-plunger and the forms with fixed core.

STOPPED-COIL ELECTROMAGNETS.

One form of electromagnets in this class consists of a plunger working into a tubular coil in which a short fixed core is inserted, closing the tube up to a certain distance. Bonelli's form of electromagnet, depicted in Fig. 30, p. 55, comes under this category. The force with which the movable part of the core is sucked into the coil is somewhat greater, owing to the concentrating effect of the fixed part, than would be the case with an open coil; and when the movable part has so entered, it is drawn in by a pull increasing greatly as the gap in the magnetic circuit closes up. The reader should refer back to a passage earlier in this Chapter, p. 255, bearing on this matter.

Another example is afforded by the plunger electromagnet employed in the Brush arc lamps. A couple of tubular coils receive each an iron plunger, connected together by a yoke; whilst above, the magnetic circuit is partially completed by the sheet of iron which forms part of the inclosing box. You have here, also, the advantage of a fairly complete magnetic circuit, together with a comparatively long travel of the plunger and coil. It is a fair compromise

FIG. 119.   FIG. 120.

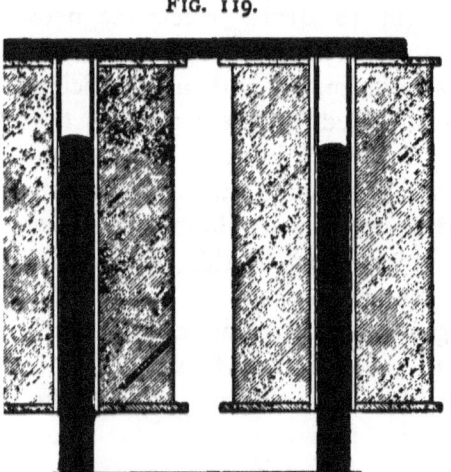
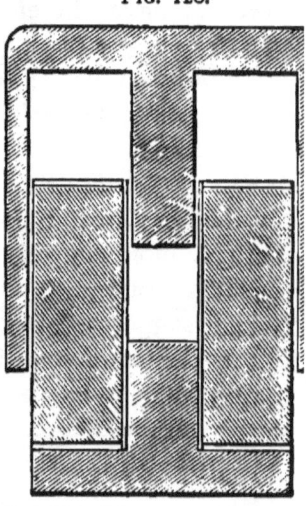

ELECTROMAGNET OF BRUSH ARC LAMP.   PLUNGER ELECTROMAGNET OF STEVENS AND HARDY.

between the two ways of working. The pull is not, however, in any of these forms, equal all along the whole range of travel; it increases as the magnetic circuit becomes more complete.

Fig. 120 represents a peculiar form of electromagnet combining some of the features of the iron-clad electromagnet with those of the movable plunger. It has a limited range of action, but is of great power over that range, owing to its excellent magnetic circuit. It was invented, in 1870, by Stevens and Hardy, for use in an electric motor for running sewing machines. A very similar form is used in Weston's arc lamp.

Long-range Electromagnets. 271

In Rankin Kennedy's arc lamp, the mechanism of which is shown in Fig. 121, a similar form of electromagnet is used. Here, however, the action of this electromagnet M, which is in the main circuit of the lamp, and is required to pull with

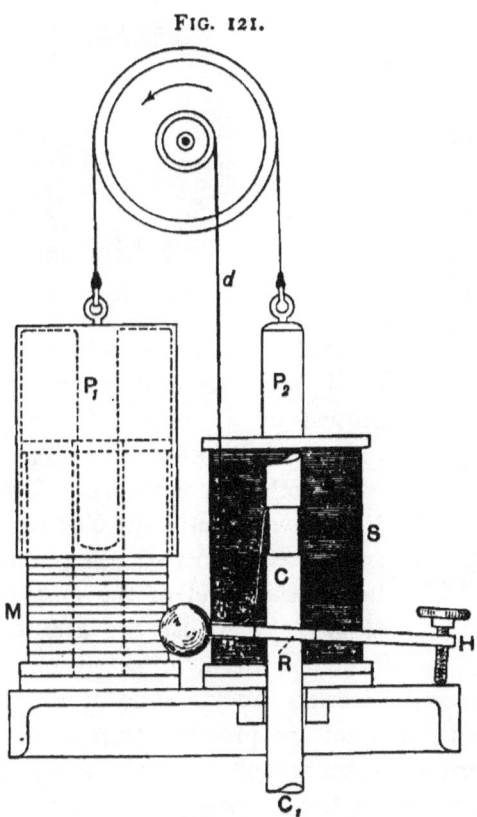

FIG. 121.

MECHANISM OF KENNEDY'S ARC LAMP.

considerable power and certainty in order to strike the arc, is partly balanced against the action of a shunt-wound coil S, which acts upon a second plunger P_2 connected by cord and pulley with the plunger P_1 of the main-circuit electromagnet The action therefore becomes a differential one from the moment when the arc has been struck.

A form of plunger electromagnet invented by Holroyd

Smith in 1877, resembles Fig. 120 inverted; the coil being surrounded by an iron jacket, whilst a plunger, furnished at the top with an iron disk, descends down the central tube to

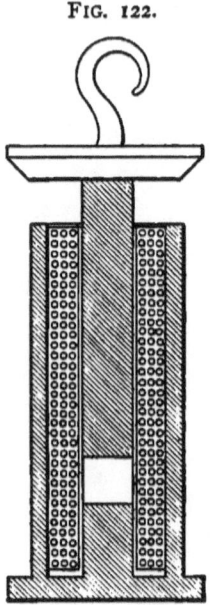

FIG. 122.

HOLROYD SMITH'S ELECTROMAGNET.

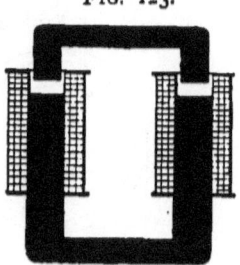

FIG. 123.

ROLOFF'S ELECTROMAGNET.

meet the iron at the bottom. It is shown in Fig. 122. A closely similar form is employed in Timmis and Currie's form of railway signal electromagnet.

Another intermediate form is due to Roloff, who made his electromagnets with iron cores not standing out, but sunk below the level of the ends of the coils (Fig. 123), whilst the armature was furnished with little extensions that passed down into these projecting tubular ends of the coils. Some arc lamps have magnets of precisely that form, with a short plunger entering a tubular coil, and met half-way down by a short fixed core inside the tube.

Fig. 124 represents a form of tubular ironclad electromagnet that deserves a little more attention, being the one used by Messrs. Ayrton and Perry in 1882; a coil has an iron jacket round it, and also an annular iron disk across the top, and an annular iron disk across the bottom, there being also a short internal tube of iron extending a little way down from the top, almost meeting another short internal tube of iron coming up from the bottom. The magnetic effect of the inclosed copper coil is concentrated within an extremely short space between the ends of the internal tubes, where there is a wonderfully strong uniform field. The range of action you can alter just as you please in the construction, by shortening

FIG. 124.

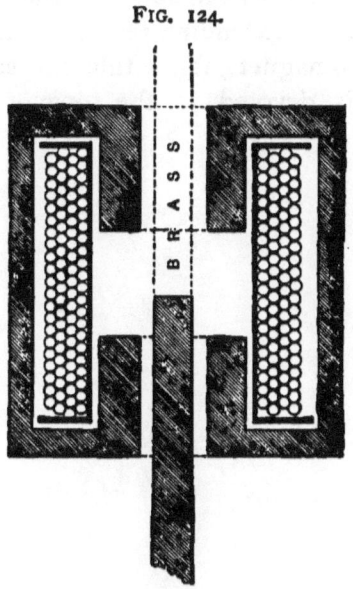

AYRTON AND PERRY'S TUBULAR IRON-CLAD ELECTROMAGNET.

or lengthening the internal tubes. An iron rod inserted below is drawn with great power and equality of pull over the range from one end to the other of these internal tubes.

FIG. 125.

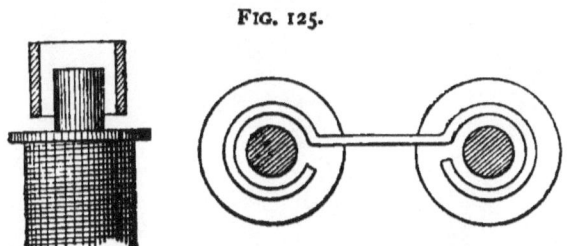

GAISER'S LONG-RANGE ARMATURE.

The reader should contrast with the last two devices, and with Fig. 133, the suggestion of Gaiser, Fig. 125, for obtaining

T

greater range of action from an ordinary electromagnet with cylindrical cores. The bobbin is made short so as not to cover the protruding end of the core. The armature consists of a strip of sheet iron bent to surround the poles.

Other devices for extending the range, and for equalizing the pull of electromagnets, fall within the scope of the next chapter, which is devoted to the topic of electromagnetic mechanism.

CHAPTER IX.

ELECTROMAGNETIC MECHANISM.

IN many ways the electromagnet lends itself to mechanical adaptations and purposes of such a special kind that its use opens out a new department in the general science of mechanism. In this new department there is great need of some classification of a systematic kind. It is half a century since Willis, at Cambridge, reduced to something like a system the subject of mechanism in general. The work which he began has now for twenty years been reduced to a well-nigh perfect system, by Reuleaux, in his great treatise on the Kinematics of Machinery. As yet it has not been found possible to reduce to systems of electro-kinematic pairs, all forms of electromagnetic mechanism. Nevertheless some sort of classification, however crude, will be helpful. It will, therefore, be convenient to preface this chapter by a simple categorical enumeration of all the various kinds of electromagnetic mechanism that are known.

ENUMERATION OF ELECTROMAGNETIC MECHANISMS.

I. *Electromagnets—*
    A. Fixed coil and core; moving armature.
        1. Short compact magnetic circuit, for traction in contact.
        2. Elongated circuit and heavier coil, for attraction at distance.
        3. Special forms: Ironclad, club-footed, laminated, consequent, multipolar, &c.
    B. Fixed tubular coil and moving plunger.
        1. Plunger longer than coil.

2. Plunger shorter than coil.
3. Special forms of plunger; coned, laminated, &c.
4. Coil constructed in sections for successive action.

C. Intermediate forms; stopped coil, &c.
D. Fixed tubular coil and movable coil.

II. *Electromagnets with Opposing Forces*—
1. Counterpoise weights.
2. Counterpoise springs.
3. Counterpoise magnets.

III. *Equalizers for Electromagnets*—
1. Electrical equalizers.
2. Set-up springs.
3. Rocking levers.
4. Linkage movements.
5. Cam equalizers with shaped polar faces.

IV. *Electromagnetic Cams.* Devices depending on the lateral approach of a shaped polar surface.

V. *Electromagnetic Linkages.* Devices depending upon the mutual interaction of two or more separate electromagnets.

VI. *Apparatus depending on Electromagnetic Repulsion*—
1. Mutual repulsion of parallel cores.
2. Elongation of jointed or tubular cores.

VII. *Polarized Electromagnetic Devices*—
1. Electromagnet with parallel polarized armature.
2. Electromagnet with transversed polarized armature.
3. Coil with polarised plunger.
4. Polarised electromagnet with counterpoise spring (Hughes's magnet).
5. Fixed permanent magnet with movable coil.

VIII. *Electromagnetic Vibrators*—
A. Non-polarized.
1. Break-circuit.
2. Short-circuit.
3. Differentially-wound.

B. Polarized.
  1. Single-acting.
  2. Moving part polarized.
  3. Fixed part polarized.

IX. *Rotatory Electromagnetic Devices—*
  1. Coil fixed, needle moving.
  2. Magnet fixed, coil moving.
  3. Coil fixed, coil moving.
  4. Electromagnet with obliquely pivoted armature.
  5. Wire rotating round magnet pole.
  6. Copper disk rotating between magnet pole.
  7. Pole rotating round conducting wire.
  8. Magnet rotating on itself when carrying current.
  9. Curved tubular coil and S-shaped plunger.
  10. Oblique approach of armature.
    a. Wheatstone's cam armature.
    b. Froment's armature constrained around centre.

X. *Electromagnetic Adherence—*
  1. Magnetic friction gear.
  2. Magnetic brakes for vehicles.
  3. Magnetic clutches.

XI. *Alternate-current Devices—*
  1. Copper conductors repelled from pole.
  2. Shading of pole by copper screen.
  3. Virtual rotation of magnetic field by two currents at angle in different phases.
  4. Virtual travelling of magnetic pole due to choking action of coils.

I. ELECTROMAGNETS IN GENERAL. GUIDING PRINCIPLE.

In all these most varied forms of mechanism there is one leading principle which will be found of great assistance, namely, that a magnet always tends so to act as though it tried to diminish the length of its magnetic circuit so as to make the flux of magnetic lines through the exciting coils a maximum. The magnetic circuit tries to grow more compact. This is the reverse of that which holds good with an electric current. The electric circuit always tries to enlarge itself, so as to enclose as much space as possible, but the magnetic circuit always tries to make itself as compact as possible. Armatures are drawn in as near as can be, to close up the magnetic circuit. Iron plungers are, for the same reason, drawn into their exciting coils.

In the catalogue given above, Section I. is a brief enumeration of the preceding chapters of this book. The remaining sections demand some notice in detail.

II. COUNTERPOISE DEVICES.

If an armature placed below the pole of an electromagnet, and at a little distance away, is not constrained by any other opposing forces than that of its own weight, it is evident that there must be one particular strength of electric current which will just suffice to magnetize the electromagnet sufficiently to overcome the downward pull of gravity, and raise the armature; but when once the armature has begun to rise, its mere approach betters the magnetic circuit and increases the upward attraction, so that it will be pulled right up to the pole. A strong current will perform no more in this case than one that is just strong enough to start the action. If, however, such an armature be controlled, not by the mere force of gravity, but by a spring so designed that, as the armature moves nearer to the poles, the opposing force of the spring shall be greatly increased, then a weak current will only produce a small motion, and a stronger current a larger motion, and there will be a determinate position of the armature corresponding to all different strengths of currents within the working limits of the apparatus. It is clear, then, that a properly designed spring, or combination of springs, may be used to regulate the movements of the armature in more or less strict proportion to the electric current. So far back as 1838, Edward Davy, in one of his telegraphic patents, proposed the use of a counterpoise spring to control the movements of the armature (Fig. 126). It has several times been suggested that the attraction of an electromagnet for its armature may be counterpoised by applying a permanent magnet of steel on the opposite side of the armature in place of a spring. It must be obvious, however, that this arrangement is open (and to a stronger degree) to the same criticism as that passed on the use of counterpoise weights, that on such a device a strong current will perform no more than one that is but just strong enough to start the action.

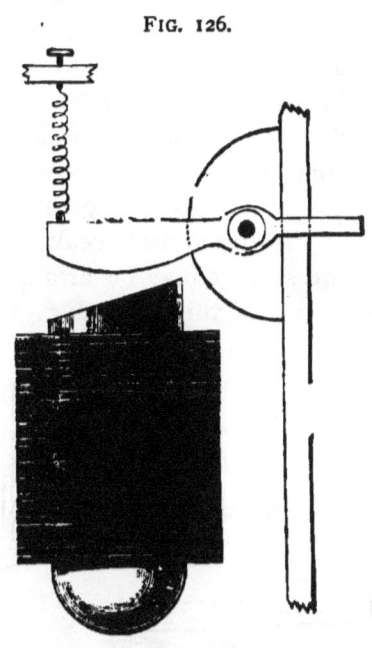

FIG. 126.

E. DAVY'S MODE OF CONTROLLING ARMATURE BY A SPRING.

III. EQUALIZERS.

Various means which have been suggested for extending the range of motion, or of modifying its amount at different parts of the range, so as to equalize the very unequable pull. There are several such devices, some electrical, others purely mechanical, others electro-mechanical. First, there is a purely electrical method. André proposed that as soon as the armature has begun to move nearer, and comes to the place where it is attracted more strongly, it is automatically to make a contact, which will shunt off part of the current and make the magnetism less powerful. Burnett proposed another means; a number of separate electromagnets acting on one armature, but as the latter approached these electromagnets were one after the other cut out of the circuit. The advantages of this method are very hypothetical.

The attraction of an electromagnet increases so greatly as the armature approaches toward contact, that many purely mechanical devices have been suggested to equalize the motion, by bringing to bear opposing forces under such conditions as to exercise great power when the armature is at small distances, and weak power when it is more remote. Set-up springs may be arranged to accomplish this function. A device of this kind, Fig. 127, consisting of a single steel

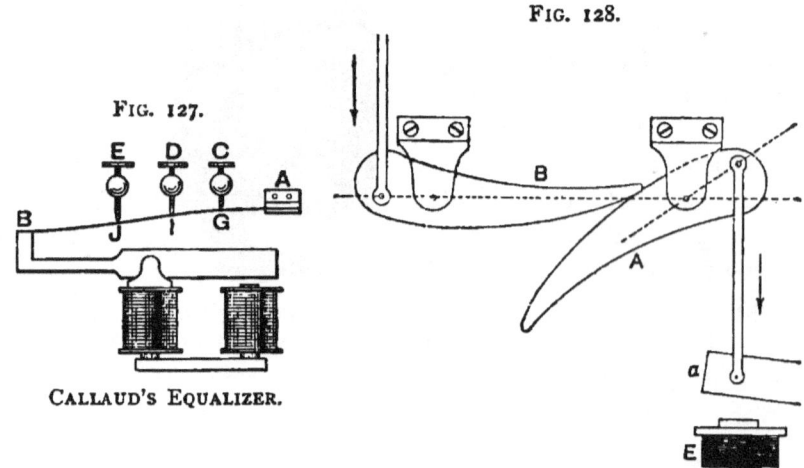

CALLAUD'S EQUALIZER.

ROBERT HOUDIN'S EQUALIZER.

spring with a number of set-screws behind it at different distances, to stiffen it as it approached, was indeed suggested by Callaud, a French engineer.

Another method is to employ, as the famous conjurer Robert Houdin did, a rocking lever. Fig. 128 depicts one of Robert Houdin's equalizers, or *repartiteurs*. The pull of the electromagnet on the armature acts on a curved lever which works against a second one; the point of application of force between the one and the other altering with their position.

When the armature is far away from the pole, the leverage of the first lever on the second is comparatively small. This employment of the *rocking lever* was adopted from Houdin by Duboscq, and put into the Duboscq arc lamp, wherein the regulating mechanism at the bottom of the lamp contains a rocking lever. In this pattern (Fig. 129) one lever, B, which is curved, plays against another, A, which is straight. A

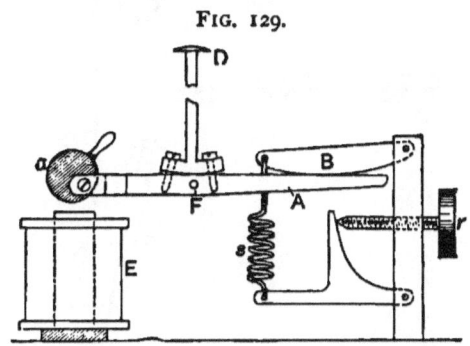

MECHANISM OF DUBOSCQ'S ARC LAMP.

similar mechanism is used for equalizing the action in the Serrin arc lamp, where one of the springs that holds up the jointed parallelogram frame is applied at the end of a rocking lever to equalize the pull of the regulating electromagnet. It is clear that by properly shaping the curve of one or both of the levers, the pull of the magnet can be redistributed in any desired proportions over a range of motion which may be of the same extension as the original motion from end to end, or greater, or less, as may be required. Du Moncel first showed how to calculate such curves, the levers of Robert Houdin having been of empirical form.

Mechanical linkages may be employed to attain, more or less perfectly, the same end. Here, Fig. 130, is a mechanical method of equalizing devised by Froment, and used by M. Roux. You know the Stanhope lever, the object of which is to transform a weak force along a considerable range into a powerful force of short range. Here we use use it backwards. The armature itself, which is attracted with a powerful force of short range, is attached to the lower end of

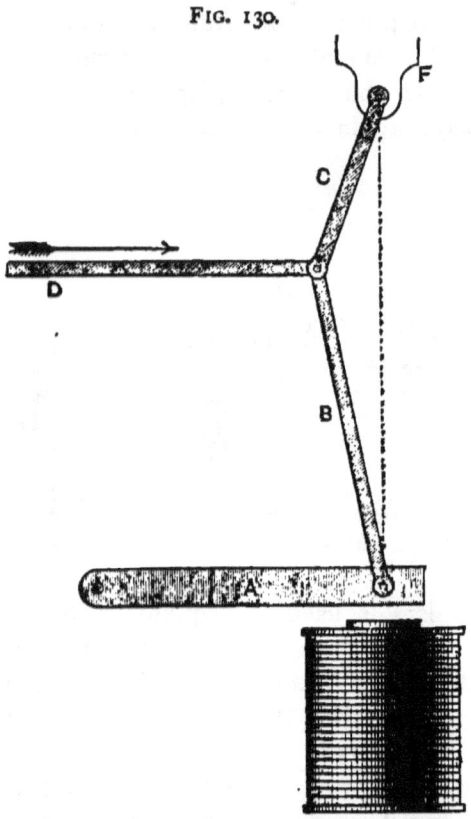

FIG. 130.

FROMENT'S EQUALIZER WITH STANHOPE LEVER.

the Stanhope lever, and the arm attached to the knee of the lever will deliver a distributed force over quite a different range.

Another device, due to Froment, consists in mounting the armature upon a sort of parallel motion which permits it to approach obliquely. The motion of attraction toward the poles is thus converted into a lateral stroke of more extended and more equable power. In the Serrin arc lamp a motion of lateral approach is used, for the armature of the electromagnet is not allowed to travel straight towards the poles of the magnet, but is pulled up obliquely past it.

In the preceding device the magnetic circuit is improved by the approach of the armature toward the poles; and the constraint imposed by the linkage upon the motion of the

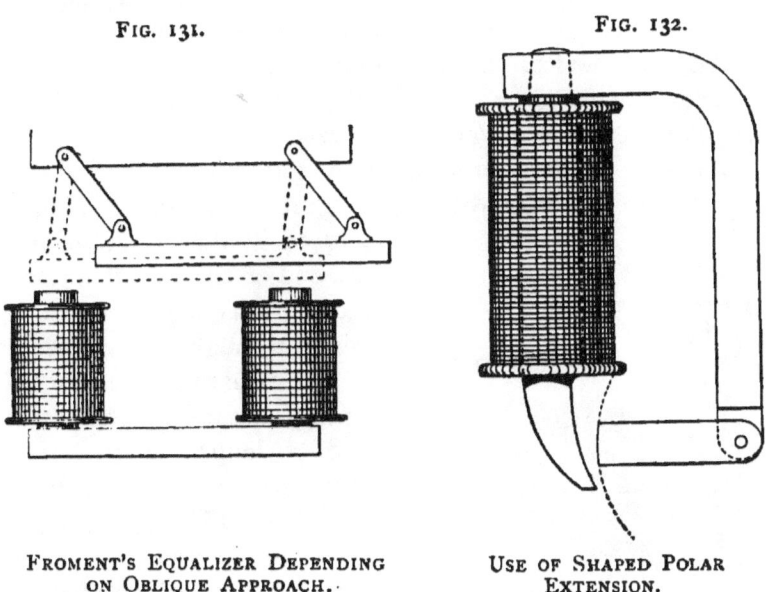

FIG. 131.

FROMENT'S EQUALIZER DEPENDING
ON OBLIQUE APPROACH.

FIG. 132.

USE OF SHAPED POLAR
EXTENSION.

armature compels this improvement of the circuit to take place more gradually than if the motion of approach were direct. The same end may be attained in various ways. It is possible, for example, by the device of shaping away the polar surfaces, or by adding pole-pieces of appropriate shape, to cause the approach of the armature to exert a very slight betterment of the magnetic circuit. One of such devices is depicted in Fig. 132, and consists in pivoting the armature to one pole of an electromagnet, the other pole of which is provided with a curved pole-piece. By varying the contour of the latter, any desired change may be made in the distribution of the pull at various points of the travel. In this particular instance, as the armature moves nearer, as a whole, it increases its distance from the curved pole-piece. An electromagnet resembling this is used in Messrs. Paterson and Cooper's Phœnix arc lamp.

There is another device for oblique approach, made by Froment. In a **V**-shaped gap in the circuit of the magnet a sort of iron wedge is put in, which is not attracted squarely to either face, but comes in laterally between guides.

All these devices may be looked upon as being the magnetic equivalents of well-known mechanical devices based on the principle of the *wedge* and the *cam*.

Another device for equalizing the pull was used by Wheatstone in his step-by-step telegraph in 1839. A hole is pierced in the armature, and the end of the core is formed into a projecting cone, which passes through the aperature of the armature, thereby securing a more equable force and a longer range. The same device, which is illustrated in Fig. 133, has reappeared in recent years in the form of electromagnet used in the Thomson-Houston arc lamp, and in the automatic regulator of the same firm. A very similar plan was used by Hjörth in 1854, for one of his electric motors, the armature being of the form of a conical iron cup fitting over the conical pole.

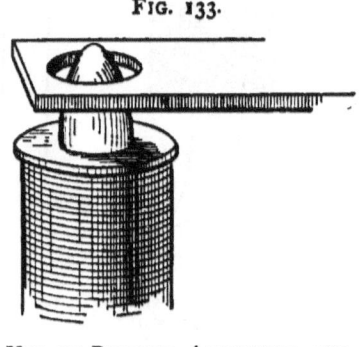

FIG. 133.

USE OF PIERCED ARMATURE AND CONED POLE-PIECE.

IV. ELECTROMAGNETIC CAMS.

It was remarked above that the devices depending on the lateral approach of an armature to a shaped polar surface presented an analogy to the mechanism of the *cam*, which is itself a variety of the inclined plane. An inverted sort of electromagnetic cam (due originally to Wheatstone) is frequently used in small electromagnetic motors. One of such forms is depicted in Fig. 134, in which it will be seen that an electromagnet, pivoted so as to revolve on its own middle, is surrounded by an iron ring, through which the magnetic lines can stream back from the north to the south pole. The inner face of this ring is cut away in an oblique manner. Applying our principle that there is a tendency of the configuration so to alter as to complete the magnetic circuit as much as possible, it is evident that the electromagnet will turn until the gaps are reduced to a minimum.

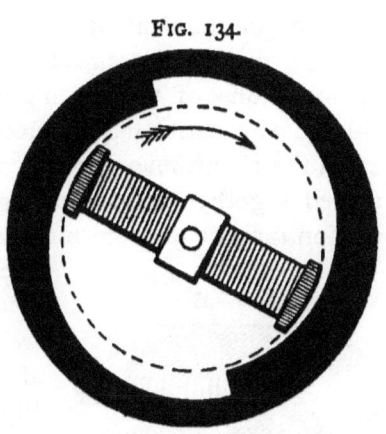

FIG. 134.

OBLIQUE APPROACH BETWEEN ELECTROMAGNET AND MASS OF IRON.

If at this point the current is cut off, the electromagnet will be carried on over the dead points, and may be again switched (automatically) into circuit, to be again attracted forward, and so forth. There is also an analogy between this piece of mechanism and that of the windmill, the oblique sails of which, acting as inclined planes, are driven round transversely to the direction of the force of the wind.

All electromagnetic devices for distributing the pull of an electromagnet, whether they consist in shaping the pole-pieces or the armatures, or whether they consist in constraining or pivoting the armatures or the electromagnet in the modes previously suggested, may be considered magnetically as either cams, wedges, or inclined planes.

V. Electromagnetic Linkages.

In the same way as a *linkage* is made by uniting together different mechanical organs, such as levers, cranks, and the like, by means of connecting rods and connecting pins, so combinations of two or more electromagnets may be considered as constituting an electromagnetic linkage. In some cases the electromagnets or their armatures may be actually united together by connecting rods; in other cases the connexion may be effected merely by mutual attraction across a gap, or simultaneous attraction of a common armature.

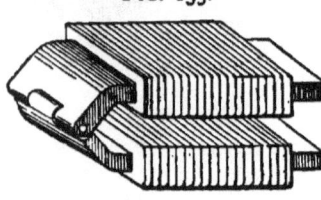

FIG. 135.

HINGED ELECTROMAGNETS.

The simplest of such cases is afforded by the example shown in Fig. 135, where two electromagnets are hinged together at one end, and attract or repel one another at their other ends, according to the directions of the currents flowing in their respective coils. A device of this kind was used by Rapieff about 1879.

Another very characteristic piece of electromagnetic linkage is found in the combined use of two electromagnets at right angles to one another. This device is excellent for producing a locking mechanism to be opened or shut at will, and is actually applied in certain forms of electromagnetic locks. It is also used in the automatic railway block system of Tyer, and in that of Professors Ayrton and Perry.* An example taken from Count Du Moncel's † apparatus for registering by the aid of electricity the notes played on the keyboard of an organ or pianoforte will suffice. The electromagnetic system is required to arrest the movement of a rotating axle, or to permit it to resume rotation, and each of

---

* See an excellent illustrated article in *La Lumière Electrique*, vol. xi., p. 345, 1884.

† Du Moncel, *Exposé des Applications de l'Electricité*, vol. ii., p. 292, and vol. iii., p. 117 (edition of 1857). See also *La Lumière Electrique*, iii., p. 339, 1881.

these actions is to be accomplished by a single electric impulse. In Fig. 136, R represents a disk on the rotating shaft, provided with a projecting tooth. This is arrested by a projection on the armature M of an electromagnet A A. When a current is sent to this electromagnet it lifts its armature, releasing the rotating shaft, and at the same time locking the armature M out of the way, a catch being provided at the end of it which latches into a notch in the armature N of a second electromagnet B B. When it is desired to arrest the rotation, a current must be sent to this

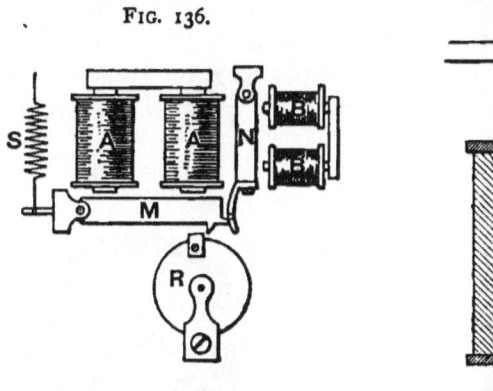

Fig. 136.

Interlocking Electromagnets.

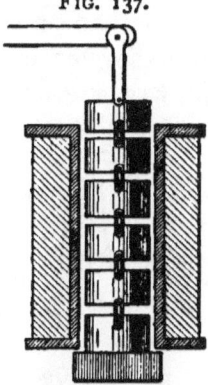

Fig. 137.

Plunger Core made in Separate Joints.

second electromagnet, which thereupon attracts N and unlatches M, so stopping the rotating disk.

Yet another variety of the combined use of two electromagnets is afforded by the mechanism of the writing telegraphs of E. A. Cowper and of Robertson. In Cowper's form the two electromagnets have separate armatures mechanically linked together. In Robertson's form they both act on the same armature, which moves in a diagonal direction according to the ratio of the two forces which act upon it at right angles to one another.

Another species of electromagnetic linkage is afforded by a device of M. Pellin, designed to give a long range of motion. A plunger core is made up of a number of short cylindrical pieces of iron, linked together in series, with the object that when they are strongly magnetized by the surrounding coil, they shall attract one another, and so draw up the lowest joint with a strong, but extended pull. The notion, which is capable of extension in various ways, is ingenious, but the expected advantage probably illusory.

VI. Repulsion Apparatus.

As we have seen above, the device shown in Fig. 135 may be used either for attraction or repulsion; but there are some forms of mechanism capable of producing repulsion only.

For example: in an apparatus due to Maikoff and De Kabath, two cores of iron, not quite parallel, pivoted at the

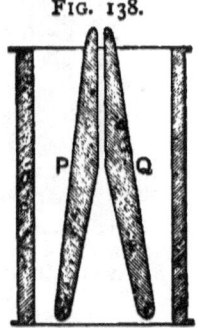

Fig. 138.

Repulsion between two Parallel Cores.

Fig. 139.

Electromagnetic Mechanism Working by Repulsion.

bottom, pass up through a tubular coil. When both are magnetized, instead of attracting one another, they open out; they tend to set themselves along the magnetic lines through that tube. The cores being wide open at the bottom tend to open also at the top.

In 1850 a little device was patented by Brown and Williams, consisting, as shown in Fig. 139, of an electromagnet which repelled part of itself. The coil is simply wound on a hollow tube, and inside the coil a little piece of iron, bent as the segment of a cylinder to fit in, going from one end to the other. Another little iron piece, also shaped as the segment of a tube, is pivoted in the axis of the coil. When these are magnetized, one tends to move away from the other, they being both of the same polarity. Of late there have been many amperemeters and voltmeters made on this plan of producing repulsion between the parallel cores.

Another example of electromagnetic repulsion is afforded by the apparatus depicted in Fig. 140. It is an electromagnet, the core of which is an iron tube about 2 inches long. There

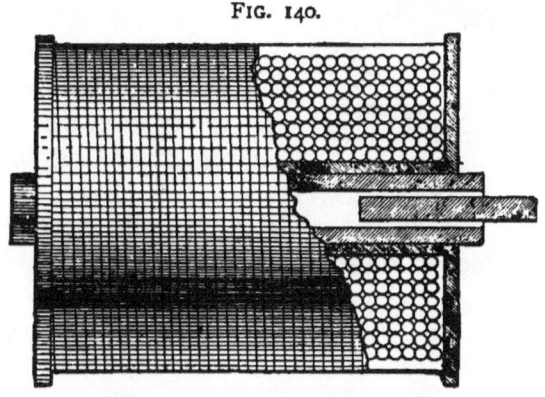

FIG. 140.

ELECTROMAGNETIC POP-GUN.

is nothing very unusual about it; it will stick on, as you see, to pieces of iron when the current is turned on. It clearly is an ordinary electromagnet in that respect. Now suppose I take a little round rod of iron, about an inch long, and put it into the end of the tube, what will happen when I turn on my current? In this apparatus as it stands, the magnetic circuit consists of a short length of iron, and then all the rest is air. The magnetic circuit will try to complete itself, not by shortening the iron, but by *lengthening* it; by pushing the piece of iron out so as to afford more surface for leakage. That is exactly what happens; for, as you see, when I turn on the current the little piece of iron shoots out and drops down. It becomes a sort of magnetic pop-gun. This is an

U

experiment which has been twice discovered. I found it first described by Count Du Moncel, in the pages of *La Lumière Électrique*, under the name of the *pistolet électromagnétique;* and Mr. Shelford Bidwell invented it independently. He gave an account of it to the Physical Society in 1885, but presumably the reporter missed it, as there is no record in the Society's 'Proceedings.'

Mr. Shelford Bidwell has also devised another model, illustrating the same principle (Fig. 141). This consists of two strips of thin springy sheet iron joined at the ends, but bowed outwards. If this is placed inside a magnetizing coil, it elongates itself when the current is turned on. If, on the contrary, you place within the coil two separate flat strips tied together at their middle, they tend to open out at their ends when the magnetizing power is applied, for by opening out they can best help the magnetic lines to find return paths through the air.

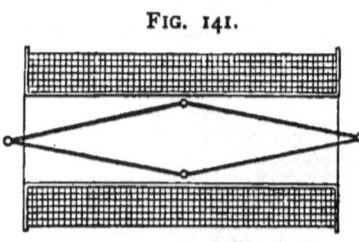

FIG. 141.

MAGNETIC ELONGATION OF DOUBLE CORE OF IRON.

VII. POLARIZED ELECTROMAGNETIC DEVICES.

We must now turn our attention to one class of electromagnetic mechanism which ought to be carefully distinguished from the rest. It is that class in which, in addition to the ordinary electromagnet, a permanent magnet is employed. Such an arrangement is generally referred to as a *polarized mechanism*. The objects for which the permanent magnet is introduced into the mechanism appear to be in different cases quite different. I am not sure whether this is clearly recognised, or whether a clear distinction has ever been drawn between three entirely separate purposes in the use of a permanent magnet in combination with an electromagnet. The first purpose is to secure unidirectionality of motion; the second is to increase the rapidity of action and of sensitiveness to small currents ; the third to augment the mechanical action of the current.

(*a.*) *Unidirectionality of Motion.*—In an ordinary electro-magnet it does not matter which way the current circulates ; no matter whether the pole is north or south, the armature is pulled, and on reversing the current the armature is also pulled. There is a rather curious old experiment which Sturgeon and Henry showed, that if you have an electro-magnet with a big weight hanging on it, and you suddenly reverse the current, you reverse the magnetism, but it still holds the weight up ; it does not drop. It has not time to drop before the magnet is charged up again with magnetic lines the other way on. Whichever way the magnetism traverses the ordinary soft iron electromagnet, the armature is pulled. But if the armature is itself a permanent magnet of steel, it will be pulled when the poles are of one sort, and pushed when the poles are reversed—that is to say, by employing *a polarized armature*, you can secure unidirectionality of motion in correspondence with the current.

Again, a reference to Fig. 20, p. 47, will show that in a mechanism containing a fixed permanent magnet and a movable conductor carrying a current, the direction of the motion will depend on the direction in which the current flows. If the current is reversed, the direction of the motion will reverse. Now, this is quite different from the action of the ordinary electromagnet of soft iron, for in that case the core pulls at the armature in the same direction, no matter which way the current is flowing around its coils. To distinguish between the two cases, it is usual to refer to those devices in which a permanent magnet comes into use as *polarized* mechanisms, whilst the ordinary electro-magnets are *non-polarized*. The single-acting mechanism is thus made into a double-acting mechanism, a point the possible advantage of which every engineer will at once grasp.

One immediate application of this fact for telegraphic

purposes is that of diplex telegraphy. You can send two messages at the same time and in the same direction to two different sets of instruments, one set having ordinary electromagnets, with a spring behind the armature of soft iron, which will act simply independently of the direction of the current, depending only on its strength and duration; and another set having electromagnets with polarized armatures, which will be affected, not by the strength of the current, but by the direction of it. Accordingly, two completely different sets of messages may be sent through that line in the same direction at the same time.

Another mode of constructing a polarized device is to attach the cores of the electromagnet to a steel magnet, which imparts to them an initial magnetization. Such initially-magnetized electromagnets were used by Brett in 1848, and by Hjörth in 1850. A patent for a similar device was applied for in 1870 by Sir William Thomson, and refused by the Patent Office. In 1871, S. A. Varley patented an electromagnet having a core of steel wires united at their ends.

It was suggested above that when a polarized armature is applied to an electromagnet it will be either attracted or repelled, according to the direction of the current. This is, of course, on the supposition that the armature is placed in a position parallel to the line joining the poles of the electromagnet. But the polarized armature may instead be placed in a transverse position between the poles.

Sturgeon was himself the inventor also of a polarized mechanism intended to be used as an electromagnetic telegraph. It is shown in Fig. 142, and consists simply of a permanently magnetized steel compass needle placed with one pole between the poles of a soft iron horse-shoe electromagnet. If the current is sent round the spiral coil from N to S, the left-hand pole will become a north pole, and the right-hand pole a south pole, with the result that the north pole of the compass needle will turn to the right. On reversing the sense of the current, the needle will of course turn to the left.

It is strange that this invention, which Sturgeon gave to the world unpatented, in 1836, was never brought into commercial use as a telegraph at a time when telegraphs were almost a monopoly in the hands of a few interested persons.

Wheatstone, in fact, patented in 1845 the use of a needle permanently magnetized to be attracted one way or the other between the poles of an electromagnet. Sturgeon had described the very same device in the *Annals of Electricity* in 1840. Gloesner claims to have invented the substitution of permanent magnets for mere armatures in 1842. In using polarized apparatus it is of course necessary to work, not with a simple current that is turned off and on, but with reversed currents. Sending a current one way will make the moving part move in one direction; reversing the current makes it go over to the other side.

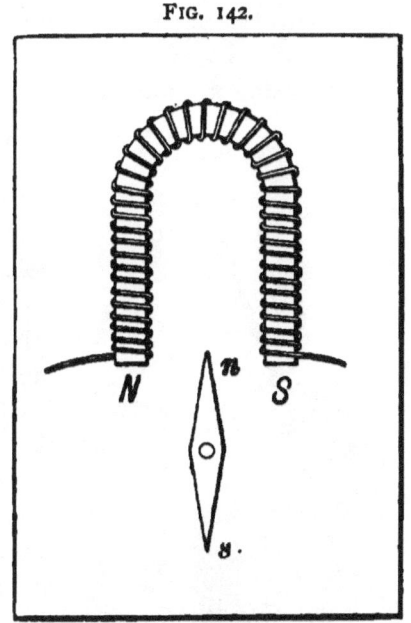

FIG. 142.

POLARIZED MECHANISM OF STURGEON'S ELECTROMAGNETIC TELEGRAPH.

According to Du Moncel,* when a soft iron electromagnet acts on a permanent magnet placed parallel to the line joining the poles to serve as a polarized armature, the force with which the armature is repelled when the current is sent in one direction around the coils is not equal to the force with which it is attracted when the current is sent in the reverse direction. The repelling force is always, for equal distances,

* *La Lumière Electrique*, vol. ii. p. 109.

less than the attractive force. This is, of course, due to the greater magnetic flux in the latter case.

In one class of electric bells—those intended to work with alternating electric currents—polarized mechanisms are employed. In one such form (depicted in Fig. 143) the two soft iron cores which constitute the electromagnet are polarized by being attached to a bent steel magnet permanently magnetized. If the bent end of this magnet be its south pole, then the upper extremities of both the cores will become north poles. These will both attract the ends of the pivoted soft iron armature, which will remain tilted indifferently in either position. But when a current is sent around the coils in the usual manner, circulating in opposite senses in the two coils, it will strengthen the flux of magnetic lines through one of the cores, and weaken that through the other. Consequently reversing the current will cause the pivoted armature to reverse its position. With these bells no battery is used; but there is a little alternate-current dynamo, worked by a crank. The alternate currents cause the pivoted armature in the bell to oscillate to right and left alternately, and so throw the little hammer to and fro between the two bells.

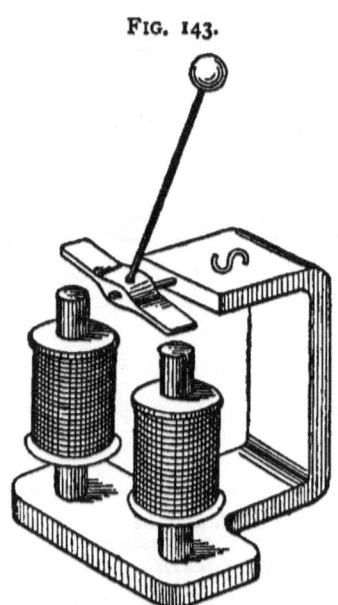

FIG. 143.

MECHANISM OF POLARIZED TREMBLING BELL.

Another form of polarized bell mechanism, due to M. Abdank, is shown in Fig. 144. Here a horse-shoe steel magnet is used to provide initial magnetism, and the alternate currents are caused to impart an alternating magnetism to a short core of iron which lies within the coil between the limbs of the horse-shoe.

Another example of a polarized device is afforded by the use of a plunger of steel permanently magnetized with a tubular coil. It is drawn in or pushed out according to the direction of the exciting current. A polarized mechanism resembling, though on a larger scale, that of a Siemens' relay, is employed in one of Hipp's forms of electric clock to drive the escapement.

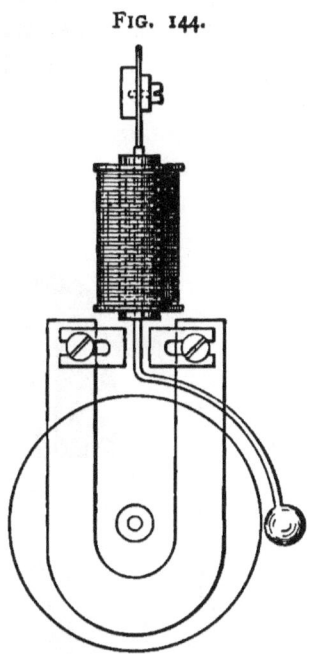

FIG. 144.

ABDANK'S POLARIZED BELL.

(*b.*) *Rapidity and Sensitiveness of Action.*—For relay work polarized relays are often employed, and have been for many years. The Post Office pattern of standard relay, a steel magnet to give magnetism permanently to a little tongue or armature which moves between the poles of an electromagnet that does the work of receiving the signals. In this particular case the tongue of the polarized relay works between two stops, and the range of motion is made very small in order that the apparatus may respond to very small currents. At first sight it is not very apparent why putting a permanent magnet into a thing should make it any more sensitive. Why should permanent magnetism secure rapidity of working? Without knowing anything more, inventors will tell you that the presence of a permanent magnet increases the rapidity with which it will work. You might suppose that permanent magnetism is something to be avoided in the cores of your working electromagnets, otherwise the armatures would remain stuck to the poles when once they had been attracted up. Residual magnetism would, indeed, hinder the working unless you have so arranged matters that it shall be actually helpful to you. Now for many years it was supposed that permanent magnetism in the electromagnet was anything but a source of help. It was supposed to be an unmitigated nuisance, to be got rid of by all available means, until, in 1855, Hughes showed us how very advantageous it was to have permanent magnetism in the cores of the electromagnet. Hughes's form of electromagnet, to which allusion has already been made on p. 186, is depicted in Fig. 145.

A compound permanent magnet of horse-shoe shape is provided with coils on its pole pieces, and there is a short

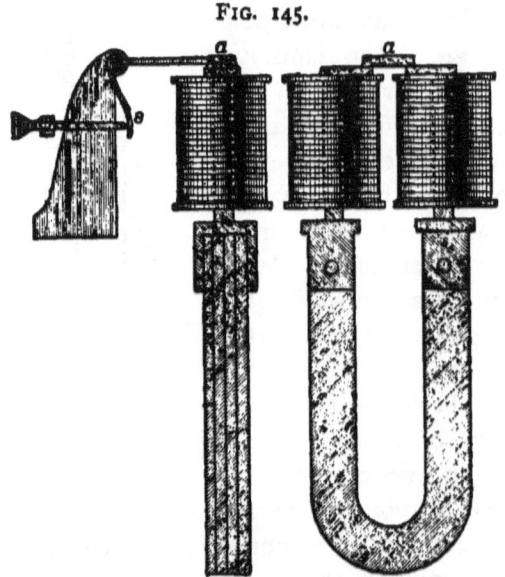

FIG. 145.

HUGHES'S ELECTROMAGNET.

armature on the top attached to a pivoted lever and a counteracting spring. The function of this arrangement is as follows:—That spring is so set as to tend to detach the armature, but the permanent magnet has just enough magnetism to hold the armature on. You can, by screwing up a little screw behind the spring, adjust these two contending forces, so that they are in the nicest possible balance; the armature held on by the magnetism, and the spring just not able to pull it off. If, now, when these two actions are so nearly balanced you send an electric current round the coils, if the electric current goes one way round it just weakens the magnetism enough for the spring to gain the victory, and up goes the armature. This apparatus then acts by letting the armature off when the balance is upset by the electric current; and it is capable of responding to extremely small currents. Of course, the armature has to be put on again mechanically, and in Hughes's type-writing telegraph instruments it is put on mechanically between each signal and the next following one. The arrangement constitutes a distinctive piece of electromagnetic mechanism.

Devices such as this, by means of which very small electric currents are enabled to perform a considerable mechanical action, may be termed electromagnetic *triggers*.

(*c.*) *Augmenting Mechanical Action of Current.*—The third purpose of a permanent magnet to secure a greater mechanical action of the varying current is closely bound up with the preceding purpose of securing sensitiveness of action. It is for this purpose that it is used in telephone receivers; it increases the mechanical action of the current, and therefore makes the receiver more sensitive. For a long time this was not at all clear to me; indeed I made experiments to see how far it was due to any variation in the magnetic permeability of iron at different stages of magnetization, for I found that this had something to do with it, but I was quite sure it was not all. Prof. George Forbes gave me the clue to the true explanation; it lies in the law of traction with which you are now familiar, that the pull between a magnet and its armature is proportional to the square of the number of magnetic lines that come into action. If we take N, the number of magnetic lines that are acting through a given area, then to the square of that the pull will be proportional. If we have a certain number of lines, N, coming permanently to the armature, the pull is proportional to N^2. Suppose the magnetism now to be altered—say made a little more; and the increment be called dN; so that the whole number is now $N + dN$. The pull will now be proportional to the square of that quantity.

It is evident that the motion will be proportional to the difference between the former pull and the latter pull. So we will write out the square of $N + dN$, and the square of N, and take the difference.

Increased pull, proportional to $N^2 + 2 N \cdot dN + dN^2$;
Initial pull, proportional to N^2 ;
Subtracting; difference is $2 N \cdot dN + dN^2$.

We may neglect the last term, as it is small compared with the other. So we have, finally, that the change of pull is proportional to $2 N \cdot dN$. The alteration of pull between the initial magnetism and the initial magnetism with the additional magnetism we have given to it, turns out to be proportional, not simply to the change of magnetism, but also to the initial number, N, that goes through it to begin with. The more powerful the pull to begin with, the greater is the change of pull when you produce a small change in the number of magnetic lines. That is why you have this greater sensitiveness of action when using Hughes's electromagnets, and greater mechanical effect as the result of applying permanent magnetism to the electromagnets of telephone receivers.

Moving Coil in Permanent Magnetic Field.—The most

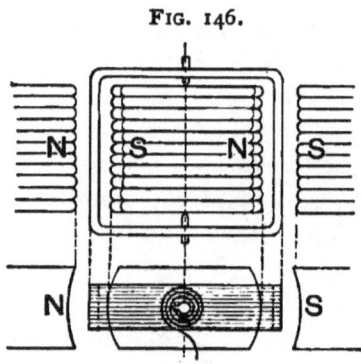

FIG. 146.

BAIN'S MOVING-COIL MECHANISM.

striking example of the use of a permanent magnet in increasing the action of a weak electric current is, however, furnished by yet another mechanism. A coil traversed by an electric current experiences mechanical forces if it lies in a magnetic field; the force being proportional to the intensity of the field. Hence, by making the magnetic field very intense, by applying powerful magnets, the effects of extremely minute currents may be shown. Sturgeon, in 1835, devised a thermogalvanometer on this principle, suspending the circuit containing the thermo-electric junctions in the field of a powerful steel magnet. Bain[*] employed this device (Fig. 146) in some of his telegraphic inventions in 1841. Of this principle also the mechanism of Sir W. Thomson's siphon recorder is a well-known example. Also those galvanometers which have for their essential part a movable coil suspended between the poles of a permanent magnet, of which the earliest example is that of Robertson (see *Encyclopædia Britannica*, ed. viii. 1855), and of which Maxwell's suggestion, afterwards realised by d'Arsonval, is a modern instance. Siemens has constructed a relay on a similar plan. The radio-micrometer of Mr. Vernon Boys is on the same principle.

A distinctive mechanism having moving coils was suggested by Doubrava for use in arc lamps. Upon a special iron framework (Fig. 147) are placed two fixed coils A, A, which are so connected to the circuit that they tend to produce consequent north poles N N at the middle of the central bar, and other consequent south poles S S at the middles of the outer bars. Between the centre bar and the outer bars there will therefore be strong magnetic fields. Two other coils, B, B, are arranged to slide up and down the two

FIG. 147.

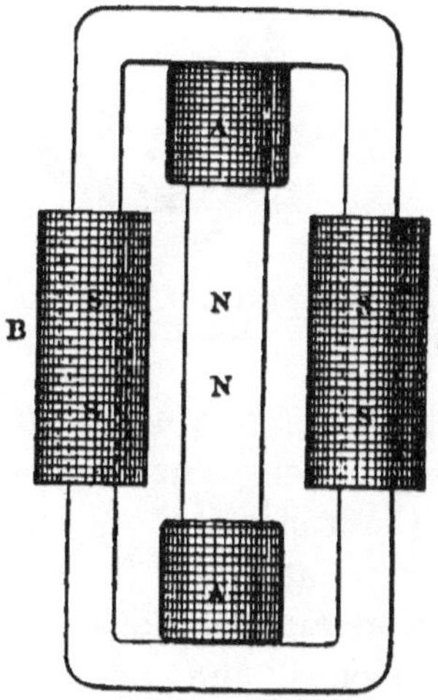

DOUBRAVA'S MECHANISM WITH SLIDING COILS.

---

[*] *Finlaison*: An account of some remarkable applications of the Electric Fluid to the useful arts, by Alexander Bain. London, 1843.

outer bars. When currents are sent through them they are urged by forces which act at right angles to the flow of the current and at right angles to the magnetic lines, and therefore tend to drive them up or down according to the sense in which the currents are circulating in them.

Amongst polarized mechanisms must be specially mentioned the ordinary receiving apparatus of the telephone. This, in the form given to it by Graham Bell, consists of a Hughes's electromagnet, (that is to say, of a permanent magnet of steel, having at its end a soft iron pole-piece surrounded by a coil) with a thin circular disk of iron as a

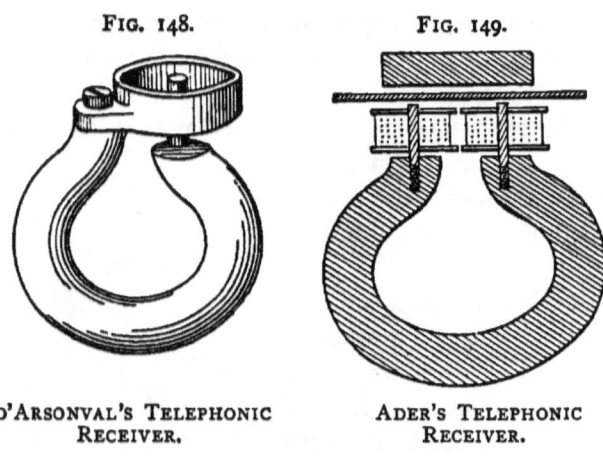

FIG. 148. FIG. 149.

D'ARSONVAL'S TELEPHONIC RECEIVER. ADER'S TELEPHONIC RECEIVER.

vibrating armature. In Bell's original form the permanent magnet was of horse-shoe shape, both poles being presented to the iron plate. The more common form with one central pole is later, and though more portable, less efficient. Two recent modifications are worthy of mention. In that of d'Arsonval the permanent magnet is curved so that an annular pole-piece may be applied around the central pole. In Ader's form a mass of iron is placed opposite the poles on the *other* side of the thin iron disk. The effect of the presence of this iron mass is to cause the disk to be more powerfully attracted by the magnet. Its presence betters the magnetic circuit as a whole, and probably concentrates the slope of the magnetic field opposite the ends of the two polar extensions.

VIII. ELECTROMAGNETIC VIBRATORS.

These form a class of mechanisms of such importance that they are separately considered in Chapter X., p. 318.

IX. ROTATORY ELECTROMAGNETIC DEVICES.

There are many devices for producing a motion of rotation by means of an electromagnet. In some of these devices a pivoted portion is turned aside through an angle of greater or less amount; in some others a motion of continous rotation is produced. The deflexion of a compass-needle by the passage over or under it, which was the discovery of Oersted in 1819, is the most elementary example of the first of these actions. The simple early electromagnetic motors of Ritchie and others, of which more will be said in Chapter XII., are examples of the second class of actions.

Of mere pivoted mechanisms not intended to generate a continuous rotation, there are many varieties, there is little resemblance between some of these forms; but they all agree in using the principle that there is a tendency to produce such a motion as will render the magnetic circuit better, or will increase its magnetism. Amongst these are several which have incidentally been already described, including the apparatus of Wheatstone, Fig. 134, p. 285, and that of Bain, Fig. 146, p. 298.

In an instrument (an amperemeter) devised by Mr. Evershed, a small piece of iron, C, pivoted upon an axis, is attracted round, parallel to itself, into a position between the curved cheeks of two other pieces of iron, A and B (Fig. 150), all being inclosed within a tubular coil of copper wire. Another way of carrying out the same idea is shown in Fig. 151, where a curiously wrought central piece of iron, R, having two wedge-shaped tongues, is placed between the poles P and Q of an electromagnet. This central piece tends so to turn as to bring into the direct line between the poles the greatest possible thickness of iron.

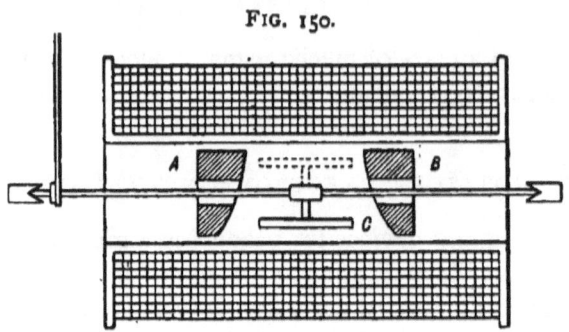

FIG. 150.

MECHANISM OF EVERSHED'S AMPEREMETERS.

One of the latest devices for procuring a rotatory movement electro-magnetically is depicted in Fig. 152. Here there are two magnetizing coils, each with an aperture through its middle. The iron cores are both coned, and

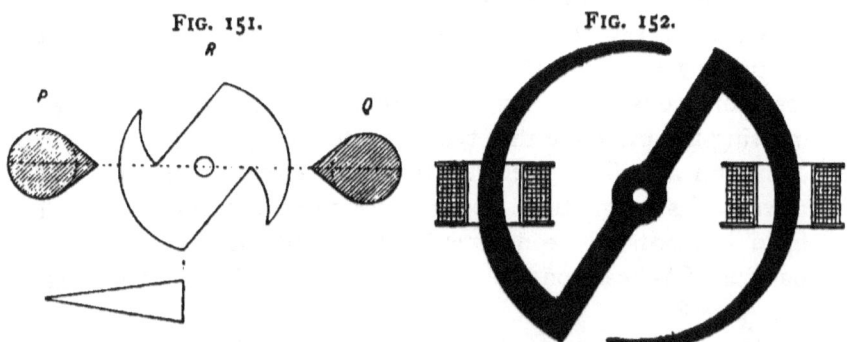

FIG. 151.

FIG. 152.

SHAPED IRON ARMATURE BETWEEN POLES OF ELECTROMAGNET.

CURVED PLUNGER CORE AND TUBULAR COILS.

joined together so as to constitute portions of the rim of a wheel. They tend to plunge deeper and deeper into their respective coils. This design is of course intended to take advantage of the peculiar property of coned plungers noticed on p. 261.

Besides the devices which I have already mentioned there are several for producing movement about or upon an axis. Some of these are designed for telegraphic purposes. For example, in Fig. 153 is depicted the method adopted by Siemens forty years ago, for pivoting an armature, A A,

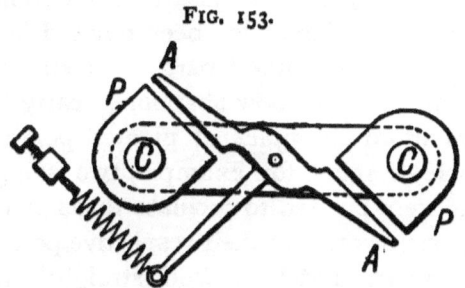

FIG. 153.

SIEMENS' FORM OF PIVOTED ARMATURE.

between the two pole-pieces, P P, attached to the end of the core, C C, of a horse-shoe electromagnet. The particular form chosen leaves very narrow gaps in the magnetic circuit, which, therefore requires little current to magnetize it,

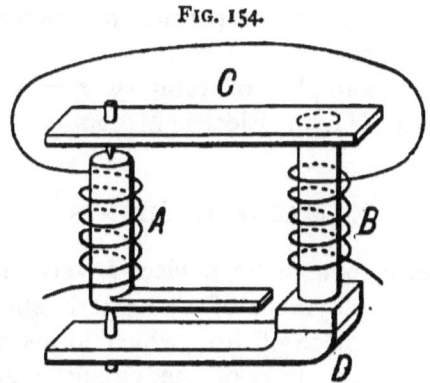

FIG. 154.

WATERHOUSE'S FORM OF PIVOTED ARMATURE.

making the arrangement very sensitive. Compare with this a device due to Waterhouse, of New York, shown in Fig. 154, who pivots one of the two cylindrical cores of the electromagnet, furnished with a projecting tongue, A. The other core, B, is provided with a large tongue, D, of brass or other non-magnetic metal, to support the lower pivot. If A is originally turned outwards, the tendency of the magnetizing current will be to turn it back into parallelism with the support D.

All the devices which have just been mentioned are non-polarized; but many have also been devised in which either the moving part or the fixed part is itself independently magnetized. Many of the now abandoned early forms of telegraphic relays afford examples of pivoted polarized mechanism. In Varley's relay, for example, two small horse-shoe steel magnets were pivoted to oscillate through a small angle, so as to bring one or other of their respective poles towards the poles of an intermediately placed straight electromagnet. But, to take full advantage of the sensitiveness afforded (see p. 297, above) by auxiliary magnets, these must be large and powerful. Therefore it is obviously inadvisable to make the magnet itself a moving part in a relay; otherwise it cannot respond rapidly to signals. In modern relays the moving part is light and is merely polarized (i.e. magnetized in a fixed direction) by auxiliary permanent magnets which are themselves stationary.

Some other examples of rotatory mechanisms will be found in Chapter XII., on Electric Motors.

X. MAGNETIC SHUNTS.

Amongst electromagnetic devices a very distinctive one consists in the employment of a magnetic shunt, that is to say, of a mass or piece of iron which forms an alternative path for the magnetic lines of the circuit. An example is afforded by the adjustable iron keeper sometimes supplied to small medical magneto-electric machines for reducing their power. When this iron keeper is placed across the poles of the steel magnet it provides a path of easy permeance for the magnetic lines, and shunts them away from their other path through the core of the rotating armature. A form of telephonic receiver containing a magnetic shunt is depicted in Fig. 155, wherein d is the ordinary thin iron diaphragm, and S N a permanent magnet of steel. Between the poles of the latter is placed a short core of soft iron surrounded by the coil. Some of the magnet's lines will pass through the iron disk, others through the core. If the current circulates around this core in such a way as to increase the magnetic flux through it, it will diminish that part of the flux which goes through the disk. If a current circulates in the opposite way it will drive out part of the flux of lines from the core, and increase those that go through the disk. This form was independently devised by M. Carpentier, and by the author.

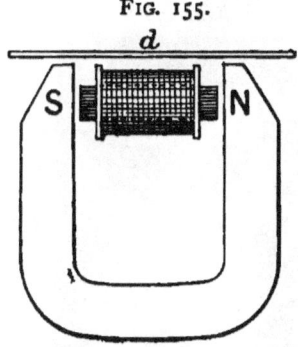

FIG. 155.

TELEPHONIC RECEIVER WITH MAGNETIC SHUNT.

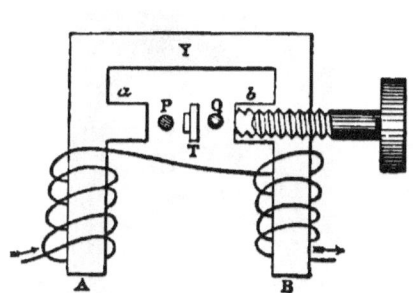

FIG. 156.

D'ARLINCOURT'S RELAY.

A magnetic shunt occurs in a peculiar and sensitive form of relay known as d'Arlincourt's, the general construction of which is shown in Fig. 156. The coils are wound on two cores not connected together by a yoke at their lower or idle ends A and B. Above, they receive two projecting pole-pieces a and b, between which lies the polarized tongue T, working between two non-magnetic stops P and Q. An iron screw, inserted in a hole in the pole-piece b, serves to adjust the action of the instrument. Beyond these two pole-pieces the cores are crossed by a yoke Y of iron, which consequently acts as a magnetic shunt. When a magnetizing current passes through the coils in the direction shown by the arrows, mag-

X

netic lines are created which pass upward through the limb A, and downward through B, some of these magnetic lines flowing across between *a* and *b*, and others going up through the shunt Y. The distribution of the magnetic flux is illustrated in Fig. 157. If the tongue T is polarized by being attached to the south pole of the polarizing magnet, it will be attracted toward the contact-stop P, when the current is turned on as described, for the pole-piece *a* will be a north pole, and *b* a south pole.

Now the tongue might be furnished with a spring, tending to keep it in the central position, which would bring it away from the contact-stop P so soon as the current was cut off,

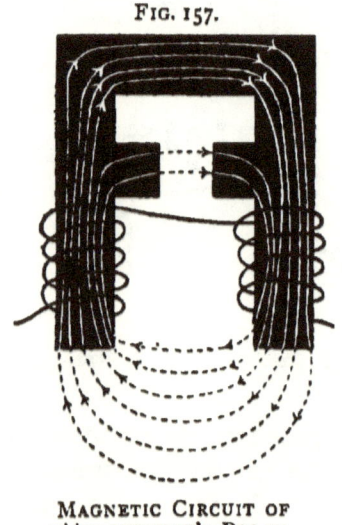

Fig. 157.
Magnetic Circuit of d'Arlincourt's Relay.

Fig. 158. 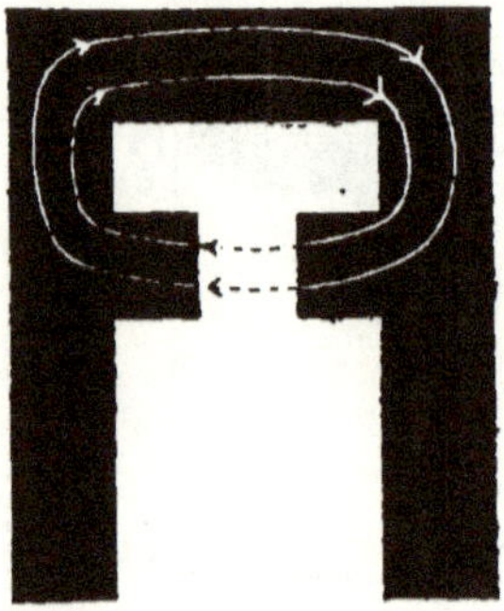
Magnetic Circuit of d'Arlincourt's Relay, after Current is Cut off.

yet its return when the current is cut off would take time. It was found by d'Arlincourt that the tongue of the relay acted more promptly when the shunt Y was present than when it was absent. The reasons hitherto given by telegraphists in explanation of the cause of this increased sensitiveness have not been altogether satisfactory.* The real

* See a paper by Mr. Brough, *Journ. Soc. Teleg. Engineers*, vol. iv., p. 418, 1875.

reason is as follows:—The more nearly the circuit of an electromagnetic apparatus is closed in form, the less readily does it lose its magnetism. It has been pointed out in Chapter III., on the Properties of Iron, that a gap in the magnetic circuit hastens demagnetization. (Compare pp. 108 and 230.) Now, there being no yoke below A and B, and the shunt Y being away from the coils, it will be the last part to retain its magnetism; and as the magnetic lines in the lower part die out, there comes a moment when, as represented in Fig. 158, the flux from *a* to *b* is reversed. At this instant the tongue will be thrown across in the other direction from P to Q.

XI. ELECTROMAGNETIC ADHERENCE.

Forty years ago it was proposed to employ the electromagnet for producing an adherence between the driving-wheels of engines and the iron rails on which they run. Experiments made by M. Nicklès at the time, on the Chemin de fer de Lyon, were not encouraging, owing to the very imperfect means then applied for magnetizing the wheels. Subsequently a better form of magnet was found. The appropriate form for this purpose is the circular electromagnet of Weber (Fig. 30, p. 55); and the proposal of M. Nicklès was to construct the driving-wheels with grooves in the periphery, as shown in Fig. 159. In this case one of the rims becomes a north pole and the other a south pole, the magnetic circuit being completed just at the one part where contact occurs with the rail.

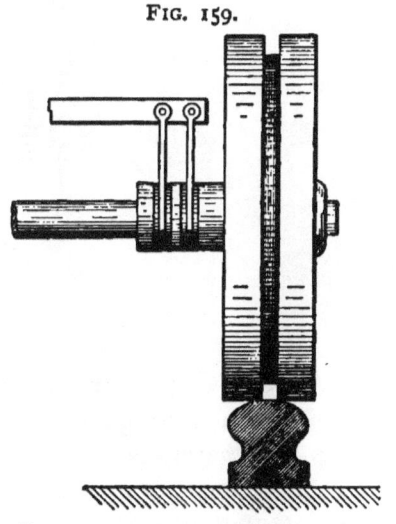

FIG. 159.

ELECTROMAGNETIC ADHERENCE OF WHEEL TO RAIL.

Another mode of employing electromagnetic adherence in the transmission of motion is represented in Fig. 160. Motion is here transmitted between two wheels without teeth by a purely magnetic gearing. The rims of both wheels being of iron, they will become strongly adherent when an electric current is made to circulate in a surrounding coil of copper.

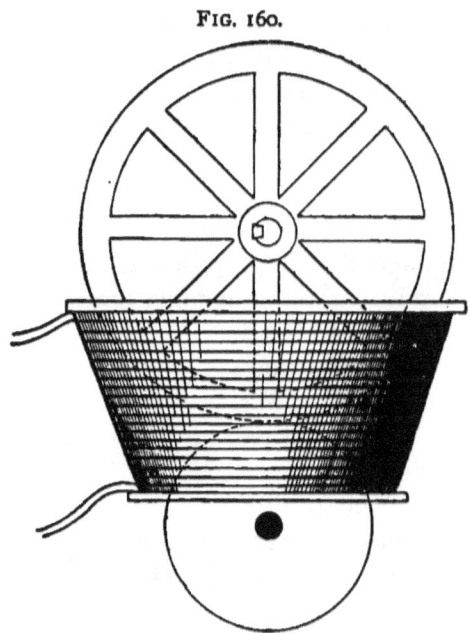

FIG. 160.

NICKLES' MAGNETIC FRICTION-GEAR.

Another important service which the electromagnet can render is in sorting or separating pieces of iron from non-magnetic materials. The earliest form [*] of electromagnetic *separator* or *sorter* is that of Arthur Wall, to whom a patent was granted in 1847. This was followed, in 1854, by the *trieuse*, or "electric sorting machine" of M. C. A. B. Chenot, in which were employed electromagnets mounted on a rotating

---

[*] It may be noted that a patent was granted in 1792 to William Fullarton for separating iron ore by the application of magnetic attraction, obviously by the use of permanent magnets.

disk to which the pulverized iron ore was brought: the portions which adhered were dropped at some distance from the non-magnetic material. Since that date many forms of magnetic separator have been devised. They are used, for example, in the porcelain industry, for preventing particles of iron from getting into the white china clay, and similarly in the white-lead industry. In engineering shops they are frequently em-

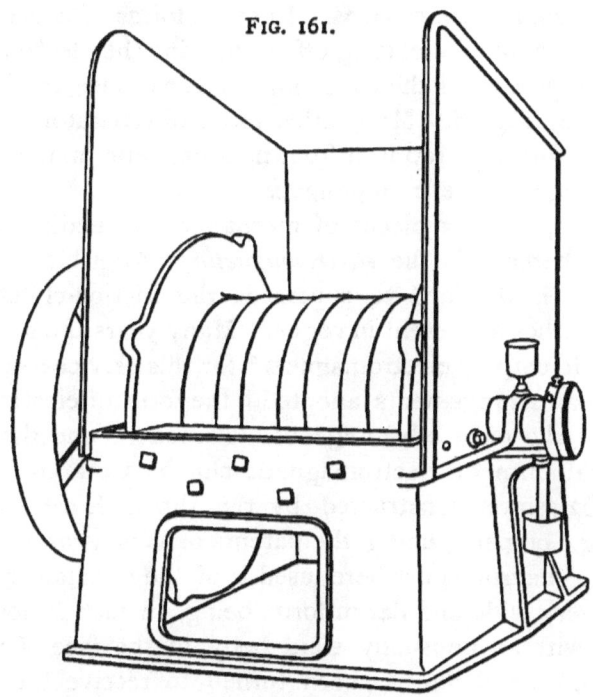

FIG. 161.

MAGNETIC SEPARATOR.

ployed to separate iron filings from brass or copper filings. The manufacturers of artificial guano, who grind up in their disintegrators all manner of refuse, chiefly dried cows' dung, employ them for catching nails and nuts and other scraps of iron, which otherwise would ruin the grinding surfaces. One form of separator, constructed for this latter service by the Brush Electrical Engineering Company, is depicted in Fig. 161. This separator consists of a revolving iron drum, in the periphery of which are cut several grooves to receive the magnetizing coils. The apparatus therefore resembles Weber's electromagnet (Fig. 32, p. 57) in its general design. The drum is revolved within a bed-plate by steam power; and the dry refuse is shot over it, any iron scraps being carried over by the drum in the opposite direction, and removed mechanically at the other side. The current is, of course, introduced through sliding connexions. In some forms of separator the device is adopted of cutting off or shunting the electromagnets during a portion of their rotation, so as to cause the adherent matter to drop off. Many other forms of separator have been used for sorting scrap iron from non-magnetic matter, and for separating iron ore from gangue.

Amongst other pieces of mechanism depending on magnetic adherence is the *electromagnetic clutch*, which for some purposes is destined to supersede the friction-clutches and claw-clutches at present in vogue. Many years ago M. Achard sought to employ electromagnets [*] for this service: but in the particular arrangements adopted [†] the form of electromagnet was not the most advantageous. A modern and thoroughly practical form of electromagnetic clutch is that depicted in Fig. 162; it is constructed by the Brush Electrical Engineering Company, under the patents of Wynne and Raworth.

The electromagnet here used is of the ironclad type, but made very wide and flat in form, being, in fact, a loose iron pulley with an unusually solid body, in the face of which a deep and broad rim has been turned to receive the coils of insulated copper wire. Opposite this, acting as the armature, is a stout disk of iron, A, firmly keyed upon the shaft. The electric current is conveyed in and out of the coils by means of two metal brushes, B B, shown on the right, which press against two gun-metal collars, C C, embedded in insulating fibre, and which are connected respectively to the two ends of the coil. When it is desired to throw the clutch into gear all that is needful is to turn on the current, when the

---

[*] See *Du Moncel's Expose des Applications* (edition 1857), vol. i., p. 310.
[†] See Specification of Patent, No. 1668 of 1855.

electromagnet-pulley at once grips the iron disk with a grip that may easily be made as much as 100 lb. per square inch of surface of contact, and the pulley forthwith turns solidly on the shaft. As the switch for turning on and off

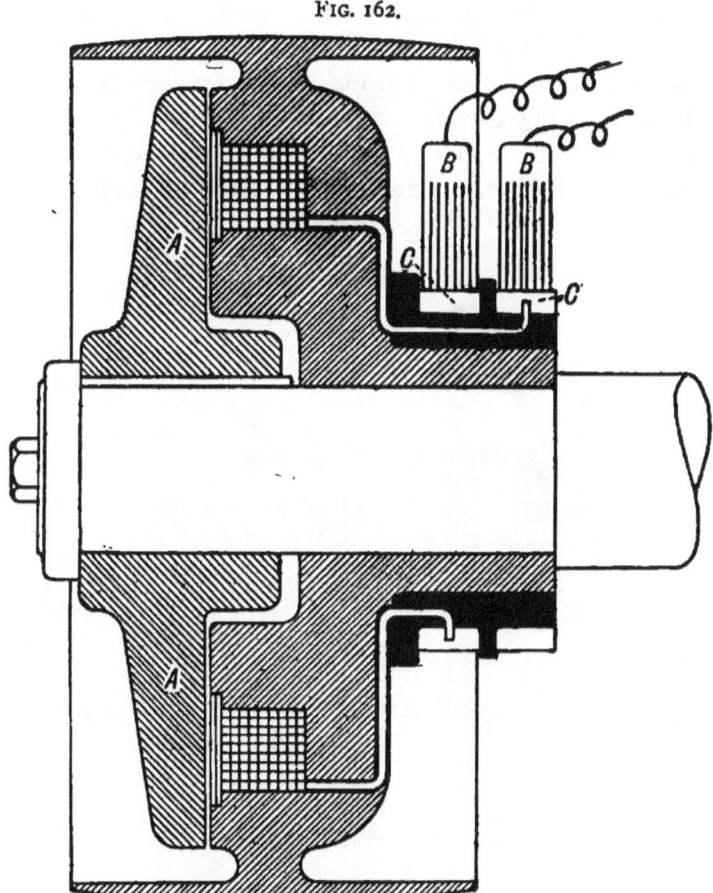

Fig. 162.

WYNNE AND RAWORTH ELECTROMAGNETIC CLUTCH.

the electric current may be placed in any convenient position, either near or far away, it is obvious that this clutch presents many advantages over the mechanical clutches — whether friction-clutches or claw-clutches — that need some kind of striking gear to be affixed to them. For clutches in inaccessible positions on a line of shafting this device is indeed superior to anything ever previously imagined.

A closely kindred invention is Willans' *electromagnetic coupling gear*, for replacing the mechanical flexible couplings frequently employed for transmitting power from an engine to a dynamo fixed on the same bed-plate. The coupling consists of a curious variety of ironclad electromagnets, made in two halves, having as polar portions an inner and an outer set of symmetrical projections which, though not touching, and admitting of play, attract one another strongly and transmit a powerful tangential driving thrust.

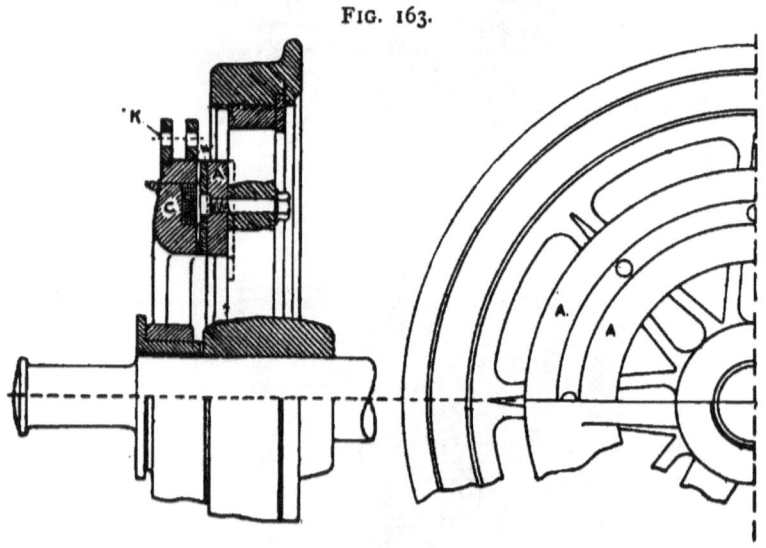

FIG. 163.

FORBES AND TIMMIS'S ELECTROMAGNETIC RAILWAY BRAKE.

Electromagnetic Railway Brakes were first suggested by Amberger, in 1850*, and they have been at various times brought into notice by Achard† and others. The most recent form of electromagnetic brake is that depicted in Fig. 163.

---

* See Specification of Patent, No. 13269 of 1850.

† For detailed notices of the inventions of Achard and others, see articles in *La Lumière Électrique*, vols. viii. and ix., 1883.

It is the joint invention of Prof. George Forbes and Mr. I. A. Timmis. The ironclad electromagnet closely resembles that described on p. 53, Fig. 27. The core C, is a ring of wrought iron, bolted securely to the axle-box by bolts passing through the holes K, and constructed with a deep groove to contain the coils. The armature A A, is another ring of iron bolted to the carriage wheel. As in this case there will be much wear of the surfaces in contact, a renewable pair of iron washers W, in two parts, are interposed to serve as pole-pieces. If every wheel of every carriage of a train were

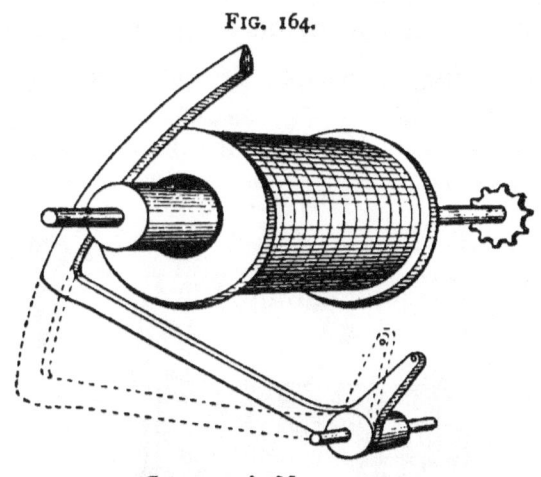

FIG. 164.

COLOMBET'S MECHANISM.

provided with such a brake, these could be simultaneously thrown into action by merely turning on the electric current to the whole of them at once. The tangential drag given by the brake is found to be proportional to the current supplied to the coil.

Two other pieces of mechanism dependent on electromagnetic adherence remain to be described. In the first of these (Fig. 164), which is due to M. Colombet, a mechanically rotating shaft of iron is caused to raise a curved iron arm, when desired, by sending a current of electricity into an exciting coil surrounding the iron shaft. The shaft then adheres magnetically to the curved arm, and maintains a rolling contact while it raises the arm.

The last instance of magnetic adherence to be mentioned is the magnetic clutch used in the arc lamp of Gülcher, in which an electromagnet, balanced on trunnions, sticks itself to an iron rod which forms the upper carbon-holder, and then raises the latter by tilting upon its bearings.

XI. ALTERNATE-CURRENT DEVICES.

These are fully described in Chapter XI., p. 331.

MECHANISMS USED IN ELECTRIC BELLS AND INDICATORS.

Before concluding this chapter it will be appropriate to refer to several varieties of electromagnetic mechanism which are used in electric bells,* and in the indicators or annun-

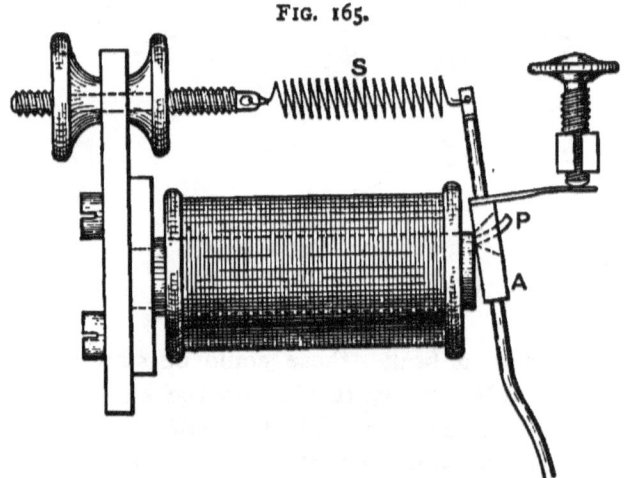

FIG. 165.

MECHANISM OF WAGENER'S ELECTRIC BELL.

ciators for bells. Many of these furnish striking instances of the universal guiding principle that there is always a ten-

* For other bell mechanisms see other parts of this book as follows:— Ordinary electric bell, Fig. 170, p. 319; Short-circuit bell, Fig. 171, p. 322; Bell for alternate currents, Fig. 143, p. 294; Abdank's alternate-current bell, Fig. 144, p. 295; Differentially-wound bell, p. 371.

dency of the configuration of the system to change in such a way as shall better the magnetic circuit and increase the flux of magnetic lines.

An excellent example is afforded by the design of an electric bell (Fig. 165) in the physical cabinet of the Finsbury Technical College, constructed by Wagener, of Wiesbaden. The electromagnet of this bell is a horse-shoe placed horizontally. From each of the polar ends of the cores there protrudes an upward-curving brass peg, P. Upon these two pegs is hung the armature A, which has two countersunk holes pierced in it. It rests against the upper edges of the pole-ends, and so in one sense completes the magnetic circuit, but is held up by a spiral spring S, so that between it and the polar surfaces there is a narrow angular gap. If it moves so as to close this gap by tilting upon the two pegs it thereby betters the magnetic circuit; and in this manner is set into

FIG. 166.

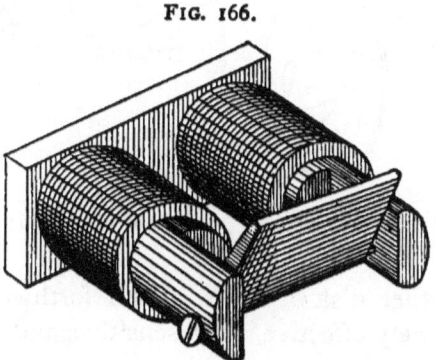

MECHANISM OF DRAW-UP INDICATOR.

vibration. Its great sensitiveness may be attributed to the way in which the initial position adopted aids the excitation of the magnetism.

Fig 166 illustrates a kindred case. A flat armature of iron is pivoted between the poles of an electromagnet which are cut away to receive it; but, when the magnet is not excited the armature falls into an oblique position, and exposes a signal-disk which is carried on an upright support on the armature (not shown in the cut) behind an aperture in the indicator board. When the current is sent around the coils of the electromagnet it draws up the armature, and moves the disk back into shadow.

Thorpe's semaphore indicator (Fig. 167) consists of a single central core surrounded by a coil, while a little strip of iron coming round from behind serves to complete the circuit all save a small gap. Over the gap stands that which is to be attracted, a flat disk of iron, which, when it is attracted,

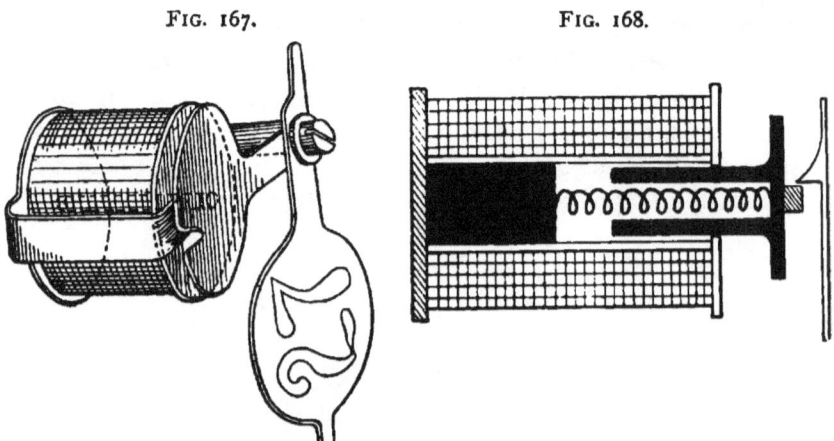

FIG. 167.   FIG. 168.

THORPE'S SEMAPHORE INDICATOR.   MOSELEY'S INDICATOR.

unlatches another disk of brass, which forthwith falls down. It is an extremely effective, very sensitive, and very inexpensive form of annunciator.

Moseley's indicator (Fig. 168) consists of a stopped coil with an iron tubular plunger, pushed out by an interior spring. When it is drawn in it withdraws a catch, and permits a signal-piece to drop.

In a common form of polarised mechanism for indicators a curved permanent magnet of steel, N S, is pivoted (as in Fig. 169) over the pole C, of a short, straight electromagnet, which projects horizontally from a back-board. The pivoted magnet bears a red signal disk D, upon a vertical support

Electrical Indicators. 317

On sending a current in one direction the magnet is thrown to one side; and, on reversing the current, it is thrown over to the other side. Hence this mechanism allows of an electrical replacement, without compelling the attendant to walk up to the indicator board. Polarised apparatus for indicators has this advantage, that it admits of electrical as distinguished from mechanical replacement. In an indicator movement by

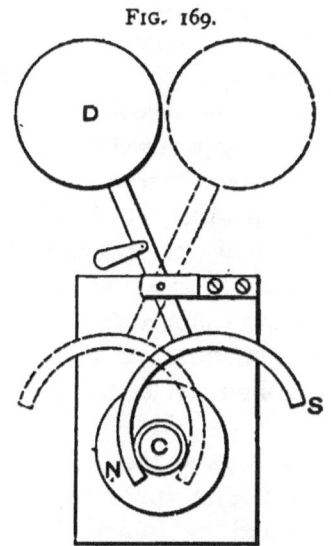

FIG. 169.
FIG. 169a.

POLARIZED INDICATOR MOVEMENT.
TRIPOLAR ELECTROMAGNET FROM GENT'S INDICATOR.

Messrs. Gent, of Leicester, a tripolar electromagnet is used, having one coil on the central pole. The rising of the armature permits a trip-lever to fall, and indicate the reception of a signal.

There are many other indicator movements in use, some with trip levers to be unlatched by the action of an electromagnet, others with pendulums to be set vibrating; but the above examples will suffice.

CHAPTER X.

ELECTROMAGNETIC VIBRATORS AND PENDULUMS.

MUCH use is made of mechanisms in which a vibratory motion is maintained electromagnetically. The armature of an electromagnet is caused to approach and recede alternately, with a vibratory motion, the current being automatically cut off from, and restored to, the coils of the electromagnet by a self-acting break. Such vibratory apparatus is to be found in every electric trembling bell, in the breaks of induction-coils, in electrically-maintained tuning-forks, in the transmitters of harmonic telegraphs; also in the slower movements of some obsolete forms of oscillating electric motors, and in those of the electromagnetically-driven pendulums of certain forms of electric clock.

The electromagnetic vibrator has gone through various phases of development. Its earliest form appeared in 1824, in the vibrating electric wire invented by James Marsh, of Woolwich, the chemist. It consisted of a wire suspended at the top by a flexible metallic joint, dipping at its lower end into a shallow mercury-cup, and placed between the poles of a powerful horse-shoe magnet. When a current of sufficient strength was sent along it, it was driven laterally across the magnetic field, and in so moving it broke contact and fell back, being thereupon again impelled. This primitive apparatus was followed by Dr. Roget's dancing spiral (which was at one time used as the interrupter for induction-coils), and later by devices of sounder mechanical construction. Amongst these were the oscillating lever movements of Dal Negro[*] and of Professor Henry;[†] the hammer of Wagner,[‡] also

---

[*] *Ann. Roy. Lomb. Venet.*, April 1834.
[†] *Silliman's Journal*, vol. xx., p. 340, July 1831.
[‡] *Pogg. Ann.*, xlvi., p. 107, 1839.

known as Neeff's hammer;* and the vibrating interruptor of Froment,† sometimes called Froment's "buzzer." This was followed by the invention of the present form of trembling electric bell, about 1850, by John Mirand; and about the same time of other forms, by Siemens and Halske, and by Lippens.

Fig. 170, depicts the mechanism of the ordinary electric trembling bell. In this apparatus, so soon as the electric

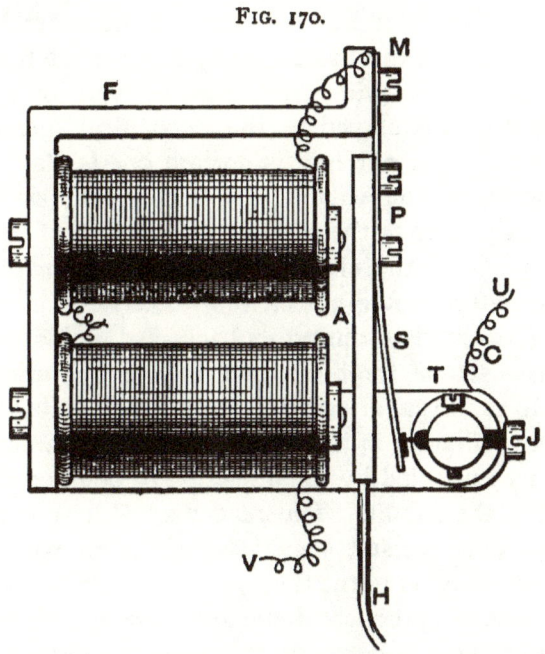

FIG. 170.

MECHANISM OF ORDINARY ELECTRIC TREMBLING BELL.

circuit is completed, the electromagnet E attracts its armature A, and draws it from its position of rest. The contact spring S behind it still maintains contact while the armature moves forward for a short distance, but as the armature increases its displacement, a breach of continuity occurs (with a spark) between the spring and the contact-pin P, so cutting off the current. From this instant the magnetism begins to disappear

---

* *Pogg. Ann.*, xlvi., p. 104, 1847.

† *Comptes Rendus*, xxiv., p. 428, 1847; also see Daguin's *Traité de Physique*, vol. iii., p. 21.

and the pull diminishes; nevertheless, the armature having attained a certain speed, continues by virtue of its kinetic energy to be carried forward, until arrested by the bending of its spring support, which at once begins to drive it backwards. Again it gets up speed, the contact spring strikes against the pin, establishing anew the circuit; but the armature does not at once stop, its inertia carries it on, and, moreover, time is required for the electromagnet to attain its full attracting power. Before it has done so, the armature has come momentarily to rest and has begun again to move towards the electromagnet. It is evident that in this cycle of operations an important part is played by the mechanical inertia of the moving part; and a not less important one is played by the electric inertia of the circuit. If the current were instantaneously responsive to the opening and closing of the contact —if the latency of the electromagnet were of zero duration— then the armature would not on the whole be more attracted during its forward than during its backward oscillation.

It must not be forgotten that in the operation of the electric bell, as in that of every electromagnetic motor, whether rotatory or oscillating in its movement, the mechanical motion is always accompanied by the induction of counter electromotive forces in the circuit. The reader who does not already know this should consult some treatise dealing with dynamo-electric machinery as to the theory of electric motors. During the movement of approach of the armature towards the polar ends of the cores it is bettering the magnetic circuit, and consequently is tending to increase the magnetic flux in the iron parts; this sets up a counter electromotive force which diminishes the current. If the motion were so rapid, and this counter electromotive force so great as at this instant entirely to dam back the current, and if at the same instant the contact at the spring were broken, there would be no spark. Again, as the armature flies back it tends to set up electromotive forces which help the current to magnetize the core; so that probably the magnetizing current is greatest at or shortly after the instant when the armature has swung back to its most distant point. In the case of the trembling bell the precise phases of these variations are of no great importance, but it is otherwise in the case of such harmonic vibrators as electromagnetically-driven tuning-forks. No vibrating organ will execute a truly harmonic motion unless the impulses are properly timed. In the ideal case of simple harmonic motion, the forces tending to restitution of the initial position are proportional at every instant to the displacement, but owing to friction such movements die out, unless sustained by compensating forces of some kind. Now it is obvious that the frictional forces called into play by the motion of the system are greatest when the velocity is greatest—namely, when the vibrating system is passing through its position of zero displacement. If at this instant a small impulse is given, the diminution of amplitude due to friction may be compensated. If, however, the impulse is imparted at any other part of the oscillation, the impulse will produce also another effect, namely it will affect the time of vibration, causing an acceleration of phase.

In clocks and other devices with pendulums, the pendulum will not beat isochronously unless the discontinuous impulses are imparted precisely at the instant when the moving part is passing through its zero or median position. We shall consider this matter again in connexion with tuning-forks.

In another disposition, sometimes adopted for the sake of avoiding the spark which occurs at break of contact, the current in the magnet coils, instead of being interrupted by the approach of the armature, is short circuited. The contact-pillar is in this case placed on the other side of the armature, so that contact is made during its approach towards the poles of the electromagnet. Some constructors save the use of a contact-pillar by making one of the poles of the electromagnet serve this function. As soon as the electromagnet is thus short circuited its magnetism disappears, and the armature, after coming to rest at the end of its movement of approach flies back. No spark occurs at the moment when the contact

322   *The Electromagnet.*

spring parts from the contact-pillar, as at that instant there is no demagnetizing set up. Short-circuit bells, therefore, need not to be provided with infusible contact-pieces of platinum. They will also work if two or more of them are joined in series, which is not the case with the ordinary bells.

A still more perfect mode of arranging the circuits of an electric bell is to use a differential winding (see Chap. XIV).

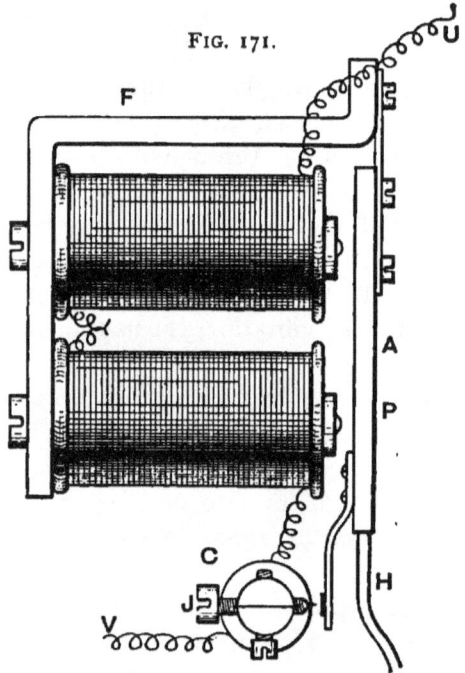

FIG. 171.

SHORT CIRCUIT BELL MECHANISM.

INDUCTION-COIL VIBRATORS.

In the break-mechanism used for ordinary induction coils a similar vibrator is used, with the addition of a device for stiffening the spring of the vibrating part at will, so as to change the rate of the vibrations. This break, depicted in Fig. 172, consists of a spring S, upon which is mounted a short cylinder of iron H, called the hammer, which has a platinum pin P at the back of it, pressing against a second platinum pin borne upon a clamped screw at the top of the upright contact-pillar U. A second screw, with a head E of ebonite, passes through an ivory collar in the upright support U, and serves to tighten the spring. When the spring presses the contact pins together only weakly, the break of current occurs at a time when the core has but feeble magnetic strength. By screwing back the contact pin and tightening the spring,

FIG. 172.

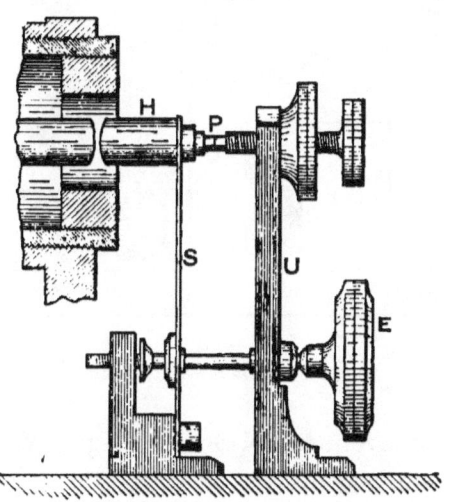

VIBRATING BREAK OF INDUCTION COIL.

so that a greater pull is required to break contact, the break does not occur until the core is more highly magnetized, thereby generating more powerful inductive effects.

Induction coil breaks of several different patterns have been devised. The extra-slow break of M. Foucault,[*] and the extra-rapid break of Mr. Spottiswoode,[†] are both worthy of study.

In Spottiswoode's break (Fig. 173) the vibrating spring C consists of a stout rod of steel rigidly clamped in a heavy

[*] Recueil des Travaux scientifiques de L. Foucault, 1878.
[†] *Proc. Roy. Soc.*, xxiii., p. 455.

mass of brass B, and maintained in vibration by the action of a special electromagnet I I. The amplitude of vibration is very small—about ·03 cm.—and the frequency about 2,500 per second. Across the electromagnet stands a strong bridge upon two pillars D; and through this bridge passes a finely-cut platinum-tipped contact screw E, provided at its upper

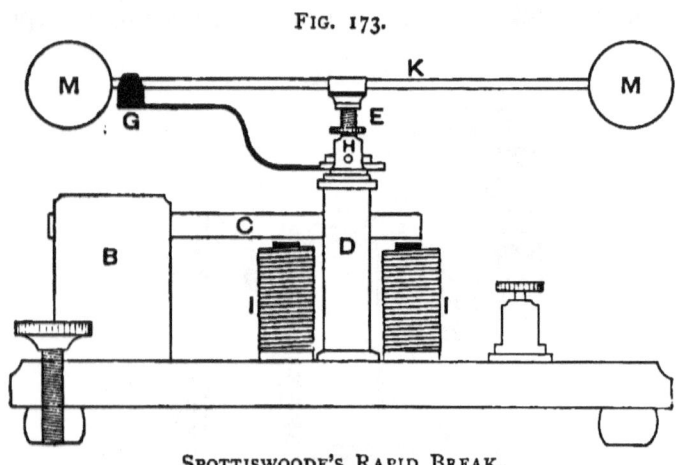

FIG. 173.

SPOTTISWOODE'S RAPID BREAK.

end with a long lever arm K so as to give fine adjustment, and with a steadying-arm G.

M. Marcel Deprez has described* a vibrating break for induction coils which he considers to have special advantage.

ELECTROMAGNETIC TUNING-FORKS.

Lissajous seems to have been the first to use an electromagnet to maintain tuning-forks in vibration, employing as a break a wire dipping into mercury. Such forks have also been used by Regnault,† von Helmholtz,‡ and others. In

---

* *La Lumière Électrique*, iii., p. 325.
† Regnault, *Rélation des Expériences*.
‡ See von Helmholtz's *Sensations of Tone* (*Tonempfindungen*), pp. 129, 176, and 604 (Ellis's Edition).

1872 Mercadier* replaced the mercurial contacts by an elastic stylus. His instrument was termed by him *electrodiapason*. One form of this apparatus, as modified by Lacour, is shown in Fig. 174, having a horse-shoe electromagnet, the prongs of which lie between the pole-pieces N S of the electromagnet, carrying two coils M M. The fork being of steel becomes a permanent magnet, having poles s, n. Upon its prongs are

FIG. 174.

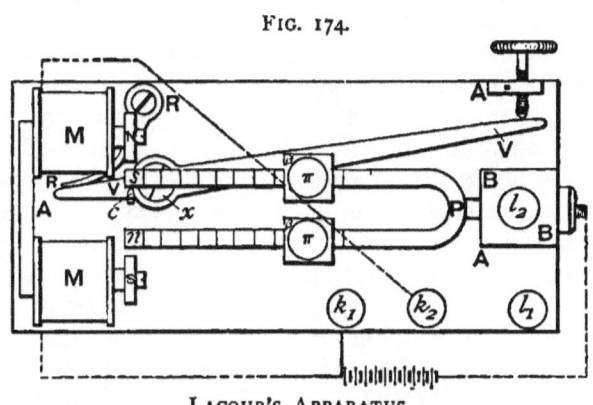

LACOUR'S APPARATUS.

supported two heavy masses, which can be slid to any of the marks engraven on the prongs, thus tuning the fork to different notes. The current enters the fork through the block B, and passes out through the contact spring c. Sometimes a single electromagnetic bobbin is placed between the prongs of the fork, instead of the external electromagnet.

With all such apparatus it is absolutely essential to consider the phase of movement at which the sustaining impulse is given. Lord Rayleigh, whose authority in acoustical matters is beyond dispute, has treated † the question with great clearness in a discussion of an early form of electromagnetic tuning-fork, having one bobbin between the prongs, and effecting the periodic interruption by means of a U-shaped rider, carried by the lower prong, and dipping into

* *Comptes Rendus*, 1873; *Journal de Physique*, 1873, vol. ii., p. 350; and *Annales Télégraphiques*, July and August, 1874.
† Rayleigh, *Theory of Sound* (1877), vol. i., p. 59.

mercury cups. The following paragraphs are taken from Lord Rayleigh's work on Sound.

"The *modus operandi* of this kind of self-acting instrument is often imperfectly apprehended. If the force acting on the fork depended only on its position—on whether the circuit were open or closed—the work done in passing through any position would be undone on the return, so that after a complete period there would be nothing outstanding by which the effect of the frictional forces could be compensated. Any explanation which does not take account of the retardation of the current is wholly beside the mark. The causes of the retardation are two : irregular contact, and self-induction. When the point of the rider first touches the mercury, the electric contact is imperfect, probably on account of adhering air. On the other hand, on leaving the mercury the contact is prolonged by the adhesion of the liquid in the cup to the amalgamated wire. On both accounts the current is retarded behind what would correspond to the mere position of the fork. But, even if the resistance of the circuit depended only on the position of the fork, the current would still be retarded by its self-induction."

"From whatever cause arising, the effect of the retardation is that more work is gained by the fork during the retreat of the rider from the mercury, than is lost during its entrance, and thus a balance remains to be set off against friction."

"Any desired retardation might be obtained, in default of other means, by attaching the rider, not to the prong itself, but to the further end of a light, straight spring carried by the prong and set into forced vibration by the motion of its point of attachment."

"The deviation of a tuning-fork interruptor from its natural pitch is practically very small, but the fact that such a deviation is possible is at first sight rather surprising. The explanation (in the case of a small retardation of current) is, that during that half of the motion in which the prongs are the most separated, the electromagnet acts in aid of the proper recovering power due to rigidity, and so naturally raises the pitch. Whatever the relation of phases may be, the force of the magnet may be divided into two parts, respectively proportional to the velocity and displacement (or acceleration). To the first exclusively is due the sustaining power of the force, and to the second the alteration of pitch."

In order to procure such a correspondence between impulses and phases of displacement as will secure true isochronism in the vibrations of forks, several devices have been suggested. The author of this work* proposed to use two forks in unison, each controlling the circuit of the other's electromagnet. Other suggestions have been made by Prof. J. Viriamu Jones,† and by Mr. Gregory.‡

Electric Pendulums.

Electrically-driven pendulums are subject to the same condition as tuning-forks, namely, that to cause them to keep true time, the sustaining impulses must be given only when the moving portion is passing through the position of zero displacement. Many inventors have essayed to construct clocks, driven not by gravity or by a spring, but by electric power, the pendulum receiving its impulses periodically from an electric battery, and itself driving the hands by a ratchet wheel and pawls. In some of these devised by Wheatstone, and in some others by Bain, the bob of the pendulum was periodically attracted electromagnetically. In others by Robert Houdin, by Froment, by Détouche, and other inventors, some other organ is periodically attracted by an electromagnet, and its motion is mechanically imparted to the pendulum. In one modern form due to Hipp, a pendulum having below its bob an iron armature, executes ten or twelve free motions, during which time its amplitude is being diminished by friction. When the arc of swing has diminished down to a certain point, contact is made to an electromagnet which gives a sudden impulse to the iron armature, and sends the pendulum flying over a wider arc. The peculiar mechanism for doing this—a device of Foucault's—consists in a small trailing tongue, hinged on the bottom of the pendulum, which passes lightly over a spring beneath it when the arc of swing is long, but catches upon the spring, and presses it down when the arc of swing is less than a certain amount.

---

* *Phil. Mag.*, Aug. 1886, p. 216.
† *Proc. Physical Society*, p. 288, 1890.
‡ *Ib.*, 1889.

328   *The Electromagnet.*

For further details of these devices the reader is referred to treatises on electric clocks.*

HARMONIC TELEGRAPH TRANSMITTERS.

In the harmonic telegraphs of Varley, Lacour, Elisha Gray, Langdon-Davies, and others, vibratory devices are used for creating the intermittent or alternating currents. Two only of these need here be described.

Elisha Gray's Transmitter.—In this apparatus a tuned bar or reed of steel is placed between two separate electro-magnets, being alternately attracted by them. Fig 175 shows

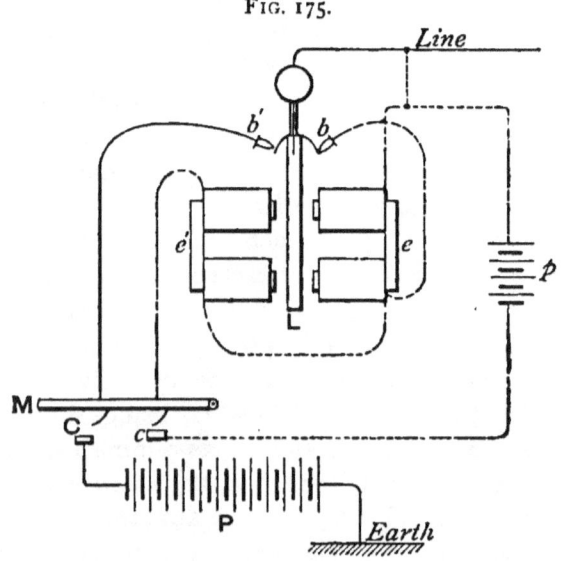

FIG. 175.

ELISHA GRAY'S VIBRATOR.

the disposition of the apparatus; the line circuit being marked in firm lines, and the local circuit for producing the vibration in dotted lines. Close to the root of the vibrating

* The most recent of these are :—Favarger's *L'Électricité et ses Applications à la Chronométrie*, Paris, 1886; Merling's *Die Elektrischen Uhren*, Brunswick, 1884; Fiedler's *Die Zeittelegraphen*, Vienna, 1889; also Du Moncel's *Exposé des Applications de l'Electricité*, vol. ii.; and Kareis's *Der Elektromagnetische Telegraph*, Brunswick, 1888.

bar L are fastened two springs which make contacts against two stops, b and b'. The electromagnet on the right e has many turns of fine wire (about 30 ohms in all in its resistance), while that on the left e' is wound with fewer turns of thicker wire (three or four ohms in resistance). If the same current is passed through both these, that with more numerous windings (on the right) will exercise the more powerful attraction. This occurs as soon as the depression of the key at M completes the local circuit at c. But when the bar is attracted to the right, it makes contact against the stop b, which has the effect of short-circuiting the electromagnet e. This electromagnet being cut out ceases to pull, and at the same instant a larger current flows through the other electromagnet e', attracting the tuned bar L to the left. It therefore takes up a rapid vibration, and imparts correspondingly rapid intermittent currents from the line-battery P to the line through the stop b'.

Langdon-Davies's Rate-Governor.—Mr. Langdon-Davies has devised a system of harmonic telegraphy in connexion with a peculiar induction-device of his invention known as the *Phonopore*. By using rapidly oscillating currents corresponding to different musical notes, broken up into dots and dashes, he is able to transmit two, three, or more separate telegraphic messages in the same wire at the same time. But to do this without a hitch it is necessary that the vibrating portions of the transmitting as well as those of the receiving instruments shall execute extremely pure vibrations. Ordinary tuning-fork interruptors, such as those above described, are not good enough for this purpose. The snarling sounds that they emit prove that their vibrations are continually changing their rate, and are not independent of the electromagnetic forces acting upon them. This is a consequence of the impulses not being given precisely at the proper phase of the motion. To remedy this defect Mr. Langdon-Davies made some hundreds of experimental devices with reeds, tuning-forks, and stretched ribbons of steel as interruptors. His final form of transmitter is called the *rate-governor*. It is depicted in Fig. 176. The

330 • *The Electromagnet.*

tuned reed A beats against a light spring V, which makes contact during precisely one half of the reed's periodic movement, being limited by a set screw S from following the

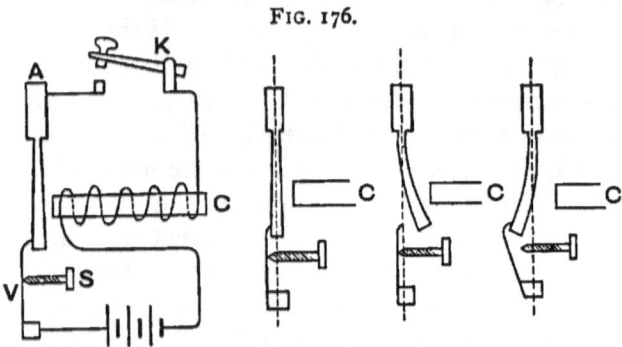

Fig. 176.

Langdon Davies' Rate-Governor.

motion of the reed. C is the magnetized core which attracts the reed when the key K is depressed. The three chief positions of the reed are shown, somewhat exaggerated, on the right of the cut.

CHAPTER XI.

ALTERNATE-CURRENT ELECTROMAGNETS.

ALL electromagnets intended for use with alternating currents must be provided with laminated cores in order to prevent the circulation within the mass of the iron of induced eddy-currents. Further, all frames, formers, or metallic bobbins, which otherwise would constitute closed metallic circuits around the cores, must be so divided or separated by non-conducting material, otherwise eddy currents will be induced in these also.

MODES OF LAMINATION.

There are several modes in which lamination may be carried out; but they must all conform to the rule that the iron shall be divided in directions at right-angles to that in which the eddy-currents would be set up. As the tendency is for these to circulate parallel to the coils that surround the cores, that is to say, transversely to the length, it follows that all lamination must be longitudinal to the cores.

For the cores of straight electromagnets a common mode of construction is to take annealed iron wires, of 1 to 2 mm. in diameter, in sufficient numbers to make up cores of the desired size. Such wires should be separately varnished, and then secured together in bundles by lightly binding them with thread or tape, which is afterward served with varnish. Metallic binding wires should not be used.

For larger electromagnets, and for those of horse-shoe pattern, thin sheets of iron are preferred. Two modes of constructing laminated cores from iron sheets are shown in Fig. 177.

In the first of these, a number of square-cornered U-shaped stampings are assembled together in sufficient numbers. In the second, the iron sheets are bent over a mould, one outside the other. In either case the separate layers of iron should be protected from metallic contact by the interposition of layers of varnished paper, or even of thin mica, though the latter is too costly for common use.

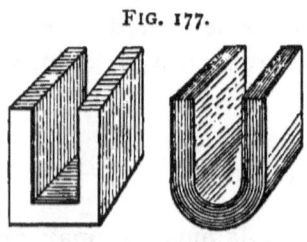

FIG. 177.

LAMINATED IRON CORES.

Some constructors simply japan or enamel the iron to prevent contact. It is important that the edges of the stampings should be trimmed from any burr which would produce contacts from lamina to lamina. To secure the parts mechanically together, bolts may be passed through the stampings, but such bolts should be insulated from contact by a surrounding tube of vulcanized fibre or of ebonite, and by washers of fibre or ebonite placed underneath the heads and nuts of the bolts. In cases where paper well varnished with shellac varnish has been interposed, it is well to consolidate the whole core into one substantial mass by serving the whole with varnish after putting it together; and then heat it in a stove to a temperature sufficient to soften the shellac, the core being then compressed until it is cold.

The cores that are to be used as plungers in coil-and-plunger mechanisms must also be laminated if they are to be used with alternate currents.

PROPERTIES OF ALTERNATE-CURRENT ELECTROMAGNETS.

In recent years much attention has been given to the properties of alternate-current electromagnets. These are very singular, and differ from anything to be observed when using electromagnets excited by means of direct currents flowing steadily in one direction. An ordinary electromagnet excited by a steady current does not attract any of the ordinary non-magnetic metals, such as copper, silver, or brass, neither does it repel them save to a very minute degree, requiring special apparatus to make the delicate effects visible. The so-called diamagnetic repulsion of the metals, due to their being less permeable than the surrounding medium, air, is so minute an effect that it does not call for any further consideration. But, as will be shown, cases arise in which an alternate-current electromagnet repels a mass of copper with an active and powerful repulsion, arising from a mutual action between the electromagnet and the currents which it sets up in the neighbouring piece of metal.

Here an important consideration presents itself in considering the phases of electric currents. Two parallel currents attract one another (or rather the conducting wires in which they flow attract one another), provided the currents in them are both flowing towards the same direction; and repel one another if the currents are flowing in opposite directions. In the case of alternating currents which flow first in one direction along a wire, then in the other, current succeeding current with enormous rapidity, it will be evident that two alternating currents flowing in two parallel wires will not attract one another unless their alternations keep step with one another. If they are exactly in step there will be attraction. If they are exactly out of step there will be repulsion. If they are in some intermediate state of phase-relation, something much more complex may happen, which we must consider.

It is usual to study alternating currents by the aid of wave-diagrams (such as Fig. 178), in which time is plotted out horizontally and current vertically. Suppose we wish to represent a current which makes 100 periods of alternation a second, and at its strongest is 10 amperes. The upper line shows the variations of the current, which, starting at time = 0, has strength = 0, but which then increases and attains its maximum of 10 amperes at time marked 1 (the 1 four-hundredth of a second). At 2 four-hundredths of a second it has died away to zero, and is going to reverse.

At 3 four-hundredths of a second it has attained its reversed (or negative) maximum; and at 4 four-hundredths of a second it has once more come back to zero, and is ready to start again. It has thus made one complete cycle of changes in one one-hundredth of a second. Its mean* strength during the time will have been 7·07 amperes. Now consider the lower curve of Fig. 178. It represents a precisely similar current in the same phase; one that keeps step with the current already considered. In the first four-hundredth of a second —the first quarter period — both currents are flowing the same way and rising to their maximum. In that quarter they will attract with a force that also goes to a maximum. In the next quarter period, from 1 to 2, they are still both flowing the positive way, and both falling in strength; so they will still both attract, but the attraction is falling to zero. In the next quarter period, from 2 to 3, both currents have reversed, but they are still in step with one another; so that they are still attracting with an attraction that becomes a maximum at 3. In the last quarter, from 3 to 4, they are still in step, but both diminishing to zero, and they still attract; but the attraction diminishes and becomes also zero at the end of the complete period. So we see that when two alternate currents are in step there is an attraction between them; but that attraction goes through periodic changes in amount, being twice at a maximum and twice falling to zero within each complete period. If the current were made to alternate very slowly—say 3 or 4 times a second only,—one might *see* the two parallel wires vibrating toward one another 6 or 8 times per second. But when the alternations come as

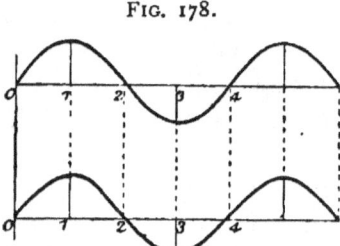

FIG. 178.

TWO WAVES IN SIMILAR PHASE.

* For the purposes of alternating currents the mean value is not the arithmetical mean but the square root of the mean square. The wire will be *heated* for example by this current that oscillates from + 10 amperes to − 10 amperes and back, as it would be by a steady current of 7·07 amperes.

Differences of Phase.

rapidly as 100 times a second (as usual when such currents are used in electric lighting) the maxima of attraction come so rapidly—200 times per second—that the wires have not time to vibrate, and are simply urged toward one another.

Consider now Fig. 179, which represents the variations of two alternate currents which are in exactly opposite phases—precisely out of step. By similar reasoning to that which we have just gone through, it will be evident that in each quarter period there will now be repulsion between the wires, for the currents start off in opposite directions, and after reversing at the same instant, are still in opposite directions. In this case

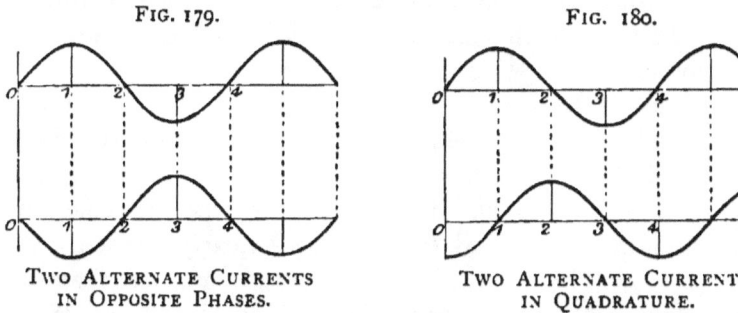

FIG. 179.   FIG. 180.

TWO ALTERNATE CURRENTS IN OPPOSITE PHASES.   TWO ALTERNATE CURRENTS IN QUADRATURE.

there will be two maxima of repulsion in each complete period.

Now let us take the case represented by Fig. 180, in which the currents have a difference of phase of one quarter of a period, one current coming to its maximum just when the other is at zero, and *vice versa*. Currents in this particular relation of phase are sometimes spoken of as "in quadrature" with one another. In the first quarter period, from 0 to 1, the first current is positive, and rising to a maximum; but the second current has in the same time been flowing in the opposite or negative sense, and is dying down from a negative maximum to zero. During this first quarter period then there will have been a repulsion, and this repulsion will have been zero when the first current was zero, and will be zero again when the second current is zero; so that it will have a maximum at some intermediate time, or at about the middle of the first quarter period. In the second quarter period there will be an attraction, because from 1 to 2 both currents are positive, and this attraction will also be at its maximum at the middle of the quarter period. In the third quarter period, from 2 to 3, since the first current has now reversed, there will be repulsion as there was in the first quarter of a period; and in the fourth quarter period, 3 to 4, there will again be attraction, as in the second quarter period. So then, in each complete cycle there will be alternately, repulsion, attraction, repulsion, attraction, and if the two currents are *exactly* in quadrature the attractions and repulsions will be of precisely equal duration and of equal amount. The net result is that the wires in this case are neither attracted nor repelled as a whole.

But now, if the currents are partly out of step, and not in exact quadrature, what will happen? In that case the durations of the two attractions and of the two repulsions of each period will not be equal. If the currents differ in phase by *less* than a quarter period (or are more nearly in step), the attractions will be greater in amount and duration than the repulsions, *and there will be on the whole an attraction.* If the currents differ in phase by *more* than a quarter of a period, the repulsions will prevail, and there will be on the whole a *repulsion.*

These things are the key to the peculiar actions produced by alternate current electromagnets.

Let us first consider the action of an alternate current electromagnet in inducing eddy currents in a copper ring hung opposite its pole as in Fig. 181.

Suppose the coil of the electromagnet to be traversed by a current in such a direction that the current is coming over the top and down the front, and to be increasing in strength. This current tends to make the end of the core nearest the ring a north pole, N, which will also be increasing in strength and sending its magnetic lines out into space in increasing numbers. This increase of the magnetic flux through the copper ring tends to induce or set up an opposing electro-

Induction of Eddy Currents.

motive force in the ring, in the direction shown by the arrow; and, of course, this electromotive force in turn tends to set up an eddy-current in the same direction around the ring. Now such induced electromotive forces are, as is well known, proportional to the rate of change of the magnetism; and the *change* of magnetism goes on at the greatest rate, not when the magnetism itself is at its maximum, positive or negative, but when it is at its zero, just reversing in sign. Hence it follows that as the magnetism of the core goes through cycles of alternations the induced electromotive force in the ring will go through cycles of alternations which are *in quadrature* with those of the electromagnet. If then, we assume that the cycles of magnetization keep step

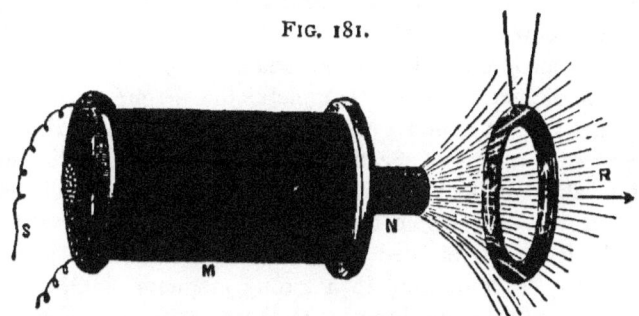

Fig. 181.

ACTION OF ALTERNATE CURRENT ELECTROMAGNET ON COPPER RING.

exactly with those of the exciting current; and if we further were to assume, *what is not true*, that the eddy-currents produced in the ring keep step exactly with the electromotive forces set up in it by the electromagnet, it would follow that the eddy-current induced in the ring would be in quadrature with the exciting current in the magnet (as Fig. 180), and in that case the ring would, on the whole, be neither attracted nor repelled. But as a matter of fact, the ring of copper possesses *a time-constant* (see p. 222) of its own. To start or stop a current in such a ring requires time. All eddy-currents lag behind the electromotive forces which produce them. Hence the currents in the ring will not be in quadrature with the changes of magnetism of the electromagnet; the difference of phase will be more than a quarter of a period, consequently *the ring will be repelled* on the whole.

ELIHU THOMSON'S EXPERIMENTS.

A most striking series of experiments have been conducted with alternate current electromagnets, by Professor Elihu Thomson,[*] of Lynn, Massachusetts. Elihu Thomson's attention was first drawn to the matter by an experiment on an electromagnet excited by a constant current; which is here described in his own words.

"In 1884, while preparing for the International Electrical Exhibition at Philadelphia, we had occasion to construct a large electromagnet, the cores of which were about 6 inches in diameter and about 20 inches long. They were made of bundles of iron rod about $\frac{5}{16}$ inch in diameter. When complete, the magnet was energized by a current from a continuous-current dynamo, and it exhibited the usual powerful magnetic effects. It was found that a disk of sheet copper of about $\frac{1}{8}$ inch in thickness and 10 inches in diameter, if dropped flat against a pole of the electromagnet would settle down softly upon it, being retarded by the development of currents in the disk due to its movement in a strong magnetic field, and which currents were of opposite direction to those in the coils of the magnet. In fact it was impossible to strike the magnet pole a sharp blow with the disk, even when the attempt was made by holding one edge of the disk in the hand and bringing it down forcibly towards the magnet. In attempting to raise the disk quickly off the pole, a similar but opposite action of resistance to movement took place, showing the development of currents in the same direction as those in the coils of the magnet, and which currents, of course, would cause attraction as a result. The experiment could be tried in another way. Holding a sheet of copper just over the magnet pole (see Fig. 182), the current in the magnet coils was cut off by shunting them. There was felt an attraction of the disk, or a dip towards the

---

[*] See "Novel Phenomena of Alternating Currents," by Elihu Thomson, in *Electrical World*, May 1887, p. 258; or *The Electrician*, May 16, 1890, p. 35 *et seq.*, in a series of articles by Prof. J. A. Fleming, to whom the author's thanks are due for permission to reproduce Figs. 181 to 187.

pole. The current was then put on by opening the shunting switch, and a repulsive action or lift of the disk was felt. The actions just described are what would be expected in such a case, for when attraction took place currents had been induced in the disk in the same direction as those in the magnet coils beneath it, and when repulsion took place the induced current in the disk was of opposite character or direction to that in the coils. Now let us imagine the current in

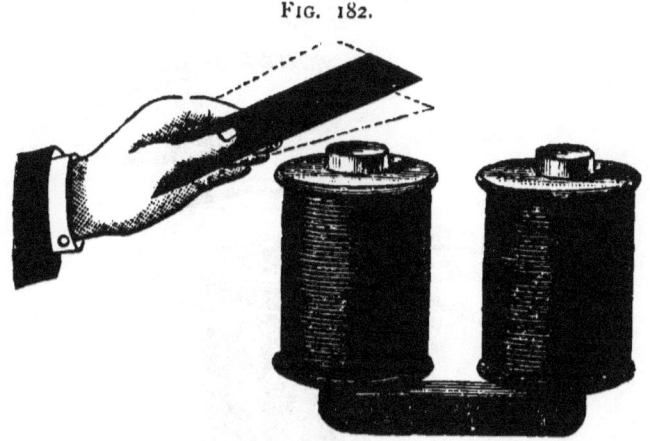

FIG. 182.

EFFECT ON SHEET OF COPPER OF TURNING ELECTROMAGNET ON OR OFF.

the magnet coils to be not only cut off, but reversed back and forth. For the reasons just given, we find that the disk is attracted and repelled alternately. The disk might, of course, be replaced by a ring of copper or other good conductor, or by a bare or insulated wire, or by a series of disks, rings, or coils, superposed, and the results would be the same."

So far there was no indication that the action of an alternating current would be to set up repulsions which were more powerful, on the whole, than the intervening attractions; for the preceding experiments had given no hint of any retarding effects or lag in the induced eddy currents. These were soon made evident when an alternating current was applied. The following striking experiments were amongst those shown by Elihu Thomson.

If a stout copper ring is held in the hand (as in Fig. 183)

over the pole of a powerful alternate current electromagnet the eddy-currents set up in it produce two effects. In the first place they warm it so rapidly that, *after a few seconds only*, it becomes unbearably hot; in the second place they cause it to be violently repelled, so that force is required to hold it down, and if let go, it flies up into the air as though shot up by an invisible spring.

FIG. 183.

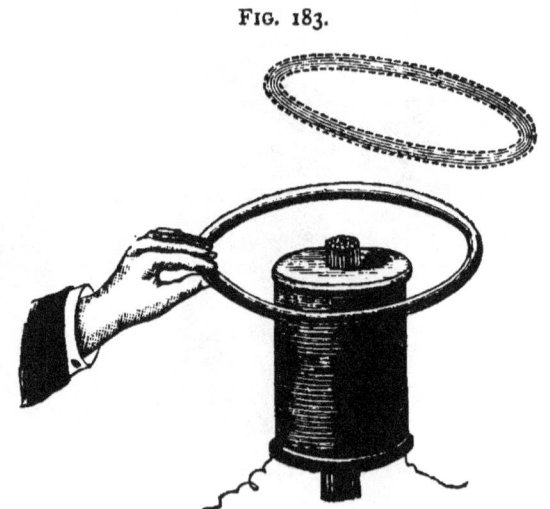

REPULSION OF A COPPER RING.

If such a copper ring be attached by thin threads to pins driven into the table, it may be held floating in the air above the pole of the electromagnet, in the manner shown in Fig. 184, like the golden aureoles pictured as floating above the heads of saints. Of course, these effects entirely vanish if the ring is cut across at any point, preventing the eddy-current from circulating. If a copper disk is hung level from the beam of a balance over the pole of such an electromagnet and carefully counterpoised before the current is turned on, then when the alternating current is turned on it is seen to rise, repelled by reason of the eddy-currents in it; but a disk slit from centre to edge is scarcely repelled at all. Other movements are possible, some of them of the most varied description. A stout copper tube or mantle placed over the upper half of an alternate current electromagnet, like that in Fig. 183, is pushed away with a steady push, resembling, but

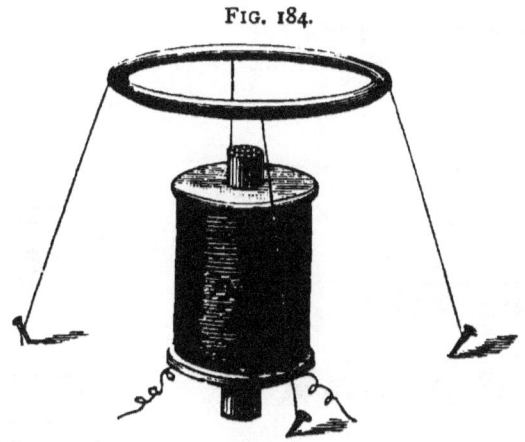

FIG. 184.

COPPER RING TETHERED TO TABLE, FLOATING ABOVE THE ALTERNATE CURRENT ELECTROMAGNET.

inversely, the manner in which a coil with continuous currents acts on an iron plunger with a steady pull. The action in this alternate current case may be modified by shaping conically either the electromagnet or the external copper mantle.

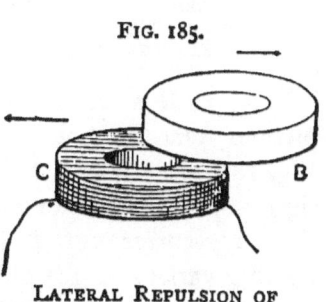

FIG. 185.

LATERAL REPULSION OF COPPER RING.

If a stout ring of copper is placed over a coil of insulated copper wire through which the alternating current is passing, the former will be repelled from the latter; but more powerfully if there be a laminated iron core in the coil. If, as in Fig. 185, the ring does not lie concentrically over the coil, it will be pushed laterally away with mutual lateral force, as indicated by the arrows, as well as repelled axially. Fig. 186 illustrates a very similar effect observed when a stout copper ring or disk, or, better still, a pile of thin rings stamped out of sheet copper, is held over one pole of a laterally-placed bar electromagnet excited by an alternating current. The eddy currents in the copper ring tend to drive it from a place where the alternating magnetic field is strongest to a place where it is weaker, namely, over the middle zone of the electromagnet.

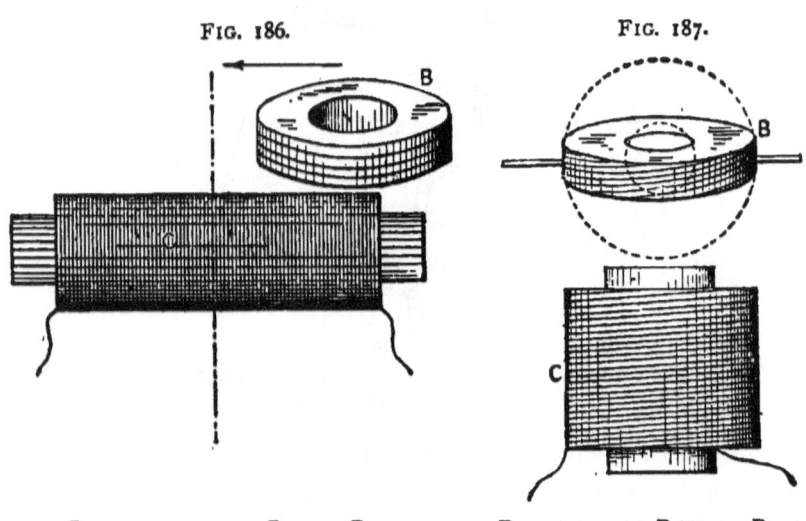

FIG. 186. DISPLACEMENT OF COPPER RING TOWARDS NEUTRAL ZONE OF ELECTROMAGNET.

FIG. 187. DEFLEXION OF PIVOTED RING BY ALTERNATE CURRENT ELECTROMAGNETS.

Again, if a copper ring or pile of copper washers is pivoted over the pole of an alternate current magnet as in Fig. 187, it is deflected from the parallel position to one at right angles. In other words it tends to move so as to reduce the eddy-currents in it to a minimum. This is really only another variety of the old principle that the configuration of any electromagnetic system tends to change in such a way as will *increase* the number of magnetic lines. In all these cases the eddy-currents tend to demagnetize—to oppose the change of magnetic field to which they are themselves due. Hence there is a tendency to such a change as shall produce the least induction of eddy-currents.

Professor Elihu Thomson has laid down four principles which govern the effects when two or more closed conductors are acted on in the same alternating field, or when masses of iron are placed in the field.

"(i.) If two or more closed circuits are similarly affected inductively by an alternating magnetic field, they will attract one another, and tend to move into parallelism.

"(ii.) Iron or steel masses placed in an alternating field give rise to shifting magnetism or lines of force moving laterally, and may, therefore, act to move closed circuits in the path of such shifting lines.

"(iii.) Closed circuits in alternating magnetic fields, or fields of varying intensity, give rise to shifting magnetism, or lines of force moving laterally to their own direction, and may, therefore, act to move other closed circuits in the path of such lines.

"(iv.) Iron or steel masses may, when placed in an alternating magnetic field, interact with other masses, or with closed electric circuits, so as to produce movement of such masses or circuits relatively, or give rise to tendencies to so move, the effects depending on continual adaptations of shifting magnetism and retained magnetism relatively."

One consequence of these interactions is a curious effect known as magnetic screening. If a sheet of copper is introduced into the magnetic field between the pole of an alternate current electromagnet and a copper ring (such as in Fig. 181), so as to intercept part of the magnetic lines, then eddy-currents will be induced in the copper sheet as well as in the ring, and these currents will be more or less parallel to one another; will, therefore, oppose one another's actions. Each will tend to stop the currents in the other. They will also attract one another. The consequence is that the presence of the interposed sheet damps down the currents in the ring; acting as if it shaded off the action of the electromagnet. Two copper rings under similar conditions are attracted together, and both are repelled from the electromagnet.

By means of such "shading" a motion of rotation can be

produced, as shown in Fig. 188. A copper disk, pivoted on a centre, is held over the pole of the alternating magnet. The edge of another disk or sheet of copper is then slid in under the first disk, so as to cover up about half the pole. The pivoted disk at once begins to rotate, the direction of movement being from the unshaded toward the shaded portion. This is readily explained by reference to the tendency of the system to re-adjust itself in such a way as to minimize the eddy-currents. The portion of the rotating disk in which the eddy-currents are is continually attracting itself up to, and trying to hide itself behind, that part of the interposed sheet in which the parallel eddy-currents are circulating. A number of very extraordinary experiments of a similar sort have been devised by Elihu Thomson. In one of these a hollow copper globe, partly immersed in water in a glass finger-bowl, rotates violently when held over the partly shaded pole of the alternating magnet. Another curious disposition is illustrated in Fig. 189, where a dissymmetrical arrangement of the magnetic field with respect to a copper conductor causes it to rotate. The conductor in this case is a cylindrical rim fastened to a laminated iron wheel;

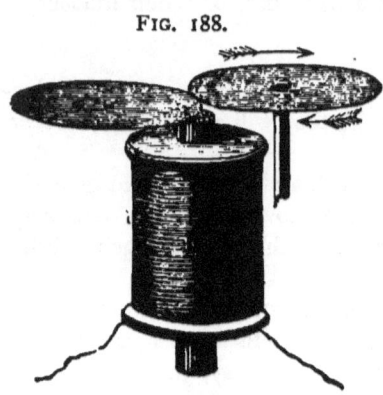

FIG. 188.

ROTATION PRODUCED BY SHADING HALF THE POLE.

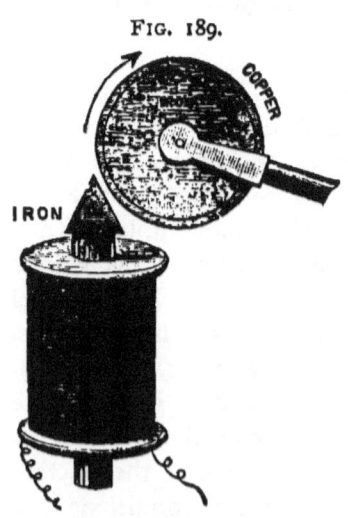

FIG. 189.

ROTATION DUE TO DISSYMMETRICAL INDUCTION OF EDDY-CURRENTS.

and the pole is furnished with a wedge-shaped iron pole-piece. The rotation occurs just as if a blast issued from the sharpened end of the electromagnet.

There is yet another way of causing an alternating magnet to produce rotation of a conductor. This is by surrounding part of the electromagnet core—usually some horn or projection of it—with a closed coil or ring of copper, in which eddy-currents are induced, and which thus retards the alternations of magnetic polarity in the projecting part. A series of such rings placed upon a protruding pole will have the remarkable result of virtually making the magnetic polarity travel continually from the root to the tip of the protruding part. The reader is referred to the chapter on Motors for some further references to this subject.

IMPEDING EFFECT OF ELECTROMAGNETS. MAXWELL'S RULE.

It was pointed out in Chapter VII., when dealing with magnets for quick action, that the electromagnet itself impedes the variations of an electric current, the magnetism in it acting with a species of inertia towards the current in the coil. This action, which is sometimes described as a result of the self-induction which occurs in consequence of the separate coils of the winding acting inductively on one another, occurs still more markedly in the case of alternating currents. The alternating electromotive force applied to the circuit by the dynamo, tends to set the current oscillating to and fro in the circuit at a high rate of frequency. If the circuit consisted simply of a plain resistance—a long thin wire without any coils in it, merely doubled back on itself—the current in it would simply depend on the (alternating) volts applied to it, and on the ohms of resistance which it offered. But if there are in it coils or electromagnets—anything possessing self-induction—anything which can act with electromagnetic inertia, then the current will be affected, damped down to some lesser strength than would otherwise be the case. This inertia-action which for short we may call the *inductance*, produces two effects on the waves of alternate current, it diminishes them in amplitude or strength, and it retards them in phase. And, as a matter of fact, the inductance is found to depend on the frequency of the alternations as well as on the self-induction of the circuit. The amount of self-induction of a circuit, or of any part of it, is usually expressed in terms of a certain unit of self-induction called the *quad*,* or *henry*.

The co-efficients of self-induction of electromagnets can be measured and specified. For example, on p. 232, it is stated that the co-efficient of self-induction of a standard relay of the British Postal Telegraphs ("C" pattern) is 26·4 quads. The usual symbol for a co-efficient of self-induction is L, meaning L quads. In order to find the inductance offered by any electromagnet of which the quads are known, when placed in the circuit in which the alternating electromotive forces go through n periods in a second, all that is needful to do is to multiply n and L together, and then multiply the product by 2π. Or in symbols :—

$$\text{Inductance} = 2\pi n L.$$

Supposing that in this way we know the inductance in a circuit, and supposing that we also know the number of ohms of resistance R in that circuit, it remains to calculate the impedance which these two causes conspire to produce. It is not sufficient merely to add them together. To calculate the impedance,† the joint effect of resistance and inductance, we have to take the square root of the sum of their squares, of

$$\text{Impedance} = \sqrt{R^2 + 4\pi^2 n^2 L^2}.$$

---

* This is short for *quadrant*, the ideal unit being called so because when comparing coefficients of self-induction with fundamental units, the length to which this unit must be compared is 10^9 centimetres, or the length of a quadrant of the earth's circumference. Another name for the *quad*, suggested by Professors Ayrton and Perry, is the *secohm*. They have devised an instrument—the secohmeter—for measuring co-efficients of self-induction. See p. 426.

† Some writers call the *impedance* the *apparent resistance*, or *virtual resistance* of the circuit.

Impedance.

Having thus ascertained the impedance of the circuit, we are ready to calculate the strength of the currents. But here we must remember that Ohm's law, which is the rule for calculating strengths of steady currents, will not help us until we have modified it to suit the case. The rule as modified is known as *Maxwell's rule*, for it was first given by Clerk-Maxwell in 1867. It may, for present purposes, be stated as follows :—

$$\text{Maximum current} = \frac{\text{Maximum electromotive force}}{\text{impedance}}$$

or as follows :—

$$\text{Mean}\dagger \text{ current} = \frac{\text{Mean}^* \text{ electromotive force}}{\text{impedance}}.$$

Neither of these two statements of the law is complete, because neither of them tells us anything about the *retarding* effect of the inductance, causing the maxima of the currents to lag behind the maxima of the impressed electromotive force. If we compare a complete period or cycle of alternations to the revolution of a point around a circle, we may express the amount of retardation of phase in terms of an angular retardation. The Greek letter ϕ is commonly used for the angle of lag. The value of ϕ is always such that its cosine is the ratio of the resistance to the impedance. So that one might re-write Maxwell's rule as follows :—

$$\text{Mean current} = \frac{\text{mean electromotive force}}{\text{resistance}} \times \text{cosine of lag}.$$

This tells us that the mean current is alway less than what it would be (by Ohm's law) if the electromotive force were a steady instead of an alternating one ; always, that is, if there is any inductance at all to cause a lag. If the inductance is very great compared with the resistance, the impedance will practically be made up of inductance alone, and

---

* The word *mean* in these cases, must be taken, as explained on p. 334, to signify the square root of the mean square of the values, not their arithmetical or geometrical mean. The above formulæ assume the alternations to follow a sine function.

the lag will be such that cosine ϕ is very nearly $= 0$; or in other words, the current will depend in no way on the resistance of the circuit but only on the inductance, and the retardation of phase will be 90°, or one-quarter of the whole period.

A well-laminated electromagnet, because of its great self-induction, offers therefore an enormous impedance to a rapidly alternating current, and such electromagnets are now frequently used as *choking coils* in (alternating) electric light circuits to damp down the currents. They do this better than a wire of high resistance would do, because self-induction does not waste energy as resistance does.

ELECTROMAGNETS FOR HEATING PURPOSES.

The alternate rapid reversals of the magnetism in the magnetic field of an electromagnet, when excited by alternating electric currents, sets up eddy-currents in every piece of

FIG. 190.

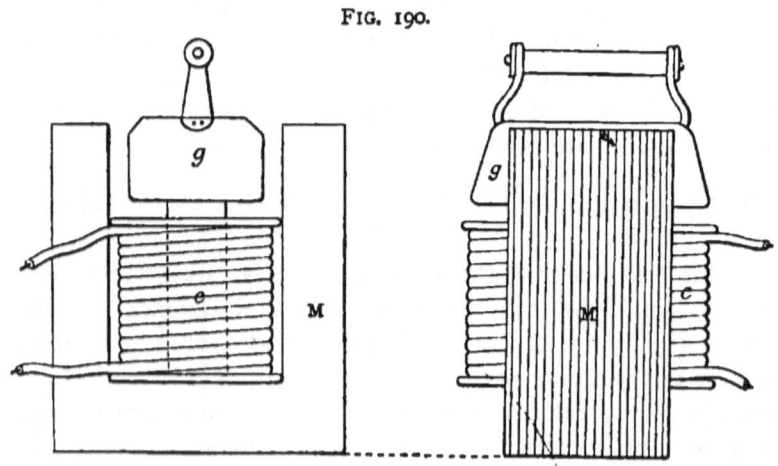

RANKIN KENNEDY'S ELECTROMAGNET FOR HEATING PURPOSES.

undivided metal within range. All frames, bobbin tubes, bobbin ends, and the like, must be most carefully slit, otherwise they will overheat. If a domestic flat-iron is placed on the top of the poles of a properly laminated electromagnet, supplied with alternating currents, the flat-iron is speedily heated up by the eddy-currents that are generated internally within it.

Seizing upon this property, Mr. Rankin Kennedy proposes to construct special forms of electromagnet on purpose to act as heating apparatus, ingeniously combining in it the transformer principle of getting large currents at a low pressure by transforming them down from small currents supplied at high pressure. His electromagnet for heating smoothing irons, copper kettles, tea-urns, &c., is of a tripolar form, built up of stampings of annealed sheet iron, the exciting coil being wound round the central limb, which is broader and shorter than the two outer limbs. Upon the top of this central limb is set the smoothing iron. Suppose the exciting coil to consist of 200 turns, and to be traversed by a current of 20 (alternating) amperes. This is 4000 ampere-turns of current; and by the principle of transformation, as the piece of metal which is acting as the secondary coil counts but as one turn, there will be a current of 4000 amperes, which will rapidly warm it.

CHAPTER XII.

ELECTROMAGNETIC MOTORS.

FROM about the year 1830, it became evident to those who were working with the electromagnet that, by suitably combining electromagnets and armatures, motive-power might be generated at the expense of the energy of electric currents supplied from a battery. Faraday's apparatus for the rotation of a magnet around a copper wire carrying an electric current had, indeed, preceded this date; likewise the rotating star-wheel of Barlow and the rotating disk of Sturgeon. Amongst those who now poposed the construction of electromagnetic motors were Henry,[*] Dal Negro,[†] and Ritchie.[‡] The motors of Henry and of Dal Negro were of the oscillatory type, the moving part being caused to vibrate to and fro alternately. Ritchie's motor was of the rotatory type, being a true progenitor of the modern machine, with fixed field-maget and revolving armature. Fig. 191 illustrates a form of Ritchie's motor as constructed by Daniel Davis, jun., of Boston, U.S., from whose book this cut is copied. The field-magnet is an inverted horse-shoe of steel about 9 inches high. Between its poles is the revolving armature, a copper coil wound on an iron core, the ends of the coil being attached to two contact pieces of silver mounted on the shaft, to serve as a commutator. This apparatus is described on page 212 of Davis's *Magnetism* (edition of 1852) as a " revolving electromagnet," and described as a motor; but on page 268 the same appa-

---

[*] *Silliman's Journal*, xx. p. 340, 1831; see also Henry, *Scientific Writings* (1886), vol. i. p. 54.

[†] *Ann. R. Lomb. Venet.*, April 1834; see also *La Lumière Electrique*, ix. p. 40, 1883.

[‡] *Phil. Trans.*, 318 [2] 1833.

ratus is again figured with a human hand turning the spindle mechanically, whilst the apparatus acts as a generator of currents, the remark being appended that "any of the electro-magnetic instruments in which motion is produced by the mutual action between a galvanic current and a steel magnet may be made to afford a magneto-electric current by producing the motion mechanically." This idea of the reversibility of function of motion and generator was still more explicitly stated by Walenn in 1860. Ritchie's motor was followed in 1834 by that of Jacobi. In this engine the field-magnet was a multipolar combination consisting of two crowns of poles between which a complex armature revolved. This motor was designed for propelling the boat with which Jacobi navigated the river Neva in 1838. The current which traversed the rotating magnets was regularly reversed at the moment of passing between the poles of the fixed magnets, by means of a multiple commutator consisting of four brass toothed wheels, having pieces of ivory or wood let in between the teeth.

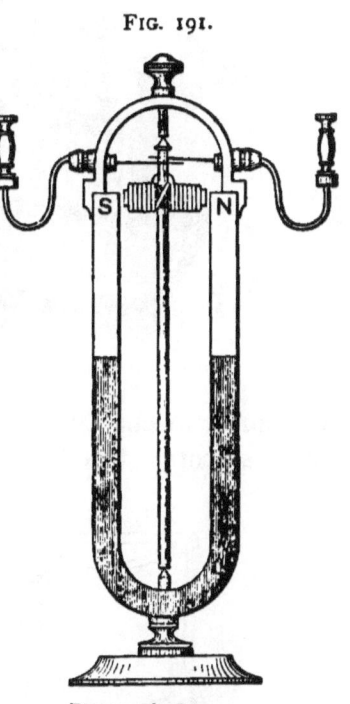

FIG. 191.

RITCHIE'S MOTOR.
(From Davis's *Magnetism*.)

Two early American inventors of motors deserve more than passing mention, viz. Davenport and Page. Davenport's [*] motor to some extent anticipated the more complete designs of Froment. Page,[†] who for nearly twenty years

---

[*] *Annals of Electricity*, vol. ii., 1838.
[†] *Silliman's Journal*, xxxiii., 1838; and [2] x. 344 and 473, 1850; and xi. 86, 1851.

continued to work at the subject, created a distinctive type of electromagnetic engine, based on the plan of using a coil-

Fig. 192.

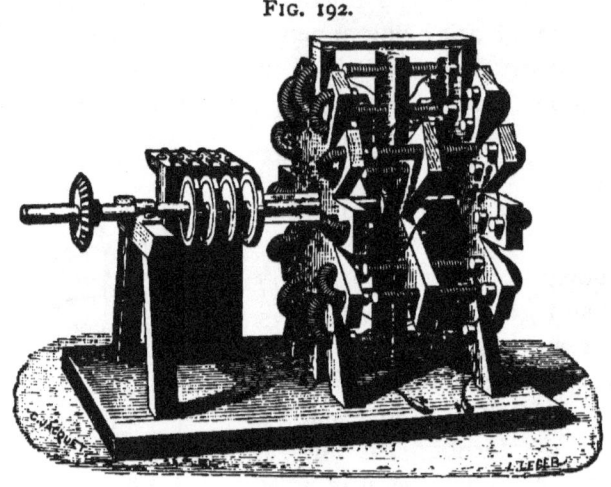

Jacobi's Motor.

and-plunger mechanism instead of an ordinary electromagnet with fixed core. One form of his motor with a double beam

Fig. 193.

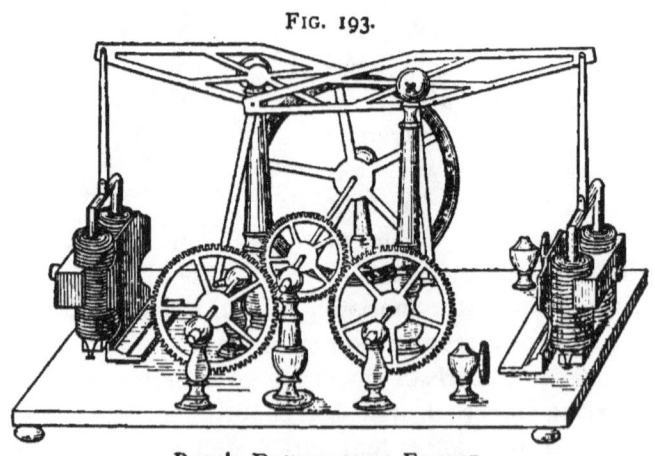

Page's Double-beam Engine.

is given in Fig. 193, which is copied from Davis's *Magnetism*. In one of Page's experiments he used a spiral coil 1 foot in diameter, and was able to raise an iron core weighing 300 lb. through a travel of 10 inches. In another experiment an iron core weighing 532 lb., with an attached load of 508 lb., in total nearly half a ton, was raised through 10 inches. He was able to apply his motors to real work, such as driving a turning lathe. But the cost of deriving electric power from primary batteries prevented these engines from coming into commercial use. Other inventors produced other forms of machine. In England, Wheatstone,[*] whose device for procuring rotation by oblique approach has already been noticed (p. 285 above), was particularly active in devising forms of motors; and Hjörth,[†] who worked at Tipton, and later at Liverpool, produced machines, both motors and generators (see p. 243 above), for which a gold medal was awarded him at the Great Exhibition of 1851. In Holland, Elias produced a motor in which both field-magnet and armature consisted of electromagnets in the form of rings wound to have several consequent poles. In France, Froment[‡] led the way with numerous forms, the best known of which were based on the plan of affixing upon the periphery of a rotating wheel a number of iron bars, as in Fig. 194, to be attracted laterally toward the poles of an electromagnet the current in which was automatically cut off just when the approaching armature had come nearly to the position of least distance. Such motors, some with multiple electromagnets, were used by Froment for many years for driving dividing-engines and other light machinery, in his instrument factory in Paris.

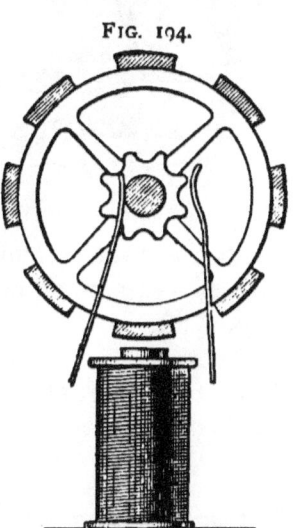

FIG. 194.

FROMENT'S MOTOR.

[*] *See* Specification of Patent 9022 of 1841.
[†] *See* Specification of Patent 12,295 of 1848, and 2198 of 1854.
[‡] *L'Institut*, lxxxii., Dec. 1834; see also *La Lumière Electrique*, ix. 194, 1883.

Another French instrument maker, Bourbouze, constructed motors such as that represented in Fig. 195, from the designs of Breton. This motor followed the plan invented by Page, but with some modifications. A stopped coil with half-long plunger (see p. 269, above), replaced the simple coil and

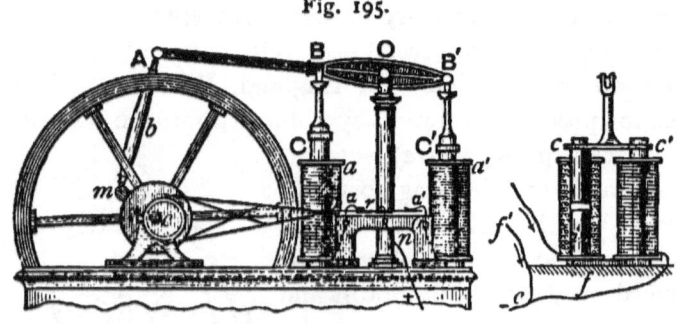

Fig. 195.

BOURBOUZE'S MOTOR.

plunger, and the commutator, constructed to imitate the slide-valve of a steam-engine, was operated by an eccentric on the crank-shaft.

Another French motor, which gained a prize at the Paris Exposition of 1856, was designed by Roux. In this motor (Fig. 196) was applied the equalizer of Froment (Fig. 131,

FIG. 196.

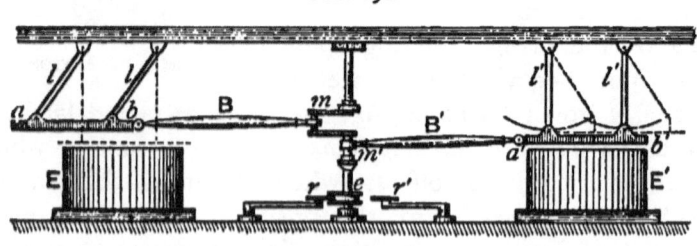

ROUX'S MOTOR.

p. 283), and the electromagnets were ironclad, a sheet of iron being folded around the outside of the coil.

In 1864 Pacinotti* produced the first machine with a true

* *Nuovo Cimento*, xix. 378, 1865.

ring-armature, the forerunner of the modern machines. This ring form was independently reinvented in 1870 by Gramme,[*] but with the modification of winding the copper wire over the whole periphery of the ring. Since that date the subject of motor design has become simply a part of the subject of the design of dynamo-electric machines in general. For, strange to say, the great advance in the design and construction of electric motors came about at the hands of inventors who were not trying to invent a motor at all, but who, in perfecting the dynamo machine for producing electric light, found themselves in possession of a far better electric motor than any that had been specially designed as such.

Modern electromagnetic motors may be classified under two heads—those intended to work when supplied with continuous currents, and those intended to work when supplied with alternate currents. In each case the machine may be described as consisting of two parts—a field-magnet and an armature—the distinction between the two parts being that in the part called field-magnet the magnetism remains constant (or nearly constant), whilst in the part called armature the magnetism is being continually reversed during the running of the machine.

In continuous current machines the field-magnet usually consists of a single stationary massive electromagnet of simple form, between the poles of which the armature revolves; the latter being usually a complex electromagnet built up symmetrically around the shaft, with a core of laminated iron (usually in the form of a ring or a drum) covered with a peculiarly wound system of insulated copper conductors through which the current circulates in a manner governed by a special commutator. Fig. 197 illustrates a modern form of electromagnetic motor with a drum armature placed between the poles of a field-magnet having a double magnetic circuit. This is a compact and powerful form; others of kindred design are made by scores of engineering firms.

[*] *Comptes Rendus*, lxxiii. p. 175, 1871; and Specification of Patent No. 1668 of 1870.

For use with alternating currents a different type of structure is adopted. The field-magnet of alternate current machines is usually multipolar, a common form resembling

FIG. 197.

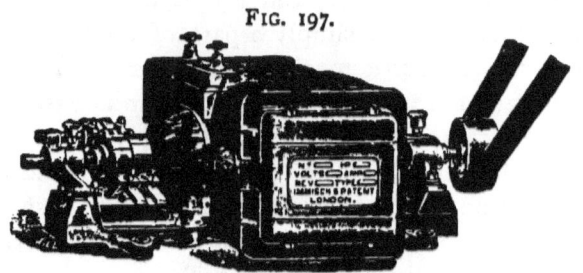

IMMISCH'S MOTOR.

that sketched in Fig. 198, consisting of two opposing crowns of poles of alternate and opposite polarity, set upon two stout frames. In the gap between these two crowns of poles rotates the armature, which, in this class of machines, usually contains no core of iron, but consists simply of a set

FIG. 198.

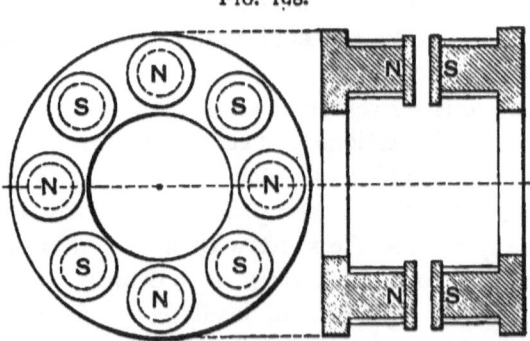

FIELD-MAGNET OF ALTERNATE-CURRENT MACHINE.

of copper coils, equal in number to the number of peripheral poles (eight in this case), wound alternately in right-handed and left-handed spirals. Alternate current motors of this type possess two important peculiarities—(1) that they are not self-starting, but require to be raised in some auxiliary way to their proper speed; and (2) that when so started they run

Alternate-current Motors.

whatever their load, in absolute synchronism with the alternations of the current, and therefore with the rotation of the dynamo which provides the current at the generating station. An important improvement has been introduced by Mr. Mordey, who substitutes for the complex field-magnet with coils wound separately around each projecting pole, a much simpler design, Fig. 199, having only one coil. The multiple

FIG. 199.

FIELD-MAGNET OF MORDEY'S ALTERNATE-CURRENT MACHINE.

magnetic field is procured by placing on the ends of the cylindrical core two massive pole-pieces furnished with projecting horns which nearly meet. In this particular machine the field-magnet revolves, and the armature coils are held in a stationary frame.

A very neat alternate-current mechanism for driving electric clocks, devised by Grau-Wagner,[*] has a polarized rotating armature with obliquely-shaped polar faces.

Other alternate-current motors have lately been devised

[*] *Zeitschrift für Elektrotechnik*, vol. iv. p. 1.

by Ferraris, Tesla, Elihu Thomson, and others, based upon the mutual actions of coils in which alternating currents in different phases are circulating. To describe them in detail would go beyond the limits of this work. The reader is referred to the author's book on 'Dynamo-Electric Machinery' for fuller account of electric motors and of the theory of the electrical transmission and utilization of power.

CHAPTER XIII.

ELECTROMAGNETIC MACHINE TOOLS, AND MISCELLANEOUS APPLICATIONS OF ELECTROMAGNETS.

MANY years have elapsed since Froment employed his electromagnetic motors to drive various light tools, watchmakers' lathes, dividing engines and the like, in his *atélier* in Paris. From this down to the present time but little has been done to develop the special capabilities of the electric motor as a machine tool. True there are many workshops and factories in those towns where the public supply of electricity is now an accomplished fact, particularly in the United States, where electric motors are employed in lieu of steam-engines to drive shafting and machinery. In all this there is no particular adaptation to special ends of the qualities by which an electric motor is distinguished from other prime movers.

The very high speed and lightness which may be given to the moving parts of an electric motor mark it out for special purposes. For a good many years electromagnetic motors have been employed for running dentists' drills. M. Trouvé, in Paris, has been prominent in devising applications of this kind; and in the United States electro-dental appliances have become widely spread. Small automatic hammers that can be held steadily in the hand, and deliver a multitude of minute blows, are found useful for inserting stopping into hollow teeth. Some account of this application of the electromagnet to dentistry will be found in the Jury Reports of the Electrical Exhibition held at Philadelphia in 1884.

Another mode of using electromagnetic power for tools consists in employing a movable core or plunger in a tubular coil wound *in sections*. If a current is introduced at any point in such a coil, it flows both ways through the coil from that point, and produces a consequent pole (see p. 55 above) at that point. Toward such a polar point in the tubular coil the end of the internal iron core is urged. Consequently, if the point at which the current is introduced into the coil is being continually shifted by sliding the contact to the circuit along the turns of the coil by a special commutator, the core can be drawn along the coil to any point as may be desired. On this plan, Page and Du Moncel constructed motors, and in 1880 M. Marcel Deprez revived the method and produced an electromagnetic hammer,* depicted in Fig. 200. The coil A B consists of 80 sections or separate coils, all connected together into one long coil, and with a branch wire connecting the junction of the end of each section and the beginning of the next down to one segment of the commutator F G. Upon this commutator press two springs C E, C D, fixed to a handle H I. The distance to which the core is drawn depends on the position to which the handle is moved. The iron core weighs 23 kilogrammes, and to its weight can be added an actual force of 70 kilogrammes when a current of 43 amperes is passed through 15 of the sections.

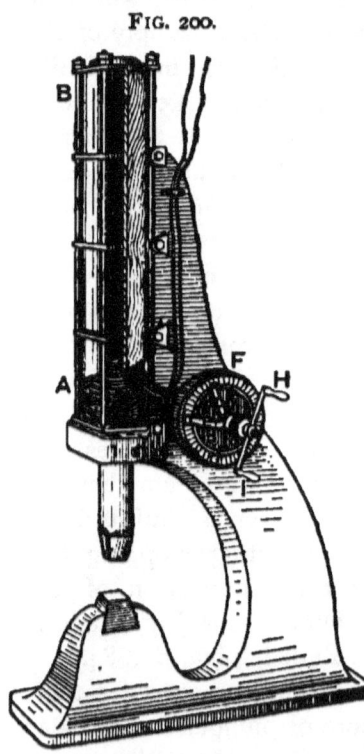

Fig. 200.

Marcel Deprez's Electromagnetic Hammer.

In 1882 there were exhibited at the Electrical Exhibition at the Crystal Palace at Sydenham, some electromagnetic

* *La Lumière Electrique*, ix. 44, 1883.

appliances for use in the salvage of ships. These were the invention of Mr. Latimer Clark. One of them consisted in an electromagnet to be used by a diver going down to repair a submerged iron vessel, or prepare her for being raised. This electromagnet was to be let down to any required depth at the side of the vessel, being connected by flexible cables to a

FIG. 201.

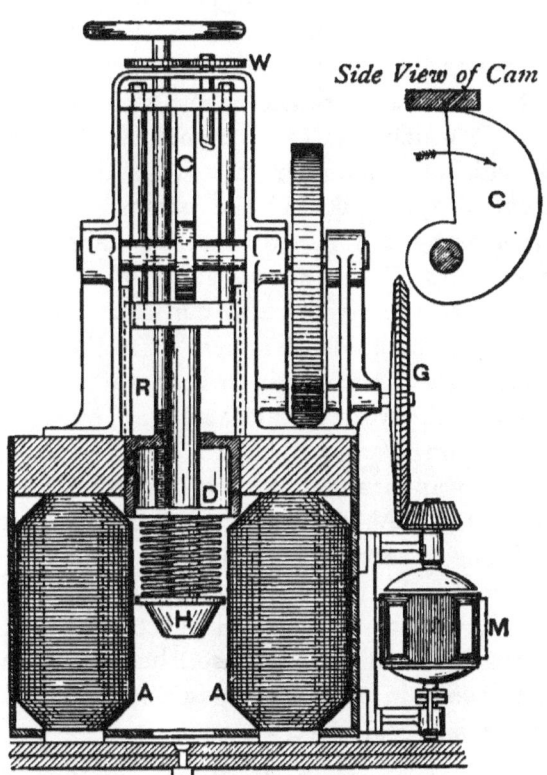

ROWAN'S ELECTROMAGNETIC RIVETER.

suitable electric source, and the diver descending with it was provided with a switch, by means of which he might at any moment excite the electromagnet and cause it to attach itself to the iron hull. It thus provided him with a seat or with a fixed point from which to begin operations, such as drilling holes in the iron plates for inserting hooks or chains. An electrically driven motor to work the drill was amongst the devices suggested.

In recent years various new forms of machine tools have been brought out by Mr. F. J. Rowan, of Glasgow. These are chiefly riveters and drillers intended for use in building iron ships. The body of the riveter is constituted by an electromagnet A A (Fig. 201), which can be applied to any part of the surface of the ship where a rivet is to be driven, and thus temporarily adhere. An electromagnetic motor M, driven by the current, is connected by gearing G to a snail-shaped cam C, which lifts the hammer H. The amount of force with which the hammer falls upon the rivet depends upon the coiled spring behind the hammer head; and the power of this spring can be adjusted by hand by means of the screw spindles R and spur-gearing W. The "bolster" or "holder-up" of the rivet is also constituted by an electromagnet which can be made to adhere on to the iron plates at the other side. Electromagnetic drillers are made to attach themselves on to the work in the same way by a holding-up electromagnet, and is also operated by an electromagnetic motor. Mr. Rowan has also constructed tools for caulking and chipping; the tool, electromagnetically operated, being set upon a frame to serve as a guide bar, which frame is itself attached to the ship by holding-up electromagnets.[*]

In February 1891, a paper on electromagnetic mining tools was read by Mr. Ll. B. Atkinson before the Institution of Civil Engineers, in which are described several electromagnetic drills and coal-cutters.

---

[*] For further information respecting Mr. Rowan's electromagnetic tools see *Proc. Inst. Mechanical Engineers*, August 2nd, 1887; and *Trans. Inst. Engineers and Shipbuilders in Scotland*, March 20th, 1888.

CHAPTER XIV.

MODES OF PREVENTING SPARKING.

EVERY one is familiar with the fact that on opening the circuit of an ordinary electromagnet a bright spark is seen. Such sparks fuse and corrode away the contact surfaces of the keys and switches employed; and in the case of vibratory apparatus, cause a rapid deterioration of the contact-pieces.

The means taken in past time to prevent such deterioration are of several kinds. The contact surfaces were made of metallic silver; and later, platinum being found superior in its power of resisting fusion and oxidation, was substituted, and has become—in spite of its high price—of general application. Also, in order to keep the contacts bright, it was found advisable to arrange the moving parts so that a rubbing-contact, rather than a mere pressing one, should be made. Lastly, snap-switches have been used; that is to say, switches arranged with mechanical devices so that on opening the circuit the moving part springs suddenly back from contact, with a rapid motion, thus breaking down the spark which otherwise would follow and burn the contact surfaces.

CAUSES OF SPARKS.

It is necessary, however, to study the phenomena of the spark a little more closely. The spark is never seen at make of contact; only at break. It is small and faint in plain circuits with no coils or electromagnets in them; but bright and sharp in circuits in which there are coils or electromagnets. As these sparks are able to leap over air-gaps of a very visible length—say from $\frac{1}{20}$ to $\frac{1}{10}$ inch,—it is evident that they are impelled by electromotive forces much greater than those of the ordinary batteries. An electromotive force of 100 volts even, will not of itself cause a spark to strike across a gap $\frac{1}{100}$ inch in width. The electromotive force that sets up the sparks is one due to self-induction between the various convolutions of the copper coil, which at the moment when the magnetism is dying away, act on one another as the primary and secondary wires of an induction coil do. The more rapid the rate at which the magnetism is decreasing, the greater the self-induced electromotive force. The magnitude of the electromotive force is also attested by the shocks which are often felt by experimenters if their hands touch the two parts of the circuit between which the spark-gap occurs.

It was the fashion at one time to speak of the current thus self-induced, as an "extra-current"; and it was usual to say that on making a circuit there was a momentary extra-current which ran in a direction opposing the main current, and prevented it rising instantly to its full value, and that at break of circuit there was another momentary extra-current which flowed in the same direction as the dying current, and manifested itself as a spark.

It has been explained in Chapter VII., p. 222, how every circuit possesses a "time-constant," which is, indeed, the ratio of its coefficient of self-induction to its resistance. It is to this same self-induction that the extra-current spark at break is due. But the time-constant at break of circuit is always less than the time-constant at make, because, though the coefficient of self-induction may be a constant, there is a great increase in resistance in the path of the spark. No one knows how much is the resistance along a spark $\frac{1}{100}$ inch long. Indeed it is doubtful whether such a spark can be said to offer a definite resistance to the flow of electricity along it. It partakes somewhat of the nature of the arc or electric flame ; and no doubt its resistance increases as it is lengthened, though probably not in any rational proportion.

An instructive way of considering the matter is to remember that when the circuit is first closed some of the energy flowing from the battery has to be employed in building up the magnetic field; or, in other words, in thrusting into the core the magnetic lines. This energy, which remains in the system as long as the field is maintained, as long therefore as the current is kept flowing, will disappear or run down as soon as the current is turned off. *The extra-current spark seen at break is the visible mode in which the energy of the magnetism spends itself.* The amount of heat developed in this spark may be taken as a measure of the energy accumulated in the system during the variable period when the current, at make, was rising to its final strength. The quantity of electricity so accumulated and subsequently restored is equal in amount to that which would be conveyed by the current if flowing at its steady value for a time equal to the time constant. If E be the *volts* applied to the circuit by the battery, R the *ohms* of resistance of the circuit, and L the *quads* of self-induction, then the quantity Q of *coulombs* of electricity virtually stored in the system is:—

$$Q = \frac{E}{R} \times \frac{L}{R} = \frac{EL}{R^2}.$$

And, as this quantity was virtually stored under an electromotive force E, while the quantity grew from zero to Q, the work done in the operation by the battery, and stored as potential energy, is:—

$$W = \tfrac{1}{2} \frac{E^2 L}{R^2}.$$

If on opening the circuit we can imagine the resistance to be abruptly increased by the interposition of a gap to a much higher value R_2, the time constant will be diminished to the corresponding value $\frac{L}{R_2}$, and consequently as the potential energy stored up will run down with a greater electromotive force E_2 such that

$$\tfrac{1}{2} \frac{E^2_2 L}{R^2_2} = \tfrac{1}{2} \frac{E^2 L}{R^2}.$$

Now it is quite conceivable that the resistance introduced by the gap may be hundreds of times greater than that of the original circuit; hence the electromotive force that forces the spark across the gap may be hundreds of times greater than that of the battery.

It must be obvious from the foregoing considerations that the most powerful electromagnets, will—*ceteris paribus*—set up the strongest sparks; for the coefficient of self-induction is proportional to the flux of magnetic lines in the core and to the number of coils which surround it. Nevertheless, there are one or two other matters that may affect the spark. If beside the copper wire coil there is any other metal circuit surrounding the core, or if the core itself is solid (not laminated), then there will, at the moment when the magnetism is disappearing, be generated in such circuit or solid core, induction currents, as well as those in the wire coil; and these eddy-currents will tend to keep up the magnetism, and thus by making it die out more slowly will prevent the induced electromotive force from rising so high. Such damping circuits therefore lessen the spark at break. Another way of stating the same facts is to say that if there is mutual induction the self-induction will be thereby virtually lessened. The presence of metal masses in which eddy-currents can be set up by mutual induction, has the effect of distributing the electromotive forces due to self-induction, diminishing them in amplitude, and extending the time over which they last.

The abrupt opening of a circuit in which there is an electromagnet, thus throws into play a sudden electromotive force of high amount, and gives rise to sudden oscillations of current, and sets up sudden tensions in the insulating coatings, which may even pierce them and injure them permanently. Higgins observed[*] that such sudden discharges are not regulated by the usual laws of steady currents, and instead of passing through a spiral conductor of 2 inches of No. 30 or No. 24 B.W.G. copper wire, will go preferably by a more

[*] *Journal of Society of Telegraph Engineers*, vi. 139.

direct route made up of two thicknesses of lacquer and an air space, though it offers perhaps a resistance of some millions of megohms.

Mechanical Devices for Suppressing Sparks.

Some of these have already been alluded to; but it will be convenient to classify the various methods.

(*a*) *Snap Switch.*—A rapid-break switch which parts the separated ends of the circuit to a really wide gap before the self-induced electromotive force of the discharge has had time to rise to its maximum, causes the electromotive force to spend itself in some other way. It produces a momentary high charge on the end of the wire, which then surges backward and forward (unless it pierces through insulation to some other path) in the wire and dies away.

(*b*) *Break under Liquid.*—If the break in the circuit is caused to occur under water or alcohol, the spark is much diminished, or even suppressed, but the conditions are awkward to carry out. If oil or other hydrocarbon is used, the contact points become clogged with carbonized matter.

(*c*) *Wiping out Spark.*—The author has suggested that a moving break should carry behind it a pad of asbestos, renewable from time to time, to wipe out the spark.

(*d*) *Blowing out Spark.*—An air blast delivered just at the right instant will blow out the spark. This is done at the commutator of the Thomson-Houston dynamo.

(*e*) *Break in Magnetic Field.*—If the break takes place at a part of the circuit that lies in a magnetic field, the spark is blown out laterally with extraordinary suddenness. If the circuit of an ordinary (large) electromagnet be broken close over its pole, or in a narrow gap between pole and armature, the snap of the spark will resound like a pistol-shot. The magnetism under these circumstance falls more rapidly, and the extra-electromotive force is much higher.

ELECTRICAL DEVICES FOR SUPPRESSING SPARKS.*

(A) *Shunting Break with a Resistance.*—If across that part of the circuit where the gap is to be made there is attached as a shunt a wire of considerable resistance (*not coiled*), the extra current due to the self-induction of the circuit will find here a path through which it can flow without having to leap the air-space as a spark. This device was first introduced by Dering in 1854, who proposed to use a resistance about forty times as great as that of the electromagnet coils in the circuit. The same was revived by Dujardin in 1864. A coiled wire is no use for this purpose, as it offers self-induction, and the discharge refuses to go around its convolutions. A very thin wire of platinum is better, or even a streak of a plumbago pencil on a rough surface. A drop of water between two platinum wires joined to either part of the circuit will also answer the purpose. The author, in 1880, proposed a special form of switch in which there is a high resistance wire across the break, this wire being subsequently itself parted after the passage of the discharge, by a further movement of the switch handle. The theory of the high-resistance shunt is very simple. Let the electromagnet have a resistance of r ohms, and let the current in it be i amperes. Suppose it to be laid down that the potential at the terminals of the break must never rise to more than E volts. Then dividing E by i we find the value R of a resistance, such that if the current were to continue full strength, the value E would be just attained. But the interposition of this resistance will itself cut down the current, which simply dies down without sparking. As an example, we may take a case where $i = 0.05$ ampere, and 300 volts is prescribed as the limit for E. Then R the high resistance to be used as shunt must be equal to or not greater than $300 \div 0.05 = 6000$ ohms. This is 60 times as great as the resistance of the coils.

(B) *Shunting with a Condenser.*—Two cases arise here.

* On this topic, see Vaschy, *Annales Télégraphiques*, 1888, p. 290.

A condenser of suitable capacity may be placed as a shunt—(i.) across the break, or (ii.) across the terminals of the electromagnet itself.

If the condenser used is of proper capacity, then it may, when placed as shunt to the electromagnet, entirely counterbalance, so far as the rest of the circuit is concerned, the impeding action of the self-induction of the coil; it does not prevent the retarding action in the coil itself. If the condenser is placed across that part of the circuit where break is to occur it has no effect on the current at make, because it is itself at that instant short-circuited; but when the circuit is broken it lowers the striking distance, because any increase of capacity of the terminal parts lowers the potentials at those parts. If the capacity is not great enough in proportion to the self-induction of the circuit it does not entirely obviate the spark. If it is as great as, or greater than, the value $C = 4 L \div R^2$, then there will be no spark, and the discharge from the electromagnet will simply oscillate along the wire circuit into and out of the condenser, and die away. A third way of using a condenser is to connect it as a shunt to a plain resistance inserted in the circuit or across the break. The first to suggest the combined use of condenser and high-resistance shunt, appears to have been von Helmholtz.*

(C) *Shunting with a Voltameter or Liquid Resistance.*—It has been suggested by d'Arsonval that the extra current might be provided with a path which, though impassable to the ordinary electromotive force of the circuit, should be readily passable by the higher electromotive force of self-induction. A set of small voltameter cells joined as a shunt across the break, may, if sufficient in number, entirely obstruct the ordinary current. For example, if the working battery is 10 volts, then a set of six small cells with leaden plates in them (capable of polarizing, like accumulators, to 2 volts each) will be permeable only to electromotive forces exceeding 12 volts.

(D) *Mutual Induction Protectors.*—In 1867, Mr. C. F.

* See Appendix VIII. of von Helmholtz's *Sensations of Tone.*

Varley proposed the use of a copper sheath surrounding the core, to lessen the effect of self-induction. The action of such a closed circuit surrounding the core has been explained at the beginning of this chapter. The suggestion was revived in 1878 by Brush, in the construction of the field magnets of his dynamo machine.

A further development of the same idea came about in 1870, at the hands of Messrs. Paine and Frost, who laid a layer of metal foil between each of the successive layers of wire. This mode of using a mutual-induction has also been adopted by Dr. Aron, of Berlin, since 1887. It has the disadvantage of increasing the liability to break down of insulation.

(E) *Short-circuit Working.*—In this method of working the circuit is never opened, but a path of very low resistance is provided as a shunt to the coils of the electromagnet. When a switch in this path is closed, the current takes the path of low resistance and no longer circulates around the electromagnet. To excite the electromagnet the switch must be opened. This is the inverse of the ordinary case, and when the switch is opened there will be no magnetism in the core to set up a discharge in its disappearance, and therefore there will be no spark due to this cause at the break. And there will be no spark at make, because only after the circuit has been thus made does the magnetism of the core die out. The method is not perfect unless the shunt path is relatively of negligibly small resistance and inductance. Neither does it obviate sparks due to self-induction in other parts of the circuit. It is excellently adapted for use with electromagnets to be used in the circuits of arc lamps, or in other constant-current circuits.

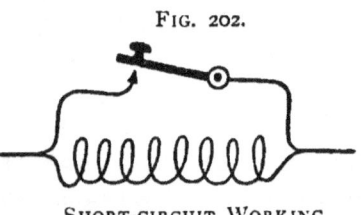

FIG. 202.

SHORT-CIRCUIT WORKING.

(F) *Differential Winding.*—In this method the electromagnet is provided with two separate coils, either wound in opposite directions or else so connected up that the current will circulate in opposite directions around the core, and with an equal number of convolutions in each coil. There are several ways of doing this, but of these the best is to wind the coil throughout with two separate well-insulated wires laid side by side. One of these two windings is connected straight into the circuit, and the second is connected as a shunt to the first, but with a switch inserted. When this switch is open the current circulates round one coil only, and magnetizes the core. When the switch is closed the current divides, half flowing through one coil, half through the other, and exactly neutralizing one another's magnetizing power.

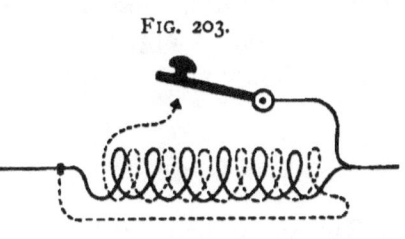

FIG. 203.

DIFFERENTIAL WINDING.

This method of operating has all the advantages of the short-circuit method, but requires a greater weight of copper. But it is even more perfect, as it eliminates the spark due to self-induction in other parts of the circuit. It has been used with great success in electric bells by Messrs. Jolin, of Bristol.

(G) *Multiple Wire Method.*—For the purpose of the transmitting electromagnets used in his system of harmonic telegraphy (the Phonopore) Mr. C. Landon Davies has used a device called *multiple wire*. The bobbin is wound with a number (from four to twenty) of separate layers of fine wire, a separate wire being taken for each layer, and all are wound in the same direction, not turning back at the end of each layer as in ordinary windings. After the bobbin is thus filled all the wires are joined up in parallel, so as to act electrically as a single

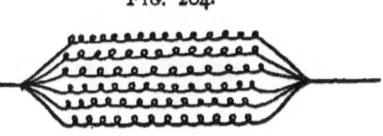

FIG. 204.

MULTIPLE WIRE WINDING.

wire of greater cross-section. In this case the time-constants of the separate circuits are different, because, owing to the fact that these coils are of different diameters, the coefficient of self-induction of the outer layers is rather less, and their resistance, because of the larger size, rather greater than those of the inner layers. The result is that instead of the extra currents running out all at the same time, it runs out at different times for these separate coils. The total electromotive force of self-induction never rises so high, and it is unable to jump a large air-gap, or give the same bright spark as the ordinary electromagnet would give.

THE COMPARISON OF VARIOUS METHODS.

The author had five bobbins constructed, all of the same size, to fit to the same core, and wound in five separate modes with about the same weight of copper on each, in order to compare the sparks produced. The first was wound in ordinary way; the second with a copper sheath around the core; the third with interposed layers of foil; the fourth differentially wound; and the fifth with multiple wire in 15 layers. The differential winding gives absolutely no spark at all; and second in merit comes No. 5, with the multiple-wire winding. Third in merit comes the coil with intervening layers of foil. The fourth is that with copper sheath. Worst of all, the electromagnet with ordinary winding.

COMPENSATING THE SELF-INDUCTION IN THE CIRCUIT.

The self-induction of an electromagnet can be entirely compensated, so far as the rest of the circuit is concerned, by the joint employment, as in Fig. 205, of a condenser and a resistance. If the coefficient of self-induction of the electromagnet be called L and its resistance R_1, and the capacity of the condenser C and the resistance with which it is shunted

R_2 then the retarding effect of the electromagnet will be precisely compensated, if the condition is fulfilled that $C R_1 R_2 = L$.

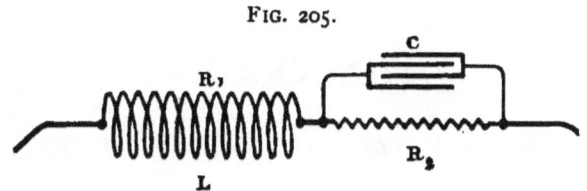

FIG. 205.

COMPENSATING THE SELF-INDUCTION OF AN ELECTROMAGNET.

CHAPTER XV.

THE ELECTROMAGNET IN SURGERY.

FOR many years past ophthalmic surgeons have used both steel magnets and electromagnets. Indeed, the use of the loadstone has been known for two centuries,[*] for the purpose of extracting from the eyeball chips of iron which have by accident lodged therein, as frequently happens in engineering workshops to fitters and metal turners. Later, in 1745, Dr. Milnes, in his "Observations of Medicine and Surgery," describes how he applied a loadstone to remove a piece of iron from the iris. Probably this was not strictly the case, as he says "immediately it jumped out." It is more likely that he meant that the fragment of iron was embedded in the cornea in front of the iris. According to Hirschberg,[†] Dr. Meyer, of Minden, was, in 1842, the first to use a magnet in extracting fragments of metal from the actual interior of the eyeball.

In the middle of this century cases of successful removal were recorded by Critchett,[‡] Dixon, White, Cooper, and others. There seems to have been little advance until 1874, when Dr. McKeown, of Belfast, recorded in the *British Medical Journal* several successful cases treated by means of a permanent steel magnet tapered at each end to facilitate its introduction. In 1880, Gruening, of New York, employed a permanent steel magnet composed of a number of "cylinders" (*sic*), united at each end and fitted at one extremity with a needle-shaped pole.

---

[*] Fabricius Hildanus, *Opera observationum et curationum*, Frankfort, 1646. Of this rare work I have not seen a copy.

[†] Hirschberg, *Der Elektromagnet in der Heilkunde*, Leipzic, 1885.

[‡] *The Lancet*, April 1854, p. 358.

The first use of the electromagnet in this connection is accredited to Hirschberg, of Berlin, in 1877.

In this country, Snell,* of Sheffield, has been the chief advocate of its use, and in his hands splendid results have been obtained. In the *British Medical Journal* for 1881 he describes his instrument and records the results of several operations. His experiments have steadily progressed since then, and in a recent number of the same journal he brings up the number of his recorded cases to 77.

The thing needed for this special service is an electromagnet which can be fitted with polar appliances of various elongated forms which can be introduced into the wound in the eyeball to a sufficient distance to enable the particle of iron to be attracted and withdrawn. It is necessary at the same time that the electromagnet should not be too heavy to be conveniently held in the hand of the operating surgeon, and that it should be capable of being excited with current from a battery of moderate size, such as an oculist might be expected to be able to manage and maintain without having to call in a skilled electrician at every occasion of its use. It is obvious that, other things being equal, a magnet of considerable projective power is desirable ; hence one with a long core, or with an iron expansion at the posterior end of its core, is preferable to a short one. Possibly in some cases the placing of a mass of iron behind the patient's head during the operation might aid the electromagnet to project its magnetic field to the desired depth. Various forms of magnets have been devised by various experimenters.

Snell's instrument, which may be quoted as the typical form commonly used by ophthalmic surgeons in this country, consists of a soft iron core surrounded by a coil of insulated copper wire enclosed in an ebonite case. At the one extremity are the terminals for the battery connections, and at the other the soft iron core is tapped to receive the variously shaped and sized needle poles. He recommends the use of a single quart bichromate cell to generate the current.

* Snell, *The Electromagnet and its Employment in Ophthalmic Surgery*, London, Churchill and Co., 1883.

Hirschberg's instrument consists of a hollow cylindrical core with two curved pointed poles of different thickness for external and internal use.

Bradford, of Boston, devised an instrument almost identical with Snell's, and simultaneously with the latter. It is, however, a much heavier and more powerful agent.

McHardy, of King's College Hospital, used a very similar instrument, but of much reduced size and weight.

The latest form is that brought out by Tatham Thompson, of Cardiff, and is a further modification of Snell's instrument, but for the same weight and bulk, additional power is claimed, as the magnetic circuit is perfected by means of a collar of iron continuous with the core projecting forwards over the insulated coil towards the manipulative pole-piece.

FIG. 206.

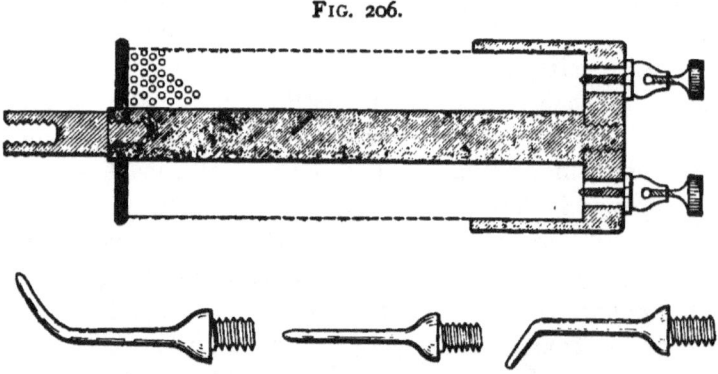

SPECIAL ELECTROMAGNET FOR EXTRACTING CHIPS OF IRON FROM THE EYEBALL.

The core is tapped at its extremity to receive the differently shaped poles or an intermediate polar extension where deeper probing is required. The battery used with this consists of one or more quart bichromate "dip" cells. This instrument is shown in the illustration in its actual size, and its weight is only about five ounces.

The most frequent use of the instrument is in removing fragments of steel or iron from the surface or embedded in the coats of the eye, and by its use much disturbance and laceration of the delicate structures is avoided, which the older methods of cutting out or withdrawing by means of forceps, spuds, or curettes, too often involved. Its most valued properties are, however, best demonstrated where it is used for the removal of foreign bodies which have penetrated the walls of the eye. It is not too much to say that in many cases it has been the only possible means of preserving sight.*

In the first place, it is of diagnostic value, for where a fragment, for instance, has perforated the cornea or anterior transparent part of the eye, it can often be proved whether it is steel or iron, by observing whether it is acted on by the close approach of the instrument. If no movement is observed, and no peculiar sensation of dragging pain felt by the patient on the approach of the magnet, it may be due either to its not being a magnetizable metal or to its being gripped by the tissues, or encysted with inflammatory material thrown out by its irritating presence. If in this case a very lightly suspended needle be magnetized and held close to the eye, it may sometimes be found to dip towards the foreign body. If it should chance to be a steel fragment, this effect is greatly increased by holding the electromagnet close to the eye for a time, so as to magnetize the offending particle.

Further still, the instrument is at times of immense use in moving the fragment from one part of the eye where it is inaccessible, and dropping it in a more favourable position for subsequent removal. For instance, suppose the fragment of steel or iron has penetrated the cornea and lodged in the crystalline lens, it has happened in several cases that by applying the electromagnet to the cornea, the metal has been attracted forward through the lens substance, across the aqueous humour contained in the outer chamber of the eye, and then dropped by opening the circuit on to the floor of the outer chamber; then a small opening being made at the

* In addition to references previously given, see Mellinger, *Ueber die Magnet Extractionen an der Basler ophtalmologischen Klinik* (Inaug. Dissert.), Basel, 1887.

most dependent part, the pole of the instrument introduced there, and the fragment readily removed.

In at least one recorded case a sharp-pointed fragment, which had penetrated deep into the lens, has been seen to force its way forward through the lens substance and lens capsule, through the iris, and actually be drawn through the cornea, and so removed completely without involving any cutting operation whatever.

It is, however, in those cases where the foreign body has penetrated the posterior or vitreous chamber of the eye that the wonderful results obtained by the electromagnet are best seen. Where only the outer chamber or the lens is the seat of the mischief, other means may possibly succeed; but when a fragment has penetrated the vitreous chamber, we require an instrument not to go blindly searching round, disturbing the delicate structures and risking the escape of the semi-fluid contents, but one which will attract to itself the offending particle, and this is where the instrument which we are considering reaches its highest achievements.

Many successful cases have been recorded by Snell, Hirschberg, Lloyd Owen, McHardy, and others, of fragments of steel thus removed from eyes which otherwise must have inevitably been lost.

The most serious and desperate cases of all are where the metal has actually traversed the inner chamber of the eye and become embedded in the delicate retina and choroid lining the posterior wall. So far as can be ascertained, only three successful cases of removal from this position have been recorded. The first by Galezowski in 1882, where a fragment of iron had passed through cornea, iris, and lens, and was lodged in the retina. Its position was noted by means of the ophthalmoscope, an instrument employed by ophthalmologists for examining the interior of the eye. An incision was made in the sclerotic or white coat of the eye over the situation, and the fragment removed by means of the electromagnet. The second, recorded by Hirschberg in 1888. The third, by Tatham Thompson, of Cardiff, in 1890. Some detailed account of this will well illustrate the method of the procedure.

The patient was a blacksmith employed at one of the South Wales collieries, and on December 8th, whilst "stamping" a new "pick," was struck on the left eye by a fragment of the new tool. The fragment struck him on the "sclerotic" or "white of the eye," about a quarter of an inch from the margin of the cornea. He felt little pain at the time, but two days later dull pain supervened, and considerable irritation and dimness of vision, not only in the injured, but in the sound eye.

He was sent to the Cardiff Infirmary on December 10th. When examined there, a small wound was found at the point where the fragment had entered on the inner side, and on ophthalmoscopic examination, the piece of steel could be seen embedded in the retina at the upper and outer part. The track which it had made in traversing the vitreous chamber could also be distinguished by slight opacities in the humour. The uninjured eye showed distinct signs of sympathetic irritation. The next day the patient was put under the influence of ether, the wound of entrance slightly enlarged to allow the introduction of the pole of the electromagnet. This was passed across the vitreous chamber in as nearly as could be calculated the direction originally taken by the fragment. On withdrawing the first time nothing resulted, but at the second attempt the piece of steel came readily through the wound "in tow" of the pole. Only a very small bead of vitreous humour escaped, and the eye was dressed antiseptically and bandaged. At the end of twenty-four hours he could count fingers at four feet distance, was quite free from pain and dread of light, and the dimness of the other eye had largely disappeared. The wound healed in three days, and the sight steadily improved, till on December 30th he could read moderately sized print. Examined by means of the ophthalmoscope, the scar formed in the retina at the point where the fragment had lodged could be distinctly made out, as well as some patches of hæmorrhage, caused either by injury from impact of the fragment or by injury of retinal vessels at the time of its removal.

On January 13th, 1891, he could read small print; his central vision, in fact, was as good as ever, but his " field of vision" was limited slightly below—the limitation corresponding to the portion of retina injured by the foreign body.

The electromagnet has been used also to some extent in general surgery for the removal of needles from the soft tissues of the body or limbs.

CHAPTER XVI.

PERMANENT MAGNETS.

RETENTION of magnetism, as exhibited by loadstone and by hard steel, was the fact which first drew the attention of mankind to magnetic phenomena. In Chapter III., on the Properties of Iron and Steel, some precise data about the retentive properties of various samples are given. For the purpose of comparing together various brands of steel to ascertain which is the best for the manufacture of permanent magnets, two things should be known: (1) the *remanence* (see p. 98), or residual value of B after the application of a powerful magnetising force; and (2) the *coercive force* (see p. 98), or amount of negative magnetizing force, H, which would be needed to reduce the remanent magnetization to zero. The force thus required to deprive any specimen of its remanent magnetization may be taken as a measure of the tendency of the steel of this particular quality to retain permanent magnetism. Hence, for the present purpose, it is more important that the specimen possess great coercive force. In the specimen of annealed steel examined by Ewing (see p. 99), the remanence was 10,500, and the coercive force 24.

The following table exhibits the remanence and coercive force of a number of specimens of steel and iron, together with some information as to the degree to which the magnetization was temporarily pushed in order to leave the respective residua. The figures are taken from the researches of Ewing and Hopkinson.

In Ewing's experiments the specimens employed were either wires of great length in proportion to their diameter, or rings.

REMANENCE AND COERCIVE FORCE.

Nature of Specimen.	Observer.	Max. H applied.	Max. B attained.	Residual B.	Coercive Force.	Residual I.	"Specific magnetization."
Steel wire, hard drawn	Ewing	57	14,300	8,200	16	652	84
,, annealed	,,	53	14,600	11,700	17·5	931	119
,, glass-hard	,,	55	9,400	6,800	39	541	69
Pianoforte steel wire, normal temper	,,	92	14,600	11,800	27	939	120
,, ,, annealed	,,	94	14,300	10,500	24	836	107
,, ,, glass-hard	,,	98	12,700	9,600	41	747	98
Cast iron	,,	16	3,700	2,600	8	207	26·5
Very soft iron wire	,,	17	13,500	11,000	1·9	875	112
Annealed wrought iron	,,	90	16,200	12,700	3	1010	129
Whitworth mild steel, annealed	Hopkinson	250	16,120	10,740	8·3	855	109
,, ,, oil-hardened	,,	,,	16,120	8,736	19·4	695	89
Chrome steel, as forged	,,	,,	14,680	7,568	18·4	602	77
,, annealed	,,	,,	13,233	6,489	15·4	516	67
,, oil-hardened	,,	,,	12,868	7,891	40·8	628	81
Tungsten steel, as forged	,,	,,	15,718	10,144	15·7	807	104
,, annealed	,,	,,	16,498	11,008	15·3	876	112
,, hardened tepid water	,,	,,	15,610	9,482	30·1	755	97
,, (French) oil-hardened	,,	,,	14,480	8,663	47·1	687	88
,, very hard	,,	,,	12,133	6,818	51·2	542	69

In Hopkinson's experiments thin rods were employed in the apparatus described on p. 73; the presence of the soft iron yoke having the effect of making the specimen act as if it were practically of indefinitely great length in proportion to its thickness.

MAXIMUM VALUES OF PERMANENT MAGNETIZATION.

Observer and Material.	Residual I.	Residual B.	σ.
Weber, common steel magnet	314	3,947	40
von Waltenhofen, tungsten steel, glass-hard	369	4,638	47
Schneebeli, sewing-needles, 2·5 to 6·6 cm. long, 0·06 cm. thick	557 / 671	7,001 / 8,435	71·4 / 86
,, knitting-needles, 19·8 to 21 cm. long, 0·083 to 0·175 cm. thick..	765 / 832	9,626 / 10,458	28 / 107
Hopkinson, tungsten steel, oil-hardened	687	8,643	88
,, ,, very hard	542	6,818	70
Ewing, steel wire, glass-hard	541	6,800	69
,, pianoforte wire, glass-hard	747	9,600	98
Perry, Jowitt's steel	1003	12,600	129
Preece, Wall's steel	120	1,519	15·5
,, Ashforth's steel	143·5	1,704	17·3
,, Saunderson's steel	114·2	1,435	14·6
,, Jowitt's steel	109·6	1,503	15·3
,, Vicker's steel	93·4	1,174	12
,, Crewe "rivet steel"	14·8	186·6	1·9
,, ,, "spring steel"	110·5	1,391	14·2
,, Clemandot steel (compressed and tempered)	170	2,264	23·2
,, ,, (compressed but untempered)	106·1	1,333	13·6
,, Marchal steel	202·2	2,540	26
,, Allevard steel (mercury-hardened)	104·6	1,315	13·4
,, ,, (water-hardened)	132·1	1,660	16·9
Gray, magnet steel, glass-hard	520	6,536	66
Evershed, Wall's and Jowitt's steels (mean) ..	318 to 398	4,000 to 5,000	41 to 51
Brown, magnet steel, glass-hard	477 to 556	6,000 to 7,000	61 to 72

In the foregoing table are given some of the values of maximum magnetization retained by actual steel magnets magnetized by usual process, and of the "specific magnetization," σ, or magnetic moment per gramme.

It is important to note, with respect to the above table, that there is a wide difference in the relative dimensions of the samples used by different observers. As was pointed out on p. 94, the ratio of length to diameter makes a great difference to the temporary magnetization of a specimen, and in the case of permanent magnetizetion of a specimen the differences are even more marked. In Schneebeli's experiments the length used was from 100 to 800 diameters; in Ewing's about 200; Hopkinson's results were with practically endless bars; Perry employed a horse-shoe form, closed with an armature of soft iron; Gray[*] used square bars, about 60 diameters long; Brown[†] used cylinders from 33 to 37 diameters; whilst Preece[‡] used short square bars 10 centimetres in length and 1 centimetre in the side. The results given by Preece are the means of several samples of each sort, not the actual best obtained with each. For example, the mean residual B for Marchal's steel is given as 2540; but the best magnet of this sort showed 2835.

The numbers given in the first column of figures of the last table are the values of the residual *magnetic moment per centimetre cube* of the substance (in C. G. S. measure), and are obtained from the residual B by dividing 4π. Many Continental writers on this subject prefer to give the values in terms of the *magnetic moment per gramme;* a quantity which is also called the *specific magnetization* of the material. If the density of the specimen is known, the specific magnetization σ can be calculated from the residual I by dividing by the density. For example, taking Gray's figure of 9804 for the residual B, dividing by 4π gives residual I, and dividing by $7\cdot8$ (which we assume as the density of the steel) gives 100 for the specific magnetization. In all the preceding table the only samples

---

[*] *Phil. Mag.*, Dec. 1885.   [†] *Phil. Mag.*, May 1887.
[‡] *The Electrician*, xxv. p. 547, Sept. 19, 1890.

showing values over 100 for the specific magnetization are those obtained by Schneebeli and by Perry. Preece's results for *short* magnets of 10 diameters show specific magnetizations varying from 2 (Crewe "rivet steel") to 26 (Marchal steel).

Ewing found, using enormous magnetizing forces which pushed the temporary value of B for Lowmoor and Swedish iron to over 30,000 (see p. 83), that the residual values of B were only 515 and 500 respectively, whilst cast iron showed only 400. This means residual I of 40, 32, and 30 respectively, and specific magnetization of between 6 and 5 only. Du Bois, also using abnormally strong fields, found the residual specific magnetization for electrolytically deposited nickel 10·6, and for electrolytic iron from 70 to 106.

RELATION OF PERMANENT MAGNETISM TO CHEMICAL COMPOSITION.

Few of the experimenters quoted above have given the chemical composition of the materials employed. Hopkinson's results[*] are of exceptional value, being accompanied by careful analyses of the tungsten steel, which gave the best results and had the following respective compositions :—

Chrome Steel :—

Iron	97·893
Carbon	0·687
Manganese	0·028
Sulphur	0·020
Silicon	0·134
Phosphorus	0·043
Chromium	1·195

Tungsten Steel :—

Iron	95·371
Carbon	0·511
Manganese	0·625
Silicon	0·021
Phosphorus	0·028
Tungsten	3·444

[*] *Phil. Trans.*, 1886, pt. ii. p. 455.

INFLUENCE OF TEMPER, GRAIN, AND FORM.

In the preparation of permanent magnets temper is almost as important a consideration as the chemical composition of the material. Most makers have their own favourite method of hardening and tempering,* and few can or will give exact particulars. Neither is it possible to give a universal rule as to the best process to follow; for the process which is best for one brand of steel is different from that which is best for another brand. Moreover, as will be seen, the temper which is best for *short* magnets is quite different from that which is best for *long* magnets and horse-shoe forms. There is, however, no doubt that it is eminently desirable to secure a fine, even uniform grain, since any lack of homogeneity tends to reduce the quality of the magnet. An unskilful smith in forging a magnet will spoil its structure by working it too much and unequally. It is commonly supposed that the harder a piece of steel is, the greater retentive power will it exhibit. But under this term two separate qualities must be distinguished. Some steels will show, after magnetization, a large residual magnetization, which, however, decays as weeks and months pass away. Some other qualities of steel will exhibit a lesser remanence, but will conserve much longer what they have thus retained. The question of constancy of retention will be separately considered hereafter.

There are two ways of hardening steel. One is by suddenly cooling it from a bright red heat; the other is by subjecting it to enormous pressure by hydraulic machinery whilst it slowly cools down. In the former method, which is the usual one, the outer crust cools suddenly, whilst the inner portion is still plastic, and probably compresses it by natural contraction. In the latter method the hardening, though less extreme, is more uniform throughout. It is also possible to harden wires by mere longitudinal traction, as is well-known;

---

* Tempering does not mean the process of hardening, though careless writers sometimes confuse their meaning by using the term in this erroneous sense. Tempering means the process of letting the steel down to a softer state by a partial re-heating subsequently to the hardening.

but this is not used in the manufacture of magnets. If thin rods of steel are heated to a brilliant redness, or even almost to whiteness, and then suddenly plunged into water, oil, or best of all mercury, they become intensely hard and brittle, and, indeed, are frequently split in the process. This state is known as *glass-hard*. If the glass-hard steel is then re-heated to near a very dull red heat for a short time, it softens somewhat, and acquires a pale yellowish tint on the surface; this temper being technically known as *straw-colour*. If it is still further let down by continuing the re-heating it becomes *blue* in tint; this temper being used for steel springs, steel pens, and other purposes requiring flexibility. If the re-heating is carried still further the steel becomes soft, being annealed by the continued exposure to a dull red heat. Now, if it were true that the hardest magnets were best, it is clear that none ought to be tempered down, but all should be left glass-hard; which is not found in practice to be by any means the best course.

The specification of the degree of hardness by tint is extremely vague. Happily it has been found that steel possesses another and more measurable quality which exists in almost exact proportion to the hardness of the specimen; namely, its electric resistance. Barus,* who examined this fact, and Fromme† have shown that this relation is very close, and Strouhal and Barus‡ have used this property to specify the hardness of various bars subjected to magnetization. As an example they give the following values for a sample of steel:—

State.	Specific Resistance in Microhms per centimetre cube.
Glass-hard	45·7
Bright straw-tint	28·9
Straw-tint	26·3
Blue-tint	20·5
Bright blue-tint	18·4
Soft	15·9

* *Wied. Annalen.*, vii. 1879, p. 411. † *Ibid.*, viii. 1879, p. 352.
‡ *Ibid.*, xx. 1883, p. 525.

In all comparisons between hardness and retentiveness, account must be taken of the form of the magnet; for, as already mentioned, short magnets and long magnets possess different properties. In almost all the experiments made on this matter the steel rods or wires used have been of square or circular section, usually the latter. Consequently, the most convenient way of describing the form of a straight magnet is to state how many times its length is greater than its diameter. In some of Ewing's experiments on soft iron (see p. 86), rods were used varying in length from 50 to 200 diameters; and the results (embodied in Fig. 44) showed that even though the temporary magnetization was forced up to about the same degree for all, yet the longer rods showed a greater remanence than the short ones. It will be convenient to use the symbol δ for the ratio of length to diameter.

Some important researches were made by Cheesman* on the effect of hardening by stretching on magnets of different dimensions. Using an English "silver-steel" wire (of Messrs. Cook) 1·28 millims. in diameter and 90 millims. long (i.e. $\delta = 70$) magnetized by contact with a large electromagnet, he found the result of hardening by stretching to *decrease* the amount of permanent magnetism which the magnet would receive. The magnetizing was repeated after each stretching.

	Stretching Weight in kilogrammes.	Specific Magnetization.
Steel soft..	0	78·1
Harder	30	77·3
Still harder	60	70·8
Still harder	70	63·7
(Wire broke)	75	55·6

* Cheesman, *Ueber den Einfluss der mechanischen Härte auf die magnetischen Eigenschaften des Stahles und des Eisens*; Inaug.-diss. 1882, Leipzig. See also *Wied. Annalen.*, xv. 204 and xvi. 712 (1882).

In another series of experiments the length was varied.

Dimension-ratio δ.	Specific Magnetization.	
	Soft.	Hard.
22·5	16·4	18·0
40·4	48·0	48·4
58·6	75·6	60·4

From these and numerous other experiments Cheesman concluded that for short magnets hardening the steel increased the retentiveness; whilst for long magnets hardening decreased it. He fixed upon 41 diameters as the critical value of δ for this kind of steel.

Messrs. Strouhal and Barus made a much more searching investigation of the matter, including the question of tempering. They used various sorts of iron and steel; but the results now quoted relate to magnets made of "silver-steel" wire of Messrs. Cook of Sheffield, of diameter 1·48 mm., and density 7·7. Of this steel five magnets were prepared, varying from 10 to 50 diameters in length. These were first made glass-hard by sudden cooling in water and then magnetized. They were then subjected to the following processes, their magnetism after being re-magnetized in a coil supplied with current of 30 amperes from a dynamo, being observed (by a magnetometer) from time to time. Heated (*a*) in steam at 100° C. for one hour, two, three, four hours; (*b*) in anilin vapour at 115° for twenty minutes, one hour, two, four, six hours; (*c*) in melting tin at 240°; (*d*) in melting lead at 330° C.; (*e*) in melting zinc at 420° (this latter temperature is a dull red heat); (*f*) to red heat so as to reduce the steel to complete softness, the magnet being laid in lime in a piece of gas-tubing, and thereafter slowly cooled down. The results, which are of highest importance, are given in the following Table.

This Table is but one of a large number, all tending to the same general result. The figures given under the column of

"Hardness" are the specific electric resistances, in microhms. It is interesting to observe how with short magnets the specific magnetization decreases, then shows a tendency to increase, then decreases again, as the softening proceeds. Another point is that in every case heating in steam slightly reduces the magnetization that a magnet will retain; but that after some five or six hours a practically constant value is attained.

Course of Tempering.	Hardness.	Specific Magnetizations. $\delta =$				
		9·9	20·3	29·2	40·5	49·8
Glass-hard	43·8	24·8	41·0	47·2	50·8	52·5
1 hour in steam .. at 100°	38·6	23·3	39·2	45·2	48·7	50·4
3 ,, ,, ,, .. ,, ,,	36·5	23·4	38·6	44·5	47·7	49·3
6 ,, ,, ,, .. ,, ,,	35·1	23·3	38·6	44·3	47·4	49·1
10 ,, ,, ,, .. ,, ,,	34·3	23·3	38·3	44·2	47·4	48·9
20 min. ,, anilin vapour ,, 185°	29·0	21·0	38·9	46·0	50·3	52·3
1 hour ,, ,, ,, ,, ,,	27·5	21·3	40·2	48·1	52·9	55·1
3 ,, ,, ,, ,, ,, ,,	25·6	21·6	42·7	52·2	57·7	60·2
7 ,, ,, ,, ,, ,, ,,	24·2	21·8	45·5	56·9	63·5	66·8
13 ,, ,, ,, ,, ,, ,,	22·9	20·9	46·2	59·7	68·1	71·8
10 min. ,, molten tin ,, 240°	22·2	20·4	46·7	61·0	70·2	74·3
1 ,, ,, ,, lead ,, 330°	19·0	19·0	46·7	68·0	83·7	90·0
1 hour ,, ,, zinc ,, 420°	16·2	14·4	40·1	72·6	93·9	103·9
Annealed	14·9	4·3	10·2	18·7	29·8	42·4

A very large number of observations with wire of the same sorts were made by these experimenters, with practically identical results. They endeavoured to fix (on the galvanic scale) the actual hardness and magnetic value of their steel when tempered to the tints recognized by workmen.

These results agree with those obtained by Ruths,* who found, using another sort of steel, that for short magnets ($\delta = 20$) the specific magnetization fell from glass-hard 51 to

---

* Ruths, *Ueber den Magnetismus weicher Eisencylinder und verschiedener harter Stahlsorten;* Dortmund, 1876.

28·4 when tempered blue; whilst for long magnets ($\delta = 70$) the specific magnetization which was 68·7 for the glass-hard state increased to 92 when tempered blue.

State of Temper.	Hardness.	Mean Specific Magnetization. $\delta =$				
		10	20	30	40	50
Glass-hard	45·7	23·5	37·6	43·6	46·5	48·3
Straw-tint	26·3	21·4	40·2	49·4	53·8	56·5
Blue-tint	20·5	19·3	45·8	67·0	80·4	87·3
Annealed soft	15·9	4·3	11·2	20·5	31·8	44·6

The researches summarized above, important as they are, do not, however, answer the question what state of temper produces a magnetization of the greatest constancy in amount. This precious quality, to which Scoresby[*] gave the name "fixidity," cannot even be pre-determined by an inspection of the values of the coercive force, such as are given in the Table on p. 382, because these are themselves subject to change with time. Scoresby suggested a method by examining the change of magnetism which was brought about by placing the magnet gently down upon the similar poles of another and more powerful one. He took as the measure of the goodness of a magnet (for the purpose of compass needles, &c.) the product of its original strength into its reduced strength after such treatment. By this treatment Scoresby (whose magnets were 6 inches long, 0·5 inch broad, and from 0·5 to 0·0625 inch thick), found the percentage of reduction to be least in cast steel when hard, and to be less when tempered blue throughout than when tempered at the extremities only. Kater[†] found that for compass needles of thin steel, it was of advantage to temper down soft the middles of the needles, leaving the ends only hard.

In relation to the question of temper and chemical compo-

---

[*] Scoresby, *Magnetical Investigations*, vol. i. p. 35.
[†] Bakerian Lecture (*Phil. Trans.*), 1821.

sition many important facts have recently been discovered. It appears that iron is capable of existing in two distinct or allotropic states, just as carbon can exist either as soft plumbago or hard diamond. In one of these two states iron is hard, in the other soft. Further, the mode in which the carbon exists in soft steel differs from the mode in which it exists in hard steel. According to Sir F. Abel,* in soft cold-rolled steel the carbon exists in the form of a carbide of iron having chemical composition Fe_3C. This is distributed in pearly masses throughout the steel, and can be detected in the microscope. When such steel is heated to bright redness and quickly cooled it becomes very hard, and now it is found that the carbon has apparently dissolved through the mass and is no longer visible in flakes. When tempered by reheating at temperatures between 200° and 400° C., the carbide again slowly separates into flakes within the mass, the degree of separation depending on the temperature, the duration of its application, and the mechanical treatment during such reheating. Chernoff, in 1868 showed that steel is not hardened by sudden cooling unless it is heated to a certain well-marked degree of temperature, about 650° C. Gore, in 1869, discovered that when a brightly red-hot iron wire is allowed to cool there arrives a moment when it suddenly elongates, after which it goes on contracting as it cools. Prof. Barrett further discovered that as the wire cools down to a certain degree it suddenly glows more brightly. To this phenomenon he gave the name of *recalescence*. Further, Tait has shown that there is a certain high temperature at which iron undergoes a curious and sudden change in its thermo-electric properties. Still more recently, Osmond, studying the cooling of hot specimens of iron and steel, has observed that there are two points in the temperature scale at which heat is evolved during cooling.

During *rapid* cooling, the carbon passes from the state in which it is combined with the iron, into a state in which it is dissolved in the iron; and during *slow* cooling this dissolved

* *Proc. Inst. Mech. Engineers,* Jan. 1883.

carbon can re-enter into combination with the iron and return to the state of softness. According to Osmond, the second point of temperature at which heat is evolved, and at which the fall of temperature is momentarily arrested, corresponds to Barrett's recalescence, and he holds that when during cooling this temperature is reached, the carbon leaves its state of solution and combines with the iron, thereby evolving heat and causing the momentary glow on the surface. But even with pure iron, one point of arrest in the fall of temperature occurs. Here there is no carbon; the phenomenon is due to something in the iron itself. Osmond accounts for this in the following way. At a brilliant red heat the atoms of the iron are so grouped in molecules that the iron is virtually *hard;* but there is a second grouping possible, which when assumed renders the iron soft. Osmond calls the first or hard sort, β iron, and the second or soft sort, a iron. If the iron is pure it will pass from the a to the β state when cooled, whether quickly or slowly. But if carbon is present a certain percentage of all the molecules are maintained in the β state, with the result that the steel is hard. According to Osmond, this molecular change occurs at the higher of the two pauses in the cooling; the second pause indicating the point at which the carbon changes from the dissolved or hardening carbon to the combined state. The higher pause is at about 770°, the lower at about 650°. Consequently, if *between these two temperatures* steel is suddenly cooled, the iron part ought to be soft but the carbon part uncombined with it. This is found to be the case, as on dissolving out the free iron no carbide of iron is found. Further, by mechanical working at temperatures below a dull redness, the a iron may be gradually converted into the β or hard iron, as indeed occurs when iron is rolled, hammered, or drawn. Now, these changes are obviously connected with the magnetic properties of the material, and researches have been made by Tomlinson, Hopkinson, Roberts-Austen, and others upon the possible connection. It would seem that at temperatures above that of the lower of the two pauses, iron is not magnetisable, and that only below the lower of the two can it receive or retain any magnetism. In any case the temperature of recalescence appears to correspond with the critical temperature for magnetization. If this were true it would follow that the β (or hard) iron cannot itself be magnetized. One fact supporting this view is that the addition of a small percentage of manganese to iron produces a so-called manganese steel, which is not only intensely hard, but is also non-magnetizable. Seven per cent. of manganese suffices to prevent the iron passing from the β state to the a state.

METHODS OF MAGNETIZING.

The ancient methods of magnetizing, including "single touch," "double touch," divided touch," "circular touch," in all their varieties, may be dismissed as obsolete. The only methods needing attention are those based upon the use of electric currents. Arago discovered this way in 1820, and magnetized steel needles by inserting them within a spiral coil connected with a voltaic battery. Elias, of Haarlem, in 1844 proposed a useful variation of this process. Instead of winding the wire in a long tubular spiral, he coiled it in a compact ring of many layers, which could be slipped loosely over the piece of steel which was to be magnetized, and then moved along it from end to end, so as to bring every part successively into an intense magnetic field. In 1846 Böttger suggested a modification for special application to horse-shoe magnets, consisting of two coils oppositely wound, to be slipped over the two polar ends simultaneously. Sinsteden used, for magnetizing horse-shoe forms, a powerful two-pole electromagnet, against the polar ends of which the steel horse-shoe was placed, and while thus, *in situ*, stroked the horse-shoe in the direction from bend to poles, with an armature of iron resting across the two limbs. Van der Willigen, in his remarkable treatise* on the Haarlem magnets forged by Van

* Van der Willigen, *Sur le Magnétisme des Aimants artificiels*. (Archives du Musée Teyler, Haarlem, vol. iv. 1878.)

Wetteren, describes the method used for these famous magnets. The steel horse-shoe is placed with its poles against the pole-pieces of a Ruhmkorff's electromagnet (like Fig. 28, p. 54), and the current from ten or twenty Bunsen cells is then turned on and off three or four times. The current being cut off, the horse-shoe is then raised into a vertical position, and the keeper is slid across its poles before these are removed from contact with the electromagnet. The magnet is then in a super-saturated state, holding on to its keeper with a pull about 30 per cent. greater than the permanent pull which it will exhibit. For larger magnets Van der Willigen made a simultaneous use of the Ruhmkorff electromagnet, as described, and of an Elias's ring, which he slid to and fro from one end to the other of the horse-shoe from 20 to 100 times, whilst the electromagnet was also in action.

For small bar magnets and compass needles it suffices to draw them over the poles of a large electromagnet, each extremity of the piece of steel being finally touched against the pole of opposite kind, and pulled off normally. For straight bars of considerable size Van der Willigen used the same electromagnet, adjusting the distances between the polar faces to correspond to the bar to be magnetized, and then applied the same process as for horse-shoes, but of course without applying a keeper.

One process, known as Hoffer's method, consists in stroking the steel horse-shoe from the poles to the bend with an iron rod laid across the limbs whilst it is in contact with the poles of another, previously magnetized, horse-shoe. Van der Willigen, who examined this method, found it faulty, and even harmful. It set up inequalities in the distribution of the magnetism, and diminished the effective power of magnets previously magnetized as described above.

Another process, which has been suggested at various times as an improvement on the ordinary processes, is to subject the magnet to powerful magnetizing forces during the act of hardening. Robinson placed the steel bars red hot against the poles of a strong magnet, and in this position applied cold water to the bar. Holtz says that thick steel bars so magnetized during sudden cooling are twice as powerful as those magnetized in the cold; and that thin bars are three times as powerful. Aimé and Hamann have also advocated this method, which, however, does not seem to have any real advantages.

Moser* made a comparative trial of various methods as applied to a parallelopipedal steel rod weighing 12 ounces. He measured the degree of magnetization by the method of oscillations, observing the time required for one complete swing. From the duration of the oscillation the relative strengths are to be calculated. The results of eight methods, each better than the preceding, are given below.

These experiments leave no doubt as to the superiority of the modern method over the ancient ones.

	Time of One Swing.	Relative Strength.
	sec.	
1. Knight's method of "divided-stroke," using two permanent magnets; 20 strokes on one side of rod only	22·13	1
2. Same, repeated on all four sides	14·87	2·21
3. Same, repeated with masses of iron laid under each end of the bar	14·63	2·29
4. Same, repeated while bar placed on top of a reversed steel magnet	12·13	3·33
5. Michell's method of "double-stroke," using two separate steel bar magnets	11·13	3·95
6. Same, using a steel horse-shoe	10·19	4·71
7. Aepinus' method of "circular stroke," using same horse-shoe magnet	8·75	6·39
8. With an electromagnet: rod laid on two iron pole pieces on the poles of the electromagnet, and while here stroked with steel horse-shoe; the current of the electromagnet then turned off, and rod then lifted	8·0	7·62

* Dove's *Repertorium der Physik*, vol. ii. p. 141, 1838.

Ewing employed a large electromagnet in his research on the magnetization of iron in very strong magnetic fields. Between the poles of this magnet he placed the specimen to be magnetized; and this consisted of a cylinder turned down thin at its girth so that it presented a narrow neck between two conical portions. In this way an enormous number of magnetic lines could be concentrated through the narrow "isthmus." It was in this way that Ewing found the residual value of I to be about 500 for wrought iron.

CONSTANCY OF MAGNETS.

It by no means follows that the magnet which shows the greatest amount of permanent magnetism, will be the most constant in its retention of the same. Magnets lose their magnetism from many causes, such as accidental shocks, contacts with other magnets or pieces of iron, changes of temperature, slow annealing of the steel, and the like. The relation of these influences to the quality and temper of the steel, and to the dimensions of the magnet are not very well known, but such information exists.

Gray,* using glass-hard magnets, found the percentage of loss of magnetism due to a small demagnetizing force (that of unit field) to vary widely for magnets of different lengths. A bar 10 diameters long lost $0 \cdot 8$ per cent.; one 20 diameters long lost $0 \cdot 6$ per cent.; 40 diameters long $0 \cdot 5$ per cent.; whilst one 100 diameters long lost about $0 \cdot 44$ per cent.

Bosanquet,† using a considerable number of magnets, made (on February 8th, 1885) from the best cast steel, as hard as they could be made, found the following fall in magnetic moments:—

February 18	12,539
March 3	11,822
March 15	11,767
April 8	11,620
September 18	11,119

---

* *Phil. Mag.*, xx. p. 484, 1885. † *ib.*, xix. p. 57, 1885.

Lamont* made an investigation lasting eleven years, as to the constancy of the magnets used in the observatory at Munich for the daily magnetic measurements. These magnets had already been repeatedly alternately immersed in warm and cold water, but had not attained by this process a state of constancy, as the following figures show. The losses are given in decimal parts of the whole magnetic moment :—

	1848.	1849.
January	0·0000	0·0000
February	0·0003	0·0001
March	0·0003	0·0002
April	0·0008	0·0005
May	0·0014	0·0007
June	0·0022	0·0011
July	0·0028	0·0016
August	0·0032	0·0022
September	0·0028	0·0022
October	0·0017	0·0013
November	0·0009	0·0007
December	0·0005	0·0001

The total decrease in various years, also in terms of the whole magnetic moment as unity, was as follows :—

1847	0·0174	1853	0·0099
1848	0·0169	1854	0·0103
1849	0·0103	1855	0·0081
1850	0·0091	1856	0·0079
1851	0·0113	1857	0·0071
1852	0·0079	1858	0·0063

According to Barus† mean atmospheric temperature, acting on freshly quenched steel for a period of *years*, produces a diminution of hardness, with consequent loss of permanent magnetism, about equal to that which would be caused by the action of a temperature of 100° C. acting for a similar period of *hours*.

* Lamont's *Handbuch des Magnetismus* (1867), p. 410.
† *Phil. Mag.*, Nov. 1888, p. 403.

' Some persons have sought to explain the slow decay of magnetism by reference to chemical changes of the surface, due to oxidation, moisture, etc., and have proposed to prevent such decay by gilding, silvering, or lacquering the surface of the magnet. Such remedies are entirely ineffectual, and the supposition on which they are based is erroneous.

Cheesman observed the percentage loss of magnetism caused by percussion, the magnets being allowed to fall from a height, with the following results:—

Brand and Temper.	Height of Fall in metres.	Percentage Loss of Magnetism.
Steel wire, soft	1·5	30
,, mechanically hardened	2·0	44
,, ,, ,,	3·0	57
Steel rod, mechanically hardened	2·0	52
,, ,, ,,	seven times, do.	81
Iron wire, mechanically hardened	2·0	84
,, ,, ,,	three times, do.	95
,, soft	2·0	83
,, ,,	3·0	99
Iron rod, soft	2·0	97
Steel wire, glass-hard	3·0	6
,, ,, .	3·0	4

W. Brown,[*] using magnets made from silver-steel, all 20 centimetres long, endeavoured to find a relation between the temper of the magnets and the percentage loss of magnetism due to percussion after the magnets had been laid aside for various periods of time. His results are embodied in the following table, from which it will be seen that the glass-hard temper appears to be the most constant in this respect. It is less certainly shown that long magnets are more constant than short ones:—

[*] *Phil. Mag.*, xxiii. p. 293, 1887.

Temper.	Diameter ratio. δ	Specific Magnetization. σ	Loss per cent. due to percussion after lying aside for				
			1 hour.	20 hours.	44 hours.	1 month.	3 months.
Glass-hard	33	41	1·98	2·0	1·95	1·04	0·8
,, ,,	50	45	2·96	3·2	1·48	1·0	0·0
Yellow ..	33	44	6·03	6·1	4·8	5·4	6·2
,, ..	50	46	4·0	3·5	3·76	2·6	4·0
Blue.. ..	33	54	11·8	10·8	9·71	11·8	7·5
,, ...	50	71	8·2	8·2	8·18	7·5	8·7

Strouhal and Barus, who have made so many researches upon the physical properties of steel, have examined the question of magnetic constancy in relation to form, temper, and exposure to fluctuations of temperature. They found that a glass-hard magnet 119 diameters long lost 30 per cent. of its magnetism when heated for six hours in steam at 100° C. ; but the same magnet, when remagnetized and again heated in steam, now only lost 5·3 per cent. The loss of magnetism on the first heating was found to vary in magnets of different lengths.

A magnet of 119 diameters' length lost 30 per cent.
,, 108 ,, ,, 28 ,,
,, 35 ,, ,, 49 ,,
,, 14 ,, ,, 67 ,,

In every case after protracted heating in steam a condition was attained in which neither the specific magnetization nor the specific resistance (i. e., the hardness as measured electrically) showed any appreciable further change. These investigators came to the conclusion that magnets so treated possess a special fixidity, and resist not only changes of temperature but mechanical shock better than magnets prepared in any other way. As an example, they took a short magnet, 2·5 centimetres long, 0·4 centimetres wide, and 0·3 centimetres thick. This was boiled in water for 4 hours, then magnetized and kept two hours longer in steam. Its magnetic moment was then observed, after which it was laid on a wooden log and violently beaten 50 times, both lengthways and sideways, with another mass of wood. It then showed a diminution of only $\frac{1}{900}$; and, after a repetition of the beating, only showed a loss of about $\frac{1}{400}$. In another experiment a tubular steel magnet was made glass-hard, magnetized, and heated for 30 hours in steam; remagnetized, and heated for 10 hours in steam. This magnet, when allowed to fall ten times, on its ends, from a height of 1·5 metres, showed a permanent diminution of only $\frac{1}{4756}$ of its magnetism. It is accordingly recommended that all magnets for use in magnetic observations should be prepared as follows:—*Make the magnets glass-hard, then place in steam at* 100° *for* 20 *or* 30 *hours, or longer for very massive magnets. Then magnetize as fully as possible, and then heat again for five (or more) hours in steam.* Such magnets will then be as constant as they can be made.

In quite another way Mr. G. Hookham* has sought to obtain magnets of constant power for use in his electrical meters. A number of bar-magnets of tungsten steel are fitted into cast-iron pole pieces which nearly meet, so constituting a nearly-closed circuit. After these are fitted on, a magnetizing current is sent round the bar magnets to saturate them as fully as can be done. They are then mechanically hammered, and a weak demagnetizing current is sent through the coils, reducing the magnetism by about 10 per cent. Then for many months the magnet will show no tendency whatever to lose; indeed, it may slightly gain.

Effects of Temperature.

Besides the effects of tempering and annealing produced on steel, heat produces various effects on magnetism.

The effects on temporary magnetization of iron and steel have been noted in Chapter III., p. 93; but it remains to mention some points respecting permanent magnetism.

Faraday† found a steel magnet to lose its permanent magnetism at a temperature a little lower than the boiling point of almond oil, and from that point onward it behaved

* *Journal Institution Electrical Engineers*, vol. xviii. p. 688, 1889.
† *Experimental Researches*, ii. p. 220.

simply like soft iron until raised to orange redness, when all magnetic properties disappeared. A fragment of lodestone kept its magnetism till just below a dull red heat.

Trowbridge,* using a severe freezing mixture of carbonic acid snow dissolved in ether, capable of cooling to about $-140°$ C., found the effect of extreme cold to diminish the magnetism of a steel magnet by 60 per cent.

Wiedemann† came to the conclusion that if the temperature of a magnet is repeatedly altered and brought back again to its initial point, the magnetism gradually attains a constant state, after which any increase of temperature will, in very hard steel bars, cause an increase, and in soft steel bars a decrease, of magnetism. A decrease of temperature produces opposite results. The phenomenon is, however, somewhat complex. On heating a magnet and again cooling it the magnetism lost during heating is only partially regained during cooling, so that at every repetition of the heating there is some loss until the steady state has been attained. But the phenomena are dependent on the prior magnetic history of the bar, and this may render the facts complicated. If, for example, a bar has been magnetized and left for a long time, or subjected to mechanical shocks, so that its magnetism may be regarded as well engrained into it, and it has then been recently subjected to some partial demagnetizing force, which has not acted for any long time, then such a bar may, on being heated and cooled, actually regain more magnetism during cooling than it lost during heating. It acts as though it possessed two independent magnetizations superposed on one another, and having different temperature coefficients.

For ordinary steel magnets it may be taken as true that the ordinary atmospheric changes of temperature produce slight alterations, of a temporary nature only. The formula for temperature corrections used at the Kew Observatory was determined by Whipple as follows:—

$$K_t = K_0 \{1 - q(t-t_0) - q_1(t-t_0)^2\} ;$$

* *Silliman's Journal*, 1881.   † *Pogg. Ann.*, ciii., p. 563, 1858.

where the mean values of q and q_1 are respectively 0·000161 and 0·00000048. Christie found for the coefficient q 0·001015, Hansteen 0·000788, Riess and Moser, for compass-needles 2 to 3 inches long, from 0·000324 to 0·000432. Cancani found that of cylindrical steel magnets 50 millimetres long to be as follows :—diameter 1 mm., 0·000312 ; 2 mm., 0·000380 ; 3 mm., 0·000539 ; 4 mm., 0·000645 ; 5 mm., 0·000869. The temperature coefficient of magnets tempered blue was nearly 50 per cent. greater than that of those tempered straw tint. Ewing found the magnetic moment of a steel bar to fall off about 18½ per cent. on heating from 10° to 100°, but this was entirely recovered on cooling. Gaugain * also made many observations on the effects of heat on magnetism.

USE OF LAMINATED MAGNETS.

Knight appears to have been the first to employ compound magnets made of bundles of steel plates separately magnetized. Since then many other experimenters have adopted this construction, notably Coulomb and Scoresby. The advantage of this construction is that in the methods adopted for hardening steel by sudden quenching, the hardening does not really penetrate far below the surface, and consequently the interior softer part of the bar adds nothing to the permanent magnetism, and may even weaken it so far as its external manifestation is concerned. Coulomb employed magnets made up of three laminæ, the central one projecting a little beyond the others, their ends being embedded in a soft-iron pole-piece. Scoresby,† who made many researches on compound magnets, showed that it was advantageous to separate the laminæ to a short distance apart. Some of his magnets, prepared from busk steel, are preserved in the Museum at Whitby. He showed that such compound magnets may be made much more powerful than any single bar, of weight equal to the combination, can be

---

* *Comptes Rendus*, 1877 and 1878.
† Scoresby's *Magnetical Investigations*, vol. i. pp. 98–329.

made; but that the absolute gain of power in the combined mass, due to each additional lamina, diminishes progressively. For example, in a set of 30 plates, the last 26 did not add to the first four more magnetic power than the first 6 had possessed alone. This is due to the tendency to mutual demagnetization; for in a powerful set the weaker plates not only add nothing, but have their polarity actually reversed by the more powerful ones. It is this state of things which is partially remedied by preventing the actual contact of the plates. Scoresby found that for compound horse-shoe magnets the best temper was not glass-hard. He found it better to temper the hard steel plates by boiling them in linseed oil at 505° F. (= 263° C.). In this way he raised the lifting power of a 5-plate Stubs' steel horse-shoe, weighing 2·91 lb., from 13–14 lb. to 25–26 lb. And another 15-plate magnet weighing 8 lb., which, when glass-hard, would only carry 26 lb., when thus tempered carried 45–50 lb.

Jamin,* who has gone over the same ground, has added little to the very complete investigations of Scoresby. Some makers of compound magnets arrange the tiers of plates in the form of steps at the extremities; the central plates being made longer than those on either side. There is no great gain in this arrangement, which is supposed to prevent the central plates from being reversed in polarity by those outside. It probably has the advantage of concentrating the magnetic lines and so adding to the lifting power. (See Law of Traction, p. 131). Van der Willigen considered one millimetre a sufficient amount of projection for the central plate.

TRACTION, OR LIFTING POWER OF MAGNETS.

Bernoulli's law that the tractive power of similar magnets was in proportion to the $\frac{2}{3}$th root of their weight, or in other words to the square of the cube root of their weight, was explained on p. 125, and shown to mean simply that for equal

---

* *Comptes Rendus*, lxxvi. 1873, and in innumerable papers in the same Journal since.

magnetic saturation the tractive power was simply proportional to the polar area. If P stands for the greatest load which the magnet will carry, and W for its own weight, then Bernoulli's rule is

$$P = a\sqrt[3]{W};$$

where a is a constant depending on the units chosen, on the quality of steel, and on its degree of magnetization. If P and W are expressed in kilogrammes, then, according to the data of the best makers, a, for horse-shoe magnets, will be between 18 and 24, or say on the average 20. This means that if we consider a magnet weighing 1 kilogramme, it will carry a load of 20 kilogrammes, or twenty times its own weight. On this reckoning a magnet weighing 10 kilogrammes should carry 92·8 kilogrammes, or 9¼ times its own weight; whilst a magnet weighing 0·1 kilogramme should carry 4·31 kilogrammes, or 43 times its own weight. If P and W are given in British pounds, the value of the coefficient for the best sort of horse-shoe magnets (corresponding with a = 20 for kilogramme units) will be 25·1, or approximately 25. That is to say a really good steel horse-shoe magnet weighing 1 lb. should carry a load of 25 lb. For example, one of Van Wetteren's magnets, examined by Van der Willigen (magnet "B"), weighed 1·074 lb., and carried, as the mean of seven observations, a load of 26·004 lb.

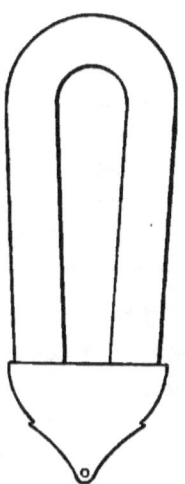

Fig. 207.

Van Wetteren's Magnet (¼ actual size).

The Haarlem magnets have been so justly celebrated for their admirable qualities that some notice of them seems to be called for. Logeman, who was continually advised by M. Elias, was the first to establish a name for the magnets made in Haarlem. By him Van Wetteren, a most accomplished master-smith, had been employed. Later, when Funckler succeeded to Logeman's business, both Van Wetteren and Funckler continued to produce magnets on Elias's plan, until in 1874 M. Van der Willigen superseded that method of magnetization by the method described above (page 395), and applied it to the magnets forged by Van Wetteren. Of the special kind of steel employed, and of the mode of hardening and tempering adopted, nothing precise is stated in the otherwise valuable memoir of Van der Willigen [*] on the subject. In this memoir a detailed account is given of about 50 magnets, and of the loads which they carried after being magnetized, in various ways and at various times. Most of these magnets were of the form shown in Fig. 207. The contact surface of the keeper is slightly rounded. The figure is exactly one quarter the natural size of the magnet "A" in the following list.

Mark on Magnet.	Weight of Magnet, in kilos.	Average Load carried, in kilos.	Value of Coefficient a.
A	0·495	13·30	21·25
B	0·487	11·78	19·03
E	0·889	19·02	20·57
3057	1·013	21·71	21·52
3053	1·521	27·90	21·17
3054	1·918	32·53	21·07
C	2·169	35·47	21·10

The magnet "A" was 0·66 centims. thick; its keeper 0·3 centims. thick; the polar ends were 2·57 centims. apart, and each of them was 2·68 centims. wide. The extreme height from the outer side of the bend to the middle point between the poles was 17 centims. The area of the polar face was 1·769 sq. centims. The greatest width between the limbs below the bend was 3·65 centims. The magnets "A," "B," "E," and "C," were from one bar of steel. The magnet "3053" was one shown in the Exhibition of Scientific

[*] Sur le Magnétisme des Aimants artificiels. *Arch. du Musée Teyler*, (Haarlem), vol. iv. 1878.

Apparatus at South Kensington in 1876, whilst "3054" was sent to the Centennial Exhibition at Philadelphia in the same year. When returned from the States in May 1877 it was remagnetized, and carried 31·44 kilogrammes.

So far as the author is aware, no British manufacturers have equalled the Haarlem makers; certainly none has surpassed them. Scoresby* mentions a nine-plate compound horse-shoe in his possession, weighing 22 lb., which would carry from 66 to 88 lb. By Bernoulli's formula, a 1 lb. magnet, proportionately magnetized, would carry from 8·4 to 11·3 lb. —less than half the load which a 1 lb. Haarlem magnet will carry. Scoresby also mentions a seven-plate steel horse-shoe, made by Dr. Schmidt, the property of the Royal Institution, which, though weighing 16 lb., would only carry 28 lb. Had this been a Haarlem magnet it would have carried at least 150 lb.

Van der Willigen gave another formula for the tractive force of the Haarlem magnets, namely,

$$P = B \cdot K \sqrt{S} \cdot \sqrt[4]{\frac{L}{S} \cdot \frac{L}{l}};$$

where P is the tractive force in kilogrammes, S the area of one polar surface (in square centim.), K the perimeter of one polar extremity, l the actual mean length of the horse-shoe, L the "reduced" length (i. e., the length between the points of maximum free magnetism observed on the flanks of the magnet when keeper is in place), all the last being given in centimetres. B is a coefficient which varies between 0·7 and 1·2, and has a mean value of 0·891.

The tractive power of bar magnets is, of course, much less than that of horse-shoes of equal weight, since the keeper is never so highly magnetized when applied to one pole. Van der Willigen states that for Van Wetteren's bar magnets the load carried by one pole is just a quarter of the load which a horse-shoe (of weight equal to that of the bar) would carry. This means that Bernoulli's coefficient a (for kilogramme units)

---

* *Op. cit.*, p. 244.

must be taken at about 5 for Haarlem bar magnets. The corresponding number for pound units will be $6\frac{1}{4}$. Has any British maker yet produced a steel bar magnet weighing one pound which will carry $6\frac{1}{4}$ pounds on one pole ?

Conservation of Magnets.

All experience shows that magnets forming a closed or nearly closed circuit are less liable to changes in strength than others, hence the provision of keepers to horse-shoe magnets, and the common arrangement of bar magnets in pairs or "magazines."

Changes of temperature and mechanical shocks are liable, as we have seen, to affect magnets.

Sudden slamming on of the keeper is liable to deteriorate the magnetism; it should always be put on gently; preferably it should be slid on across the limbs near the bend and then drawn downward toward the poles.

Sudden detaching of the armature is, on the other hand, of advantage to a horse-shoe magnet—though there is a popular superstition to the contrary. The electric eddy-currents induced in the polar masses on the sudden removal of the keeper circulate in a direction tending to augment the magnetism. It is possible to improve the power of a horse-shoe magnet several per cent. by gently sliding on the keeper as mentioned above, and then suddenly detaching it a number of times in succession. On the other hand, it is easy to deteriorate the magnetic quality by several per cent. by suddenly slamming the keeper on and then gently sliding it off at the bend for a number of times.

Bar magnets that are to be used in magnetic measurements ought never to be allowed to touch against one another or against any other magnet or piece of iron.

Unipolar Magnets.

It is impossible to make magnets having but one pole each, but it is easy to attain to a result which is virtually the same.

Let one pole, say the south pole, of a magnet be arranged to lie in the axis of rotation, the magnet being balanced by a counterpoise c. It will then act as though it possessed a north pole only. The counterpoise may be conveniently made of a piece of leaden tube fitted with a cork, which can be slid along a brass wire. Two forms of unipolar magnets are shown in Fig. 208.

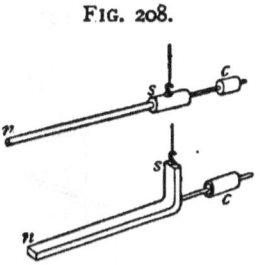

FIG. 208.

UNIPOLAR MAGNETS.

ASTATIC ARRANGEMENTS.

By suitable arrangements a suspended magnetic needle, which naturally would turn and point under the directing effect of the earth's magnetism, may be so disposed that it exhibits no such tendency, being freed, for the time being, from the control of the earth's magnetic force. Such an arrangement is described as an *astatic* arrangement. There are several varieties of such.

(1) *Use of Compensating Magnet.*—A bar magnet parallel to the needle, broadside on, placed eastwards or westwards, as in Fig. 209, or end on, and placed northwards or southwards, may be used to compensate the earth's field. If the magnet is far away it will not act strongly enough. A point P may be found such that if the magnet is brought any nearer the needle will completely turn round. When the compensating magnet is at this point a very small displacement will produce a great effect on the position of the needle. It is better to compensate with a large magnet at some distance than with a small magnet close to the needle.

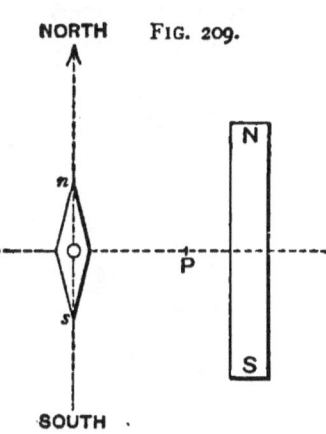

FIG. 209.

COMPENSATING MAGNET.

(2) *Screening by Iron Shell.*—Another method is to screen by surrounding the needle by an iron shell of adequate thickness.

(3) *Pivoting in Dip-line.*—Arago pivoted a compass-needle so that the axis of rotation was parallel to the line of dip. In this case there is no component of force acting on the needle in the plane in which it is free to move; it is therefore astatic.

FIG. 210.   FIG. 211.

ASTATICALLY BALANCED MAGNET.   LEBAILLIF'S SIDEROSCOPE.

(4) *Astatic Balancing.*—A bent magnet, as in Fig. 210, may be balanced so as to become astatic, if its poles have equal moments around the axis of suspension. This form is very sensitive, and easily arranged.

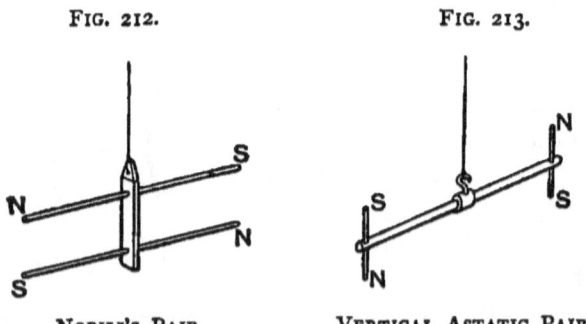

FIG. 212.   FIG. 213.

NOBILI'S PAIR.   VERTICAL ASTATIC PAIR.

(5) *Lebaillif's Pair.*—A pair of needles, of equal length and weight, and equally magnetized, may be made into an astatic pair by mounting them end to end, as shown in Fig. 211. By adjusting the poles to have greater or less leverage,

an excessively delicate apparatus is obtained. It was with such an apparatus, termed by him a *sideroscope*, that Lebaillif discovered the so-called diamagnetic repulsion of antimony.

(6) *Nobili's Pair.*—The commonest mode of arranging an astatic pair is that depicted in Fig. 212, where two needles are mounted together in reversed positions. This does not, however, give absolute astatism (*a*) unless the needles have exactly equal magnetic moments, (*b*) unless the needles are absolutely parallel. Neither condition is usually fulfilled. To secure parallelism it is better to hang the lower needle from the ends of the upper one with two small slings of aluminium foil or wire, rather than by the usual rigid central support.

(7) *Vertical Pair.*— Yet another way of arranging an astatic pair, devised by the author in 1886, is shown in Fig. 213. Like Fig. 210, it has the advantage over the combinations of Nobili and Lebaillif, that the degree of perfection of its astatism is not affected by the unequal decay of magnetism in the two needles. There are several other possible astatic combinations, but none that are in use in instruments.

APPENDIX A.

WILLIAM STURGEON.

WILLIAM STURGEON was born in 1783, at Whittingham, in Lancashire, about two miles from Kirkby Lonsdale. His father was an idle shoemaker, who neglected his family whilst poaching fish and rearing gamecocks. Young Sturgeon was apprenticed to the trade of shoemaker, under a master who starved and ill-used him. In 1802, to escape the position of degradation in which he found himself, he enlisted at the age of nineteen into the Westmoreland Militia. Two years later he enlisted as a private soldier into the 2nd Battalion of the Royal Artillery, thus gaining, though amidst the uncongenial atmosphere of the barracks, the leisure in which to pursue an absorbing passion for reading and for making chemical and physical experiments. It was during his connexion with the artillery (as he himself informed the late Dr. Leigh *) that his attention was drawn to electrical subjects, through the occurrence of a terrific thunderstorm which occurred while he was stationed in Newfoundland. He determined to study natural science, but finding himself unable to understand books on these subjects, he set himself, under all the disadvantageous circumstances surrounding the life of a private soldier, to acquire a knowledge of reading, writing, and grammar. A sergeant in the artillery lent him books, which, when he came off guard at night, he used to take from his knapsack to study. He thus devoted, as we learn from a memoir of him by Dr. J. P. Joule,† a considerable time to mathematics, and to both dead and modern languages, optics, and natural philosophy in various branches. He had, during his apprenticeship, acquired proficiency in sundry mechanical arts, being an adept at cleaning watches and clocks; and to these he now added those of a lithographic draughtsman, whilst still keeping up his old trade of shoemaking. He quitted the Royal Artillery in 1820, and for a time resumed his old avocation in Lancashire, but returned

* See obituary notice in the *Manchester Examiner and Times*, Dec. 14, 1850.
† *Memoirs of the Literary and Philosophical Society of Manchester*, vol. xiv. p. 53, 1857.

shortly to Woolwich, where he resided at 8, Artillery Place. Here, during his leisure time, he turned his attention to the construction of scientific apparatus, purchased an old lathe and taught himself turning. His earliest efforts at constructing scientific apparatus were devoted to electricity, and he showed a perfect passion for chemical and electrical experiments. He was an intimate associate of James Marsh, the chemist, and was brought into frequent contact with Barlow, Christie, and Gregory, who all interested themselves in his work. Owing to their influence he was appointed lecturer on Experimental Philosophy to the Hon. East India Company's Military Academy at Addiscombe, an appointment which he held till his removal to Manchester in 1838. His first original contribution to science was the production of a modified form of Ampère's rotating cylinders, described in the *Philosophical Magazine* for 1823. This was followed in the next year by four papers on thermo-electricity. In 1825 he presented to the Society of Arts the set of improved apparatus for electromagnetic experiments, including his first soft-iron electromagnet (see pages 2 and 3). He was at this time forty-two years of age. It is an interesting coincidence that in the volume of the Transactions of the Society of Arts for 1825 the plate of figures in which Sturgeon's electromagnet is depicted was engraved by Cornelius Varley, himself an electrician, and father of the celebrated Cromwell F. Varley. According to Dr. Joule, Sturgeon appears to have discovered the soft-iron electromagnet, and to have constructed it both in the straight and horse-shoe shape as early as 1823, though he did not publish it until 1825. In 1826 he was busied with the difficulties arising in the firing of gunpowder by electric discharges, and in the means for overcoming them. In 1830 he published an incomplete pamphlet, entitled "Experimental Researches in Electro-Magnetism, Galvanism, &c.," dealing chiefly with voltaic piles and cells. In this work he describes for the first time the now well-known process of amalgamating the zinc plates of a battery with a film of mercury. A year or two later he began to experiment on the phenomena of magnetism of rotation discovered by Arago, and had come to the conclusion that the effects were probably owing to a disturbance of the electric fluid by magnetic action, "a kind of reaction to that which takes place in electromagnetism," when the publication of Faraday's brilliant research on magneto-electric induction, in 1831, forestalled the complete explanation of which he was in search. In 1832 he constructed an electromagnetic rotary engine, the first contrivance, according to Dr. Joule, by means of which any considerable mechanical force was developed by the electric current. He had in 1823 produced the revolving "Sturgeon's disk," a modification of the pendulum of Marsh and the star-wheel of Barlow. In 1836 he communicated a paper to the Royal Society, containing *inter alia* descriptions of a magneto-electric machine, having a longitudinally-wound armature and a commutator consisting of half-disks of metal; also of a second form of magneto-electric generator, in which the revolving armature is provided with an iron core. In describing his commutator Sturgeon expressed the opinion that the magneto-electric machine as thus improved would ultimately entirely supersede the use of the voltaic battery.

For some reason this memoir was not admitted to the *Philosophical Transactions*, and was returned to him, after its reading, by the Royal Society. He afterwards printed it in full, without alteration, in his volume of *Scientific Researches*, published by subscription in 1850. In 1837 he produced his electromagnetic coil-machine for giving shocks, employing a bundle of thin iron wires as the core of the fine-wire coil. In the same year he examined the cause of the frequent fracture of Leyden jars by electrical explosions, and discovered an effectual way of preventing this species of accident, by adopting the simple device of connecting the rod which supports the ball to the upper edge of the inner coating by cross-strips of metal. So effectual did he find this contrivance, that during twelve years of active experimenting with heavy charges and discharges he did not break a single jar of his battery. In 1838 he discovered the unequal heating effects found at the two poles of the voltaic arc. In an investigation of the magnetic characters of alloys, which he communicated to the Literary and Philosophical Society of Manchester, he showed that though nickel and iron are both separately magnetic, their alloys with other metals are frequently destitute of this property. An alloy of iron and zinc was found non-magnetic.

The subject of atmospheric electricity was one to which Mr. Sturgeon devoted a great deal of attention, from the commencement of his scientific career to within a short period of his decease. Not satisfied with the ordinary apparatus in use at the electrical observatories, he elevated exploring kites into the atmosphere, and in all seasons and weathers, and even in some instances at considerable risk to his life, did he pursue this important branch of meteorology. The results of more than *five hundred* kite observations, in one of which he was nearly killed, established the important fact, that the atmosphere is in serene weather uniformly positive with regard to the earth, and that the higher we ascend, the more positive does it become; so that if the strata in which the kites are immersed are at altitudes corresponding to the series, 1, 2, 3, 4, 5, their relative states of positive electricity would be conveniently represented by those numbers. With Sir Wm. Snow Harris, Sturgeon had an active dispute on the subject of lightning-conductors for ships. He urged that these should not follow the mast down into the hold, but pass over the sides outside the shrouds, the vessel being more or less enclosed in a network of conductors. In the course of this discussion Sturgeon stoutly maintained that the so-called lateral effects of lightning flashes in neighbouring bodies were not due, as Harris maintained, to imperfect neutralisation in the discharge, but to the actual generation of induction-currents; a view now amply accepted.

In 1838, Sturgeon quitted Woolwich for Manchester, whither he had been called to act as Superintendent of the Royal Victoria Gallery of Practical Science, an institution intended, like its ill-fated predecessors in London—the Adelaide Gallery, and the Royal Polytechnic Institution—as a centre for the dissemination of popular science. Unfortunately this institution was too greatly in advance of its time to prove financially successful, and it came, after about four years, to an end. Sturgeon endeavoured bravely to establish another institution of a similar character, but met with little support. Thenceforward he had to depend for his livelihood on the precarious remuneration of an itinerant scientific lecturer. In this profession he attained, indeed, to considerable reputation, as his expositions were clear and accurate, and his experimental illustrations uniformly successful. He was, however, sorely constrained in his resources, having no fixed appointment. It was during these years that his literary activity was greatest. In the year 1836, he had established a new monthly periodical, the *Annals of Electricity*, the first journal really devoted to electrical subjects. This magazine Sturgeon conducted with immense industry and great ability through ten octavo volumes; a large proportion of the articles being from his own pen. To this journal Joule contributed various papers; and though the lack of support compelled its discontinuance in 1843, it remains a valuable work of reference. Amongst other works which Sturgeon published were: "A Course of Twelve Elementary Lectures on Galvanism;" "Lectures on Electricity, delivered in the Royal Victoria Gallery in 1841-42;" "A Familiar Explanation of the Theory and Practice of Electro-gilding and Electro-silvering." He also brought out in facsimile a reprint of a famous and rare old work, Barlowe's

"Magneticall Aduertisements." Only a few weeks before his death he completed in one large handsome quarto volume a reprint of his own original contributions to science, under the title of "Scientific Researches." This volume was published by subscription and was illustrated by a number of finely engraved plates.

Sturgeon had married, soon after entering the Royal Artillery, a widow named Hutton, who kept a shoe shop in Woolwich. They had three children, who all died in infancy. In 1829, he was married again to Mary Bromley, of Shrewsbury, who survived him. Their one daughter also died an infant; and they adopted as their daughter Ellen Coates, a niece of Sturgeon's, who died in 1884.

In the years from 1845 to 1850, Sturgeon, who was now over sixty years of age, felt keenly the pinch of poverty. After many exertions, Dr. Lee, first Bishop of Manchester, and Dr. Binney, F.R.S., President of the Literary and Philosophical Society of Manchester, succeeded in obtaining for him, from Lord John Russell's Government, a grant of 200*l.* to relieve the straits into which he had been brought, and an annuity of 50*l.* per annum, which pension, however, he only enjoyed for about eighteen months. He died on Sunday, December 4th, 1850, at Prestwich, near Manchester, and was buried in the graveyard of Prestwich Church. Over his remains stands a stone with the following inscription :—

> WILLIAM STURGEON,
> THE ELECTRICIAN,
> Born 1783, died 1850, aged 67.
>
> Also,
> MARY STURGEON,
> Died October 2nd, 1867, aged 77.
>
> ELLEN, wife of Luke Brierley,
> Died January 19, 1884, aged 51.

In the church of Kirkby Lonsdale, from which the birth-place of Sturgeon is distant about two miles, there is a marble tablet which bears the following inscription :—

"In memory of William Sturgeon, who was born at Whittington, A D. 1783, and buried at Prestwich, Lancashire, on the 8th day of December, 1850. He was the son of parents in humble life, and served as a private in the Royal Artillery for nearly twenty years. After completing his term of service, he successfully devoted himself to the study of the physical sciences, with powers of originality and industry rarely equalled. Besides contributing numerous works to the scientific literature of his country, he was the discoverer of the soft iron electromagnet, the amalgamated zinc battery, the electro-magnetic coil machine, and the reciprocating magnetic electrical machine—inventions of the highest value, and which along with many others he freely gave to the world. His name will be perpetuated as long as the science he cherished continues to exist."

Of his personal appearance and qualities Dr. Joule, in the notice previously referred to, thus speaks :—

"Mr. Sturgeon was of a tall and well-built frame of body; his forehead was high, and his features were strongly marked. His address was animated, and his conversation, as it generally is when the mind is stored with knowledge, pleasing and instructive.

"In friendship he was warm and steady, in domestic life affectionate and exemplary. He had a noble mind and a generous heart. . . . He was a close and sagacious reasoner, and an unsparing exposer of error. He detested quackery and false pretension, sought diligently for truth, and loved it for its own sake."

A portrait of Sturgeon, a fine oil-painting, which was formerly in the possession of his adopted daughter, Mrs. Brierley, is believed still to exist. According to those who have seen this, the only known portrait of this distinguished man, it fully bears out the description given by Dr. Joule.

A summary by Dr. Joule of Sturgeon's scientific claims, in the form of a letter to Dr. Angus Smith, was inserted by the latter in his work, "A Centenary of Science in Manchester." It fitly closes this record of Sturgeon's life and works.

"My Dear Sir,—

"I have sifted Mr. Sturgeon's claims to the utmost. I have examined all the periodicals likely to throw light on the history of electromagnetism, and find that Mr. Sturgeon is, without doubt, the originator of the electromagnet, as well as the author of the improved electromagnetic machine. The electromagnet described by Mr. Sturgeon in the 'Transactions of the Society of Arts for 1825' is the first piece of apparatus to which the name could with propriety be applied. Arago, and Ampère, and Davy, had already, it is true, magnetised steel needles by passing currents of electricity along spirals surrounding them, but it does not appear that they observed the phenomena with iron needles, nor that they had any knowledge of the suddenness with which the polarity of soft wrought iron might be reversed by a change in the direction of the current. It appears, therefore, quite clear that to Mr. Sturgeon belongs the merit of producing the first electromagnet constructed of soft iron, as well as of that of ascertaining its peculiar and most remarkable properties. Hence it was that M. Jacobi, of St. Petersburg, claimed for Mr. Sturgeon, in conjunction with Professor Oersted, the discovery of the electromagnetic engine. Mr. Sturgeon's claims with regard to the magneto-electrical machine appear to me to be equally well established. He was the first who devised and executed an apparatus for throwing the opposing currents into one direction, thus accomplishing for this machine exactly what Watt accomplished for the steam engine. Besides this, he is beyond dispute the author of the systems of solid brass disks and insulators, going by the name of 'commutator' on the continent, and 'unitress' in America, an apparatus now universally employed in every magneto-electrical machine. Mr. Sturgeon was without doubt the constructor of the first rotary electromagnetic engine.

"The use of amalgamated zinc plates in the voltaic battery was originated by Mr. Sturgeon. It is an improvement of such value that it has been universally adopted ever since, although all other arrangements of equal date have been superseded.

"Mr. Sturgeon's discoveries in the thermo-electricity and magnetism of homogeneous bodies are very important, and have placed his name higher than that of any other philosopher who, after Seebeck, has cultivated thermo-electricity.

"The above is only a very imperfect abstract of a small part of Mr. Sturgeon's discoveries and improvements in magnetism, electricity, and the kindred sciences. Though not himself the author of extensive generalisations, he has been signally useful in preparing the way for them, and in carrying them out practically; and I know not of one individual who, under equal or even less disadvantages, has contributed so eminently to the advancement of these highly interesting and useful sciences.

(Signed) "JAMES P. JOULE."

APPENDIX B.

Electric and Magnetic Units.

The principal units employed by practical electricians, by international agreement, are :—

The *ampere*, or unit of current (formerly called the weber).
The *volt*, or unit of electromotive-force.
The *ohm*, or unit of electric resistance.

These three practical units are based upon certain abstract units, derived by mathematical reasoning and experimentally proven laws, from the three fundamental units :—

The *centimetre*, as unit of length.
The *gramme*, as unit of mass.
The *second*, as unit of time.

The system of "absolute" units derived from these is often denominated the "C.G.S." system of units* to distinguish it from other systems based on other fundamental units.

Every system of measurement is based upon some experimental fact or law. We can only measure an electric current by the effects it produces. An electric current can (1) cause a deposition of metals from their chemical solutions; (2) heat the wire that it flows through; (3) attract (or repel) a parallel neighbouring current; (4) accumulate as an electric charge that can repel (or attract) a neighbouring charge of electricity; (5) produce in its neighbourhood a magnetic field, that is to say, can exert a force upon the pole of a magnet placed near it, as, for example, in galvanometers. Now any one of these effects *might have been* chosen as a basis for a system of units of measurement, and all of them have been proposed by one authority or another. As a matter of fact, the fifth of them is made the basis in the system now adopted by international agreement; and it is the best because, firstly, it connects the electrical units with the magnetic ones, and, secondly, it is closely connected with the mechanical units, enabling the mechanical values of the electrical quantities to be readily calculated.

* The reader who may desire fuller information about the C. G. S. system of units is referred to Professor Everett's *Units and Physical Constants*.

Taking then the experimental fact that an electric current flowing in a wire, can exert a force upon the pole of a magnet placed near it, we have next to define the conditions with the utmost precision. It is found by experiment that the force which is exerted upon the magnet-pole by the current, depends on several other things beside the strength of the current: the force is proportional (*ceteris paribus*) (1) to the length of the conducting wire, (2) to the inverse square of the distance between an element of the wire and the pole, (3) to the strength of the magnet-pole. To be very precise, then, we ought to take (1) a wire one unit in length, (2) bent into an arc of unit radius so that each element of the wire is at unit distance from the pole, (3) and take a magnetic pole of one unit strength. If these things were done, and there was made to flow through the wire a current so strong that it acted on the pole with one unit of force, then a current of such a strength might be taken as a standard of comparison; for a current that was twice as strong would exert two units of force on the pole, and so forth. But in order to be exact we have yet to define what is meant by "one unit of force" and "a magnet-pole of one unit of strength." Here again we have to go to experimental facts, and choose such as will best suit for the purpose of making a consistent system of units.

A force must be measured by one of its effects, such for example as these: that it can (1) raise a given mass against the downward pull of the earth; (2) elongate a spring; (3) impart motion to a given mass, or in other words accelerate it. The first of these, which would seem the most natural to select, is rejected because the downward pull of the earth is different at different places, the second because it would require awkward definitions of the elastic properties of springs. So the third is chosen; and to make the definition precise, it must be remembered that experiment proves that the velocity of motion which a force imparts to a mass is proportional (1) to the force, (2) to the time during which it is applied, (3) inversely, to the mass acted upon. If, therefore, one could get such a force that, if it lasted one second and was made to act on one gramme, it imparted to that mass a velocity of one centimetre per second, then such a force ought to be called the unit of force. This unit has received the name of "one *dyne.*" It may be remarked that the downward pull of the earth on a mass of one gramme is sufficient to give it at the end of one second a velocity of about 32 feet per second, or, more exactly, 981 centimetres per second (in the latitude of London); hence it is clear that the pull of the earth on one gramme (what is commonly called the gramme's weight) is equal (at London) to 981 dynes. The pull of the earth on a pound (commonly called the pound's weight) is 444,971 dynes (at London). (A pound at the Pole would weigh 445,879, and at the Equator only 443,611 dynes). One dyne is a pull equal to 0·0157, or about $\frac{1}{63}$ of the weight of a grain (at London). Now, as to the unit strength of the magnet pole or unit of magnetism: a magnet pole can (1) lift a piece of iron; (2) repel (or attract) another magnet pole at a distance. The first of these two effects is rejected as a basis for a definition of a unit because the load that a magnet pole will lift does not depend only on the amount of magnetism at the pole, but on the shape and quality of the piece of iron lifted. For precise definition of the second effect upon which the definition is based, it must be remembered that experiment has shown that the repulsion of one magnet pole by another is proportional (1) to the product of the strengths of the two poles, (2) inversely to the square of the distance between them. If, therefore, we choose two similar and equal poles of just such a strength that when placed at unit distance apart they repel each other with unit force, then such poles will possess that amount of magnetism that ought to be called the unit quantity of magnetism.

We may now retrace our steps and build up systematically the units of the C.G.S. system.

The absolute unit of force ("dyne") is that force which, if it acts on one gramme for one second, gives to it a velocity of one centimetre per second.

The unit of magnetism, or unit magnet pole is one of such a strength that when placed at a distance of one centimetre (in air) from a similar pole of equal strength it repels it with a force of one dyne.

The absolute unit of current is one of such a strength that when one centimetre length of its circuit is bent into an arc of one centimetre radius, the current in it exerts a force of one dyne on a unit magnet pole placed at the centre.

The last definition is difficult to realise in practice. and a complete circle of one centimetre radius is more easy to work with than an arc one centimetre long only. If the radius be more than one centimetre and there be more than one turn of wire, as in most tangent galvanometers, then a formula is necessary. Writing r for the number of centimetres of the radius, the length of circumference will be equal to $2\pi r$. Then writing S for the number of turns of wire in the coil, and i for the strength in absolute units of the current, the formula connecting these with the force (in dynes) exerted by the current on a unit pole at the centre is :—

$$\frac{2\pi r S i}{r^3} = f;$$

whence

$$\frac{2\pi S i}{r} = f$$

In the case of the tangent galvanometer, the force, instead of being measured directly, is ascertained indirectly, by knowing the value (at the place of observation) of the horizontal component of the magnetic field due to the earth's magnetism, commonly represented by symbol H, and measuring the tangent of the deflexion produced on a magnetic needle hung at the centre when the coil lies parallel to the magnetic meridian. In this case $f = H \times \tan \delta$; whence

$$\frac{2\pi S i}{r} = H \tan \delta.$$

From this it follows that if S, r, H, and the tangent of deflexion are known, the strength of the current i will be reckoned by making the following calculation :—

$$i = \frac{r H}{2\pi S} \tan \delta.$$

(The value of H may be taken as 0·18 at London, and of the following values at other places:—Glasgow 0·17, Boston 0·17, Montreal 0·147, Niagara 0·167, Halifax, N.S., 0·159, New York, Cleveland, and Chicago, 0·184, Philadelphia 0·194, Washington 0·20, Berlin 0·178, Paris 0·188, Rome 0·24, San Francisco 0·255, New Orleans 0·82, Mexico 0·31, Bombay 0·33.)

Now, the current that is so strong as to fulfil the above definition is far stronger than anything used in telegraphic work, being about as great in quantity as the current in an arc-light circuit. Accordingly *the practical unit* of current is fixed at one-tenth part of the absolute unit, and it is called "one *ampere*." It follows that the above equation, when i is to be given in amperes, must be altered to

$$i = \frac{10 \, r \, H}{2\pi S} \tan \delta.$$

It follows that a simple tangent galvanometer to read as an *ampere-meter* can be made as follows :—Take a piece of insulated copper wire, of a gauge not less than No. 10 B.W.G., or say, than three millimetres in diameter, and of this wire wind five turns only, so as to have a mean radius exactly as below; then such a coil will, when traversed by 1 ampere, deflect the needle exactly to 45°, that is to the angle whose natural tangent is = 1, and the natural tangents of the deflexions will therefore read amperes directly. The radius has to be inversely proportional to the intensity of the horizontal component of the earth's magnetic force at the place where the amperemeter is to be used. For use at London, where H is 0·18, the radius of the coils must be 17·45 centimetres, or $6\frac{7}{8}$ inches. Other values are stated below.

Place.	Horizontal Component of Magnetic Intensity.	Radius of Coil.	
		Centimetres.	Inches.
Montreal	·147	21·37	8·41
Halifax, N.S.	·159	19·75	7·76
Glasgow and Boston	·170	18·50	7·28
Berlin	·178	17·65	6·95
New York, Cleveland, and Chicago	·184	17·07	6·72
Paris	·188	16·76	6·60
Philadelphia	·194	16·19	6·37
Washington	·200	15·70	6·18
San Francisco	·255	12·32	4·85
New Orleans	·280	11·22	4·42
Bombay	·330	9·52	3·75

It may further be noted that the current of one ampere strength will cause the deposition in 1 hour of 1·174 grammes, or 18·116 grains of copper in a copper electrolytic cell. It will in 1 hour deposit 4·024 grammes, or 60·52 grains, of silver in a silver cell.

The other electrical units also require definition. The *electromotive-force* of a battery, or of a dynamo, is only another name for the power which it possesses to drive electricity through a circuit. (Formerly the electromotive-force of a battery was called its "intensity," as a distinction from the "quantity" of current it would furnish). It is also sometimes called the electric "pressure." As a basis for a unit of electromotive-force any one of the following experimental facts might have been selected. The electromotive-force is proportional, (1) to the current that it sets up in a circuit of given resistance; (2) to the quantity of electricity that it will force as a charge into a condenser of given capacity; (3) to the number of lines of magnetic force cut per second by a conductor moving in a magnetic field. The first of these would do if the unit of resistance were given, but it is more convenient to make this fact the basis of definition of that unit rather than of the unit of electromotive-force; the second is useful for defining the unit of capacity; the third is selected for defining the unit of electromotive-force, and is extremely appropriate for the purpose, as it is the very principle of the dynamo machine. Clearly, that electromotive-force ought to be reckoned as of unit value which is produced by the motion of a conductor cutting across one line of magnetic force in one second. But this involves the preliminary definition of the unit line of magnetic force. This is as follows. The so-called magnetic lines of force represent by their direction, the direction of the resultant magnetic force in the space through which they pass: the space traversed by magnetic forces, and lines of force being called a magnetic "field." To make the number of magnetic lines represent *numerically*, as well as in mere direction, the intensity of the magnetic forces, the following device is adopted. Remembering that experiment shows that the pull (or push) which a magnetic pole experiences when placed in a magnetic field is proportional to the intensity of that field, let there be drawn as many lines to the square centimetre as there are dynes of force exerted on the unit pole. For example, if at any point it was found that the magnetic pull on a unit pole was 40 dynes, then at that place we should draw, or imagine to be drawn, 40 magnetic lines all packed within one square centimetre of sectional area. As the earth's horizontal component at London is only 0·18 (dynes on the unit pole) it follows that there would be only 18 lines passing through an area of 100 square centimetres set up vertically east and west. Returning to the definition of electromotive-force, we see that if the moving conductor cuts one magnetic line in one second, the electromotive-force engendered will be of unit value in this absolute C.G.S. system of measurement. But such a unit would be ridiculously small—far too small for practical use. Measured in such units the electromotive-force of a single Daniell's cell would be represented by the enormous number of 110,000,000, and a Latimer-Clark standard cell by 143,400,000 units. Hence practical electricians adopt, as their working unit, an electromotive-force one hundred million times as great as the absolute C.G.S. unit; and they call the practical unit "one *volt*." Hence the definition of "one volt" is that electromotive-force which would be generated by a conductor cutting across a hundred million (10^8) magnetic lines per second. The electromotive-force of a Daniell's cell is about 1·1 volts; that of Clark's standard cell 1·434 volts. The appropriate instrument for measuring volts is called a *volt-meter*.

We come then to the unit of electrical *resistance*. It is found by experiment that the current which is produced in a circuit by applying a given electromotive-force depends on the resistance offered by the circuit to the flow of electricity, the current being less as the resistance is greater, in accordance with the famous law discovered by Dr. Ohm.

Ohm's law in fact states that the current is directly proportional to the electromotive-force that is exerted in, and is inversely proportional to the resistance of, the circuit. If the symbol E stands for the number of units of electromotive-force, and R for the number of units of resistance of the circuit, and i for the current that results, then Ohm's law will be written :—

$$\frac{E}{R} = i$$

or, the resulting current can be calculated by dividing the number of units of electromotive-force by the number of units of resistance. Another way of writing Ohm's law, which is useful when it is desired to calculate the electromotive-force that will drive any prescribed current through a given resistance is

$$E = R i.$$

Now suppose we had an electromotive-force equal to one absolute C.G.S. unit, and we required to produce by its means a current of unit strength as previously defined in the absolute system, it would be requisite to adjust the resistance of the circuit to a definite value; and that value would be extremely small, otherwise such a minute electromotive-force could not maintain so large a current. Nevertheless this very minute resistance would be rightly taken as the unit in the absolute C.G.S. system, for then Ohm's law would be numerically fulfilled as,

$$\frac{\text{one unit of electromotive force}}{\text{one unit of resistance}} = \text{one unit of current.}$$

But as there are already practical units of electromotive-force and of current, so there is required a practical unit of resistance to correspond. And reflection will show that the practical unit must be a thousand million times as great as the absolute unit. For then, again, Ohm's law will be fulfilled as

$$\frac{\text{one hundred million C.G.S. units of electromotive-force}}{\text{one thousand million C.G.S. units of resistance}}$$
$$= \text{one-tenth C.G.S. unit of current.}$$

The name of "one *ohm*" is given to this practical unit of resistance; and many researches have been made to determine its working value. The British Association Committee produced standard wire coils, which were long accepted as being exact *ohms*, but they are now known to be all slightly too low in resistance. In 1882 the International Congress fixed upon the *value of the ohm* as being *a resistance equal to that of a column of mercury one square millimetre in cross-section, and* 106 *centimetres long* (measured at the freezing-point of water). According to Lord Rayleigh's most careful measurements, the true ohm ought to be, not 106, but 106·3 centimetres long. It is the intention of the Board of Trade to adopt in Great Britain this more exact value, and make it the legal definition of the ohm.

The resistances of wires and circuits are measured in practice by comparing them with certain standard "resistance coils," sets of which are often employed arranged in "resistance boxes;" the particular instruments employed in making the comparison being of two kinds, namely, the differential galvanometers and the Wheatstone's bridge. For further information the reader must refer to the textbooks on electric testing.

A rough and ready idea of the resistance called "one ohm," may be obtained by remembering that a mile of ordinary iron telegraph line offers from $13\frac{1}{2}$ to 20 ohms resistance.

One other unit is required by electricians, namely, a unit of *power*, in which to express the quantity of work per second done in any electrical system.

To measure the work done by a current in a wire, or in a lamp, or other thing supplied with electric power, we must measure both the *amperes* of current that are running through it, and the *volts* of electromotive-force that are actually applied at that part of the circuit, and having found the two numbers we must multiply them

Appendix B. 427

together. For just as engineers express power mechanically as the number of "foot-pounds" expended in a given time, so the electrician expresses electric power as the number of "volt-amperes." The more convenient name of "one *watt*" is given to the unit of electric power. Calculation shows that one "watt" or "volt-ampere" is equal to one seven-hundred-and-forty-sixth part of a horse-power.

As an example of calculation of electric power, the following may be taken. It was required to ascertain the power expended in maintaining a certain arc lamp. The voltmeter showed an electric pressure of 57 volts between the terminals of the lamp, and the amperemeter showed a current of 10·5 amperes running through it. The product is 598·5 watts. Dividing by 746 to bring to horse-power, we get 0·80, or eight-tenths of a horse-power. The name *kilowatt* is given to 1000 watts. One kilowatt is slightly more than $1\frac{1}{3}$ horse-power.

As an example it may be mentioned that the power required to excite the great electromagnet mentioned on p. 31 as capable of sustaining a load of 46 tons, is 2500 watts, or about $3\frac{1}{3}$ horse-power.

The unit of self-induction, called by various names, *secohm*, *quad* or *quadrant*, and *henry*, is a derived unit of modern origin. Whenever a current is varying in strength, it will, if carried round a coil, set up magnetic lines which as they alter in number act inductively on the convolutions of the conductor, and set up induced electromotive forces which tend to oppose the change in value of the current. The symbol used for a coefficient of self-induction is usually L, and bears the meaning that if unit current were suddenly turned off or on in the circuit in question, the resulting amount of cutting of magnetic lines by the convolutions of the circuit would have the value L. The practical unit to correspond with volt, ohm, &c., is taken as 10^9 C.G.S. units, or one ohm-second; or a coil will be said to have as the value of its coefficient one *secohm* (or *quadrant*, or *henry*) if when unit current is turned on in it, the cutting of its own magnetic lines which results, is as much as if 10^9 magnetic lines had been each cut once by a single convolution. If the rate of change of current at any time be expressed as $\frac{di}{dt}$, then the resulting self-induced electromotive-force opposing the change will be :—

$$= - L\frac{di}{dt}.$$

For a given form and volume of coil, the coefficient of self-induction is proportional to the square of the number of convolutions. The presence of an iron core vastly increases the self-inductive effects, but renders the coefficient of self-induction a variable quantity, because of the variations in the permeability of the core. For further details the reader is referred to treatises on the theory of electricity.

APPENDIX C.

APPENDIX TO CHAPTERS IV. & V.

Calculation of Excitation, Leakage, Etc.

Symbols used.

N = whole number of magnetic lines (C.G.S. definition of magnetic lines being 1 line per sq. centim., to represent intensity of a magnetic field, such that there is 1 dyne on unit magnetic pole) that pass through the magnetic circuit. Also called the *magnetic flux*.

B = the number of magnetic lines per square centimetre in the iron; also called the *induction*, or the internal magnetization.

$B_{\prime\prime}$ = the number of magnetic lines per square inch in the iron.

H = the magnetic force or intensity of the magnetic field, in terms of the number of magnetic lines to the square centimetre that there would be in air.

$H_{\prime\prime}$ = the magnetic force, in terms of the number of magnetic lines that there would be to the square inch, in air.

μ = the *permeability* of the iron, &c.; that is its magnetic conductivity or multiplying power for magnetic lines.

A = area of cross section, in square centimetres.

A'' = area of cross section, in square inches.

l = length, in centimetres.

l'' = length, in inches.

S = number of spirals or turns in the magnetizing coil.

i = electric current, expressed in amperes.

v = coefficient of allowance for leakage; being the ratio of the whole magnetic flux to that part of it which is usefully applied. (It is always greater than unity.)

Appendix C. 429

Relations of Units.

1 inch = 2·54 centimetres;
1 centimetre = 0·3937 inch.
1 square inch = 6·45 square centimetres;
1 square centimetre = 0·1550 square inch.
1 cubic inch = 16·39 cubic centimetres;
1 cubic centimetre = 0·0610 cubic inch.

To calculate the value of B or of $B_{\prime\prime}$ from the Traction.

If P denote the pull, and A the area over which it is exerted, the following formulæ (derived from Maxwell's law, see p. 118) may be used:—

$$B = 4{,}965 \sqrt{\frac{P \text{ kilos}}{A \text{ sq. cm.}}};$$

$$B = 1{,}316 \cdot 6 \sqrt{\frac{P \text{ lbs.}}{A \text{ sq. in.}}}; \text{ or}$$

$$B_{\prime\prime} = 8{,}494 \sqrt{\frac{P \text{ lbs.}}{A \text{ sq. in.}}}.$$

To calculate the requisite cross-section of Iron for a given Traction.

Reference to p. 122 will show that it is not expedient to attempt to employ tractive forces exceeding 150 lb. per square inch in magnets whose cores are of soft wrought iron, or exceeding 28 lb. per square inch in cast iron. Dividing the given load that is to be sustained by the electromagnet by one or other of these numbers, gives the corresponding requisite sectional area of wrought or cast iron respectively.

To calculate the Permeability from B or from $B_{\prime\prime}$.

This can only be satisfactorily done by referring to a numerical Table (such as Table III. or IV., p. 76), or to graphic curves, such as Fig. 39 or 40, in which are set down the result of measurements made on actual samples of iron of the quality that is to be used. The values of μ for the two specimens of iron to which Table IV. refers may be *approximately* calculated as follows:—

$$\text{For annealed wrought iron, } \mu = \frac{17{,}000 - B}{3 \cdot 5};$$

$$\text{For grey cast iron, } \mu = \frac{7{,}000 - B}{3 \cdot 2}.$$

These formulæ must not be used for the wrought iron for tractions that are less than 28 lb. per square inch, nor for cast iron for tractions less than $2\frac{1}{4}$ lb. per square inch.

To calculate the Total Magnetic Flux which a core of given sectional area can conveniently carry.

It has been shown that it is not expedient to push the magnetization of wrought iron beyond 100,000 lines to the square inch, nor that of cast iron beyond 42,000. These are the highest values that ought to be assumed in designing electromagnets. The total magnetic flux is calculated by multiplying the figure thus assumed by the number of square inches of sectional area.

To calculate the Magnetizing Power requisite to force a given number of Magnetic Lines through a definite Magnetic Reluctance.

Multiply the number which represents the magnetic reluctance by the total number of magnetic lines that are to be forced through it. The product will be the amount of magneto-motive force. If the magnetic reluctance has been expressed on the basis of centimetre measurements, the magneto-motive force, calculated as above, will need to be divided by $1\cdot 257$ $\left(\text{i.e., by } \frac{4\pi}{10}\right)$ to give the number of ampere-turns of requisite magnetizing power. If, however, the magnetic reluctance has been expressed in the units explained below, based upon inch measures, the magnetizing power, calculated by the rule given above, will already be expressed directly in ampere-turns.

TO CALCULATE THE MAGNETIC RELUCTANCE OF AN IRON CORE.

(a) *If dimensions are given in centimetres.*—Magnetic reluctance being directly proportional to length, and inversely proportional to sectional area and to permeability, the following is the formula:—

$$\text{Magnetic reluctance} = \frac{l}{A\mu};$$

but the value of μ cannot be inserted until one knows how great B is going to be; when reference to Table III. gives μ.

(b) *If dimensions are given in inches.*—In this case we can apply a numerical co-efficient, which takes into account the change of

Appendix C. 431

units (2·54), and also, at the same time, includes the operation of dividing the magneto-motive force by $\frac{4}{10}$ of π ($= 1·257$) to reduce it to ampere-turns. So the rule becomes

$$\text{Magnetic reluctance} = \frac{l''}{A''\mu} \times 0·3132.$$

Example.—Find the magnetic reluctance from end to end of a bar of wrought-iron 10 inches long, with a cross-section of 4 square inches, on the supposition that the magnetic flux through it will amount to 440,000.

To Calculate the Total Magnetic Reluctance of a Magnetic Circuit.

This is done by calculating the magnetic reluctances of the separate parts, and adding them together. Account must, however, be taken of leakage; for when the flux divides, part going through an armature, part through a leakage path, the law of shunts comes in, and the nett reluctance of the joint paths is the reciprocal of the sum of their reciprocals. In the simplest case the magnetic circuit consists of 3 parts (1) armature, (2) air in the 2 gaps, (3) core of the magnet. These three reluctances may be separately written, as in the Table below.

If the iron used in armature and core is of the same quality, and magnetized up to the same degree of saturation, μ_1 and μ_3 will be alike. For the air-gaps $\mu = 1$, and therefore is not written in.

If there were no leakage, the total reluctance would simply be the sum of these three terms. But when there is leakage, the total reluctance is reduced.

	For Centimetre Measure.	For Inch Measure.
1. Armature	$\dfrac{l_1}{A_1 \mu_1}$	$\dfrac{l''_1}{A''_1 \mu_1} \times 0·3132$
2. The Gaps	$2\dfrac{l_2}{A_2}$	$2\dfrac{l''_2}{A''_2} \times 0·3132$
3. Magnet Core ..	$\dfrac{l_3}{A_3 \mu_3}$	$\dfrac{l''_3}{A''_3 \mu_3} \times 0·3132$

To Calculate the Ampere-turns of Magnetizing Power requisite to force the desired Magnetic Flux through the Reluctances of the Magnetic Circuit.

(a) *If dimensions are given in centimetres* the rule is :—

Ampere-turns = the magnetic flux, multiplied by the magnetic reluctance of the circuit, divided by $\frac{4}{10}$ of π ($= 1 \cdot 257$).

Or, in detail, the three separate amounts of ampere-turns required for three principal magnetic reluctances are explained as follows :—

Ampere-turns required to drive N lines through iron of armature $= N \times \dfrac{l_1}{A_1 \mu_1} \div \dfrac{4\pi}{10}$,

Ampere-turns required to drive N lines through the two gaps $= N \times \dfrac{2 l_2}{A_2} \div \dfrac{4\pi}{10}$,

Ampere-turns required to drive vN lines through the iron of magnet core $= v N \times \dfrac{l_3}{A_3 \mu_3} \div \dfrac{4\pi}{10}$,

And, adding up :—

Total ampere-turns required $= \dfrac{10}{4\pi} N \left\{ \dfrac{l_1}{A_1 \mu_1} + \dfrac{2 l_2}{A_2} + \dfrac{v l_3}{A_3 \mu_3} \right\}$.

(b) *If dimensions are given in inches*, the rule is :—

Ampere-turns = magnetic flux multiplied by the magnetic reluctance of the circuit. Or, in detail :—

Ampere-turns required to drive N lines through iron of armature $= N \times \dfrac{l''_1}{A''_1 \mu_1} \times 0 \cdot 3132$.

Ampere-turns required to drive N lines through two gaps $= N \times \dfrac{2 l''_2}{A''_2} \times 0 \cdot 3132$,

Ampere-turns required to drive vN lines through iron core of magnet $= v N \times \dfrac{l''_3}{A''_3 \mu_3} \times 0 \cdot 3132$;

And, adding up :—

Total ampere-turns required $= 0 \cdot 3132 N \left\{ \dfrac{l''_1}{A''_1 \mu_1} + \dfrac{2 l''_2}{A''_2} + \dfrac{v l''_3}{A''_3 \mu_3} \right\}$.

It will be noted that here v, the coefficient of allowance for leakage, has been introduced. This has to be calculated as shown later. In the meantime it may be pointed out that, in designing electromagnets for any case where v is approximately known beforehand, the calculation may be simplified by taking the sectional area of the magnet core greater than that of the armature in the same proportion. For example, if it were known that the waste lines that leak were going to be equal in number to those that are usefully employed in the armature (here $v = 2$), the iron of the cores might be made of double the section of that of the armature. In this case μ_3 will approximately equal μ_1.

To Calculate the Coefficent of Allowance for Leakage, v.

v = total magnetic flux generated in magnet core $\div$ useful magnetic flux through armature. The respective useful and waste magnetic fluxes are proportional to the permeances along their respective paths. *Permeance*, or magnetic conductance, is the reciprocal of the *reluctance*, or magnetic resistance. Call useful permeance through armature and gaps u, and the waste permeance in the stray field w; then

$$v = \frac{u + w}{u}$$

w may be estimated by the Table XIV. given on p. 177, or other leakage rules, but should be divided by 2, as the average difference of magnetic potential over the leakage surface is only about half that at the ends of the poles.

Rules for Estimating Magnetic Leakage.

(I. to III. adapted from Professor Forbes' Rules.)

Prop. I. Permeance between two parallel areas facing one another. Let areas be A_1'' and A_2'' square inches, and distance apart d'' inches, then :—

$$\text{Permeance} = 3 \cdot 193 \times \tfrac{1}{2}(A_1'' + A_2'') \div d''.$$

Prop. II. Permeance between two equal adjacent rectangular areas lying in one plane.—Assuming lines of flow to be semicircles, and that distances d_1'' and d_2'' between their nearest and farthest edges respectively are given, also a'' their width along the parallel edge :—

$$\text{Permeance} = 2 \cdot 274 \times a'' \times \log_{10} \frac{d_2''}{d_1''}$$

2 F

Prop. III. Permeance between two equal parallel rectangular areas lying in one plane at some distance apart.—Assume lines of leakage to be quadrants joined by straight lines.

$$\text{Permeance} = 2 \cdot 274 \times a'' \times \log_{10}\left\{1 + \frac{\pi (d_2'' - d_1'')}{2 d_1''}\right\}$$

Prop. IV. Permeance between two equal areas at right angles to one another.

Permeance (if air angle is 90°) = double the respective value calculated by II. or III.

Permeance (if air angle is 270°) = $\frac{2}{3}$ times the respective value calculated by II.

If measures are given in centimetres these rules become the following :—

I. $\frac{1}{2} (A_1 + A_2) \div d$;

II. $\dfrac{a}{\pi} \log_e \dfrac{d_2}{d_1}$;

III. $\dfrac{a}{\pi} \log_e \left(1 + \dfrac{\pi (d_2 - d_1)}{2 d_1}\right)$.

Prop. V. Permeance between two parallel cylinders of indefinite length.

The formula for the reluctance is given on p. 175 above : the permeance is the reciprocal of it. Calculations are simplified by reference to Table XIV., p. 177.

INDEX.

A.

	PAGE
Abdank, polarised bell	295
Abel's researches on composition of steel	392
Ader, telephone receiver magnet	300
Air-gap, effect of, in magnetic circuit	85, 163
„ „ on magnetic reluctance	163, 174
Alternate-current electromagnets	331
„ „ wave diagrams	333
André, equalising the pull of a magnet	279
Ampère, researches of	1
Ampere-turns, calculation of	129, 198, 432
Annunciators (see Indicator movements).	
Arago, researches of	1
Arc-lamp mechanism—	
Brockie-Pell	117
Brush	270
Duboscq	281
Gaiffe	263
Gülcher	313
Kennedy	271
Menges	267
Paterson and Cooper	283
Pilsen	265
Serrin	282
Thomson-Houston	284
Weston	271
Armature, effect of, on permanent magnets	212
„ effect of shape	188
„ length of cross section of	186
„ position and form of	178, 188
„ pulled obliquely	36
„ round *versus* flat	189
Aron, sheath for magnet coils	370
Astatic combinations	408

	PAGE
Ayrton, distribution of free magnetism	138
Ayrton and Perry's coiled ribbon voltmeters	268
„ „ tubular electromagnet	272
„ „ on magnetic shunts	x

B.

	PAGE
Bain, electric pendulums	327
Bain's moving coil mechanism	299
Bar electromagnet	50
„ magnets of steel	384, 395
„ „ proper temper of	390, 401
Barlow, magnetism of long bars	179
Barlow's wheel	1
Barrett, recalescence	392
Barus, researches	387, 398, 400
Battery grouping for quickest action	226
„ resistance for best effect	227
„ used by Sturgeon	3
Becquerel's electromagnet	30
Bell (A. G.), iron-clad electromagnet	52
„ „ telephone receiver	52, 300
Bells, electric, devices in	294, 295, 314, 319, 322, 371
Bernouilli's rule for traction	127, 405
Bidwell, electromagnetic pop-gun	289
„ measurement of permeability	79
Bosanquet, investigations of	70, 120, 397
„ magneto-motive force	70
„ measurement of permeability	70
Brett, polarised magnets	292
Brisson, method of winding	202
Brockie-Pell differential coil-and-plunger	266
Brown and Williams, repulsion mechanism	288
Bruger, coils and plungers	259
Bruger's researches	259
Brush Company's magnetic separator	309
Burnett, equalizing the pull of a magnet	279

C.

	PAGE
Callard's equalizer	280
Camacho's electromagnet	214
Cance's electromagnet	214
Cannons, electromagnet made of two	31
Carpentier's magnetic shunt	305

Index. 437

	PAGE
Cast iron, magnetization of	75, 76, 83, 385
Cheesman, hardness of steel	389, 399
Chemical composition of permanent magnets	385
Chenot's magnetic separator	308
Chronographs, electromagnets for	237, 238
Clark, electromagnetic tools	361
Cloisons, winding in	201, 267
Coercive force	96, 381
Coil-and-plunger coil	267
„ „ diagram of force and work of	251
„ „ differential	265
„ „ electromagnet	54, 242
„ „ modifications of	54, 264
Coil moved in permanent magnetic field	213
Coils, effect of position	187, 206
„ effect of size	205
„ how connected for quickest action	187, 225
Colombet's mechanism	313
Coned plungers, effect of	261
Constancy of permanent magnets	397
Cooke's experiments	18
Cores, determination of length	123, 253
„ effect of shape	139, 184
„ effect of shape of section	184
„ hollow *versus* solid	183, 262
„ lamination of	219, 331
„ of different thicknesses	257
„ of irregular shapes	261
„ proper length of	123, 234
„ square *versus* round	179, 189, 209
„ tubular	183, 262
Coulomb, law of inverse squares	156
„ two magnetic fluids	vi
Cowper, range of action	244
„ writing telegraph	287
Cumming, magnetic conductivity	vii
„ galvanometer	1
Curved plunger core and tubular coils	302
Curves of hysteresis	75, 97, 101
„ of magnetization and permeability	68, 72, 77, 78, 85, 87, 90

D.

Dal Negro's motor	350
D'Arlincourt's relay	305

Index.

	PAGE
D'Arsonval, galvanometer	299
„ telephone receiver	300
Davenport's motor	351
Davy (E.), mode of controlling armature	279
„ (Sir H.), researches of	2
De la Rive, floating battery and coil	1, 49
„ magnetic circuit	vii
Demagnetize iron, how to	107
Deprez, chronograph	237
„ electric hammer	268, 360
„ induction coil break	324
Diamagnetic action	268
Differential winding	265, 371
Doubrava's sliding coil mechanism	299
Dove, magnetic circuit	vi
Dub, best position of coils	206
„ cores of different thicknesses	257
„ distance between poles	186
„ flat *versus* pointed poles	135
„ magnetic circuit	vii
„ magnetism of long bars	144
„ polar extensions of core	143
Duboscq's arc lamp mechanism	281
Du Moncel, best position of coils	206
„ club-footed electromagnet	51, 216
„ distance between poles	186
„ effect of polar projections	146, 189
„ effect of position of armature	178, 188
„ electromagnetic pop-gun	290
„ experiments with pole-pieces	189
„ hinged armatures	217
„ interlocking electromagnets	287
„ on armatures	188
„ tubular cores	183

E.

Electric bells	294, 295, 314, 319, 322, 371
„ „ invented by Mirand	319
Electric and magnetic units	416
„ indicators	314
„ motors (see Motors).	
Electrodiapason	325
Electromagnet, Ayrton and Perry's	272

Index. 439

		PAGE
Electromagnet, bar		50
,,	Becquerel's	30
,,	Camacho's	214
,,	Cance's	214
,,	club-footed	51
,,	coil-and-plunger	55, 241, 242
,,	coils, resistance of	209
,,	design of, for various uses	214, *et seq.*
,,	Du Moncel's	51, 216
,,	Fabre's	52
,,	Faraday's	29
,,	Faulkner's	52
,,	first publicly described	2
,,	for heating purposes	348
,,	for rapid working	187, 219, 225
,,	Gaiser's	273
,,	Guillemin's	52
,,	Henry's	14, 17
,,	Hjörth's	243
,,	Holtz's	28
,,	hollow *versus* solid	183, 262
,,	Holroyd-Smith's	272
,,	horse-shoe	50
,,	Hughes's	187, 296
,,	in Bell's telephone	52
,,	invented in 1825	2
,,	iron-clad	52, 148
,,	Jensen's	218
,,	Joule's	21, 24, 25
,,	law of	27
,,	long *versus* short limbs	181, 234, 253
,,	of Brush arc lamp	270
,,	Plücker's	28
,,	Radford's	24
,,	Ricco's	214
,,	Roberts's	25
,,	Roloff's	207, 272
,,	Romershausen's	52
,,	Ruhmkorff's	54
,,	Smith's (Rev. F. J.)	238
,,	Stevens and Hardy	270
,,	Sturgeon's	3, 4, 8
,,	surgical	376
,,	Varley's	214
,,	Von Feilitzch's	28
,,	without iron	215

440 Index.

		PAGE
Electromagnetic adherence		307
„ clutch		53, 310
„ engines		244, 350
„ inertia		219
„ linkages		286
„ mechanism		275
„ motors		350
„ pendulums		327
„ pop-gun		289
„ repulsion		268, 288
„ separators of iron		308
„ tools		359
„ tuning-forks		324
„ vibrators		318
Electromagnets, alternate current		332
„ diminutive		22, 129
„ for alternating currents		331
„ for arc lamp (*see* Arc lamp mechanism).		
„ for lifting		121
„ for maximum range of attraction		215
„ for maximum traction		125, 214
„ for minimum weight		215
„ formulæ for		27, 119, 129, 428
„ for quickest action		187, 225
„ for traction		121, 214
„ heating of		102, 193
„ in telegraph apparatus		187, 231
„ saturation of		67
„ specifications of		92, 193, 231
„ to produce rapid vibrations		235
„ with iron between the windings		214
„ with long *versus* short limbs		181
Elphinstone (Lord), application of magnetic circuit in dynamo design		ix
Enamel, use for insulating		63, 201
Equalizing the pull of an electromagnet		279
Evershed amperemeter		301
Ewing, curves of magnetization		68
„ hysteresis		96
„ isthmus method of magnetizing		397
„ maximum magnetization		82, 385
„ measurement of permeability		69
„ on effect of joints		88
„ theory of magnetism		110

F.

	PAGE
Fabre, iron-clad electromagnet	52
Faraday, lines of force	viii
„ rotation of permanent magnet	1
Faulkner, iron-clad electromagnet	52
Fireproof insulation	63
Forbes, electromagnetic brake	53, 312
„ formulæ for estimation of leakage	433
Foucault, induction coil vibrator	323
Frölich, law of the electromagnet	27
Froment's equalizers	212, 283
„ vibrating mechanism	319

G.

Gaiffe's arc lamp	263
Gaiser's electromagnet armature	273
Galvanometer coils	200
Gauss, magnetic measurements	158
Gloesner, polarized magnets	293
Gray (Andrew), on decay of magnetism	397
„ (Elisha), harmonic transmitter	328
Grove, range of action	244
Guillemin, iron-clad solenoid	256

H.

Haarlem magnets	405
Häcker's rule for traction	127
Hankel, magnetism of long bars	179
„ working of coil-and-plunger	256
Hardness and tempering	386
Harmonic telegraph transmitters	328
Heating of magnet coils	193
Heaviside, magnetic reluctance	113
Helmholtz, law regarding interrupted currents	220
Henry, the, unit of self-induction	220, 346, 427
Henry's first experiments	10
„ motor	318, 350
Hipp's electric pendulum	328
Hirschburg, use of electromagnet in surgery	375
Hjörth's electromagnet	243
„ polarized magnets	292
„ motors	352

	PAGE
Hookham, on magnet of constant power	401
Hopkinson, coercive force of steel	382
„ curves of magnetization	75, 77
„ design of dynamos	27
„ maximum magnetization	82
„ measurement of permeability	73, 78
Horse-shoe electromagnet	30
„ steel magnet	405
Houdin's equalizer	280
Hughes, on distance between poles	186
„ magnetic balance	69
„ polarized magnet	186, 296
„ printing telegraph magnets	186, 296
Hunt, range of action of electromagnets	243
Hysteresis	96
„ viscous	106

I.

Iron-clad electromagnet	52, 273
„ range of action	148
Iron, electrolytic, magnetic qualities of	385
„ magnetic qualities affected by hammering, rolling, etc.	92, 399
„ maximum magnetization of	82, 382, 385
„ permeability of	76, 77
„ „ compared with air	66
„ the magnetic properties of	65, 382
„ residual magnetism of	94, 382, 385
Indicator movements	314
Induction coil vibrators	322
Isthmus method of magnetizing	397

J.

Jacobi's motor	351
Jensen's electromagnet	218
Joints, effect of, on magnetic reluctance	88
Joule, experiment with Sturgeon's magnet	23
„ law of mutual attraction	20
„ law of traction	20
„ length of electromagnet	20
„ letter on Sturgeon	416
„ magnetic saturation	21
„ maximum magnetization	121

Index. 443

	PAGE
Joule, maximum power of an electromagnet	26
,, range of action	244
,, researches	19
,, results of traction experiments	20, 22, 121
,, tubular cores	19, 183

K.

Kapp, design of dynamos	27
,, maximum magnetization	82
Keeper, effect of position on tractive power	188
,, ,, removing suddenly	213, 408
,, ,, slamming on	408
Kennedy, arc lamp	271
,, electromagnet for heating purposes	348
Kirchhoff, measurement of permeability	69
Krizik, coned and cylindrical plungers	261

L.

Lamination of electromagnets	331
,, of permanent magnets	402
Langdon-Davies, rate governor	329
,, suppression of sparking	371
Law of inverse squares	157
,, ,, a point law	157
,, ,, apparatus to illustrate	159
,, ,, defined	158
Law of the electromagnet	27
Law of the magnetic circuit, applied to traction	118
,, ,, explanation of symbols	114
Law of Helmholtz for sudden currents	220
Law of Maxwell for alternating currents	347
Law of Ohm	425
Law of traction	118, 405
,, verified	120
Leakage of magnetic lines	164, 433
,, reluctances	173, 433
Lamont, experiments on distribution	137
,, law of electromagnet	27
,, on slow decay of magnetism	398
Lenz and Jacobi's law	27
Lenz, magnetism of long bars	179
Leupold, winding for range of action	263

		PAGE
Lines of force		40, 65, 424
Long *versus* short cores		234, 253
Lyttle's patent for winding		202

M.

Magnetic adherence			307
,,	balance of Professor Hughes		69
,,	brake		53, 312
,,	centre of gravity		158
,,	circuit		112, 431
,,	,,	application of, in dynamo design	ix
,,	,,	for greatest traction	125
,,	,,	formulæ for	114, 432
,,	,,	tendency to become more compact	278, 289
,,	,,	various parts of	174, 432
,,	conductivity		66
,,	field, action of, on small iron sphere		154, 268
,,	flux, calculation of		114, 429
,,	gear		308
,,	insulation		115
,,	leakage		168, 433
,,	,,	calculation of	173, 433
,,	,,	calculation of coefficient	432
,,	,,	coefficient of, "v"	173, 433
,,	,,	due to air-gaps	162
,,	,,	estimation of	177, 433
,,	,,	lines	40, 44, 65, 424
,,	,,	measurement of	169
,,	,,	proportional to the surface	180
,,	,,	relation of, to pull	118
,,	memory		95
,,	moments		183
,,	motors		351
,,	output of electromagnets		203
,,	permeability		68, 429
,,	polarity, rule for determining		37
,,	pole of the earth		158
,,	reluctance, calculation of		114, 430
,,	,,	of divided iron ring	162
,,	,,	of iron ring	162
,,	,,	of waste and stray field, formulæ for	173
,,	resistance		113
,,	saturation		67
,,	screening		343, 410

Index. 445

	PAGE
Magnetic shunts	304
„ units	416
Magnetism, free	136
„ of long bars	179
„ permanent	94, 381
Magnetization and magnetic traction, tabular data	119, 405
„ calculation of	114
„ defined	44, 168
„ internal	38, 41
„ „ distribution of	41, 72
„ of different materials	68
„ specific	384
„ surface	40, 136
Magnetizing permanent magnets	393
Magnetometer	144, 159
Magnetomotive force	44, 430
„ „ calculation of	45, 430
Maikoff and De Kabath, repulsion mechanism	288
Marsh, first vibrating mechanism	318, 413
„ vibrating pendulum	318, 413
Materials of construction	59
Maxwell, galvanometer	299
„ law of the electric circuit stated	49
„ law of traction	118
„ law regarding circulation of alternating currents	347
„ magnetic conductivity	viii
Mercadier's electro-diapason	325
Mirand, inventor of electric bell	319
Mitis metal, magnetization of	83
Moll's experiments	6, 131
Motors, Bourbouze	354
„ Davis, book on	350
„ Froment	353
„ Henry	350
„ Hjörth	353
„ Immisch	356
„ Jacobi	352
„ Page	352
„ Ritchie	351
Mordey's alternate current machine	357
Moseley's indicator	316
Müller, law of the electromagnet	27
„ magnetization of long bars	179
„ measurement of permeability	68

N.

	PAGE
Neef, vibrating mechanism	319
Newton's signet-ring loadstone	128
Nicklès, classification of magnets	50
,, distance between limbs of horse-shoe magnets	186
,, magnetic adherence of driving wheels to rail	307
,, ,, gear	308
,, magnets	186
,, ,, distance between poles	186
,, traction affected by extent of polar surface	132
,, tubular cores	183
Nobilis, astatic pair	411

O.

Oblique approach	282, 353
Oersted's discovery	1
Ohm's law	113, 116, 220, 425
Osmond's researches	392

P.

Pacinotti's motor with ring electromagnet	353
Page, electric motor	268
,, electromagnetic engine	352
,, sectioned coils	267, 351
Paine, sheath for magnets	370
Pellin, jointed plunger	137
Permanent magnets contrasted with electromagnets	211
,, table of remanence and coercive force	381
,, uses of	290, 374, 408
Permanent magnetization, maximum values of	382
Permeability, calculation of	66, 79, 429
,, methods of measuring	68
Permeameter	81
Perry, magnetic shunts (see also Ayrton)	x
Pfaff, tubular cores	183
Plungers, coned *versus* cylindrical	261
,, of iron and steel	257
Point poles	159
,, action of single coil on	246
Poisson, on alleged magnetic fluids	vi
Polar distribution of magnetic lines	43
,, region defined	39

Polarized apparatus for indicators	316
„ mechanism	290
Pole-pieces, convex *versus* flat	132, 406
„ Dub's experiments with	141
„ Du Moncel's experiments with	189
„ effect of position on attractive power	142, 190
„ effect on lifting power	142
„ on horse-shoe magnets	189
Poles, effect of distance between	186
„ flat *versus* pointed	134
Preece, self induction in relays	231
„ winding of coils	230

R.

Radford's electromagnet	24
Range of action of electromagnets	242
Rapieff, hinged electromagnets	286
Rate-governor	329
Raworth, electromagnetic clutch	311
Rayleigh (Lord), electromagnetic tuning-forks	326
Reluctance	113
Remanence	95, 381
Repulsion experiments	34, 288, 337
„ mechanism	288
Residual magnetism	94, 381
Resistance of electromagnet and battery	203
„ of insulated wire, rule for	209
Ritchie, electromagnetic motor	350
„ magnetic circuit	vii
„ steel magnets	182
„ winding in sections	202
Robert's electromagnet	24
Robertson, galvanometer	299
„ writing telegraph	287
Roloff's electromagnet	207
Romershausen, iron-clad electromagnet	52
Rowan, electromagnetic tools	362
Rowland, analogy of magnetic and electric circuits	viii, 112
„ first statement of the law of the magnetic circuit	viii, 112
„ magnetic permeability	69
„ maximum magnetization	82
„ measurement of permeability	69
Ruhmkorff's electromagnet	53, 395
Ruths, on hardness of steel	390

S.

	PAGE
Saturation, curve of	72
,, distribution of	83, 151
,, effect of, on permeability	66, 78
Schweigger's multiplier	1
Screening, magnetic	343, 410
Sectioned coils	201, 267
Self-induction, effect of	220 et seq., 326, 345
,, in telegraph magnets	231
Shaped iron armature between poles of electromagnet	302
Shunt, magnetic	304
Siemens' differential coil and plunger	265
,, relay	233
,, pivotted armature	303
Siphon recorder	299
Smith (Holroyd), plunger electromagnet	272
Smith (Rev. F. J.), electromagnet for chronograph	238
Snell, surgical electromagnet	374
Sparking, suppression of	201, 321, 363
Specific magnetization	384
Spottiswoode's extra rapid induction-coil break	323
Steel, hardening and tempering	386, 392
,, magnetization of	68, 96
,, permeability of	94, 384
Stephenson, electric motors not practicable	244
Stevens and Hardy, plunger electromagnet	270
Stoletow, measurement of permeability	69
Strouhal and Barus on hardness and tempering	388, 400
Sturgeon, biographical sketch of	411
,, experiments on bar magnets	6
,, experiments on leakage	186
,, first description of electromagnet	2
,, magnetic circuit	vii
,, polar extensions	146
,, polarized apparatus	292
,, telegraph apparatus	292
,, thermogalvanometer	299
,, tubular cores	183
Sturgeon's apparatus lost	4
,, first electromagnet	2
,, first experiments	2
Surface magnetism	136, 151
Surgery, electromagnet used in	374

T.

	PAGE
Temperature, effect of, on magnetism	93, 400
Tempering	386
Theory of magnetism, Ewing's	110
Thomas, wire gauge table	194
Thompson (J. Tatham), surgical electromagnet	376
Thomson (Elihu), electromagnetic experiments	338
,, (J. J.), on effect of joints	88
,, (Sir Wm.), current meters	268
,, ,, polarized magnets	292
,, ,, range of action	244
,, ,, rule for winding electromagnets	200
,, ,, siphon recorder	299
,, ,, winding galvanometer coils	200
Thorpe's semaphore indicator	316
Time-constant of electric circuit	222
Traction, formula for	118, 405, 407
,, in terms of weight of magnet	127, 405
Tractive power of magnets affected by surface contact	131, 406
,, integral formula for	119
Trève, iron wire coil	263
Tubular coils, action of, on unit pole	246
,, attraction between	267
,, winding of	267
Two magnetic fluids, doctrine of	vi
Tyndall, range of action	244

U.

Unipolar magnets	408
Units	418

V.

Van der Willigen, on Haarlem magnets	393, 404
Van Wetteren, maker of Haarlem magnets	393, 404
Varley (C. F.), copper sheath for magnet coils	370
,, ,, iron-clad electromagnet	52
,, (S. A.), electromagnet	214
,, ,, polarized magnets	292
Vaschy, coefficients of self-induction	229
Vibrators	322
Vincent, application of magnetic circuit in dynamo design	ix
Viscous hysteresis	109

2 G

Volt, the (see Units).	
Vom Kolke, distribution of magnetic lines	151
Von Feilitzsch, plungers of iron and steel	257
,, magnetism of long bars	179
,, measurement of permeability	69
,, tubular cores	183
Von Waltenhofen, attraction of tubular coils	257

W.

Wagener's electric bell	314
Wagner, vibrating mechanism	318
Wall's magnetic separator	308
Walmsley, magnetic reluctance of air	176
Waterhouse, pivoted armature	303
Watt, the (see Units).	
Wave diagrams, alternate current	334
Wheatstone, Henry's visit to	18
,, equalizer for telegraph instruments	284
,, oblique approach	284
,, polarized apparatus	293
Willans' magnetic coupling gear	311
Winding a magnet in sections	201, 267, 360
,, calculation of	197
,, coils in multiple arc	201, 371
,, differential	265, 371
,, effect of, on range of action	215, 235
,, for constant pressure and for constant current	199
,, iron *versus* copper wire	214
,, of tubular coils	263
,, position of coils	206
,, size of coils	205
,, thick *versus* thin wire	193, 205
,, wire of graduated thickness	200
Wire gauge and amperage table	194
Wrought iron, magnetization of	75 *et seq.*

www.ingramcontent.com/pod-product-compliance
Lightning Source LLC
Chambersburg PA
CBHW022107300426
44117CB00007B/622